Topics in Applied Physics   Volume 57

Topics in Applied Physics Founded by Helmut K. V. Lotsch

Strong and Ultrastrong Magnetic Fields

and Their Applications

Edited by F. Herlach

With Contributions by
K. Dransfeld   J. Hajdu   F. Herlach   G. Landwehr
G. Maret   N. Miura   D. B. Montgomery
M. Motokawa   L. W. Roeland   C. J. Schinkel

With 199 Figures

Springer-Verlag Berlin Heidelberg GmbH

Professor Dr. *Fritz Herlach*

Katholieke Universiteit Leuven,
Laboratorium voor Lage Temperaturen en Hoge-Veldenfysika,
Celestijnenlaan 200D,
B-3030 Leuven, Belgium

ISBN 978-3-662-30933-9   ISBN 978-3-540-38962-0 (eBook)
DOI 10.1007/978-3-540-38962-0

Library of Congress Cataloging in Publication Data. Main entry under title: Strong and ultrastrong magnetic fields and their applications. (Topics in applied physics; v. 57) Includes index. 1. Magnetic fields. I. Herlach, F. (Fritz), 1932–. II. Dransfeld, K. (Klaus) III. Series. QC754.2.M3S77   1985   538   84-25474

© Springer-Verlag Berlin Heidelberg 1985

Originally published by Springer-Verlag Berlin Heidelberg New York Tokyo in 1985
Softcover reprint of the hardcover 1st edition 1985

2153/3130-543210

Preface

In one way or another, magnetic fields have been involved in most of the important discoveries in modern physics. As new effects are often discovered after the extension of the range of a basic experimental parameter, continued efforts are made to design and build stronger magnets. For generating fields above 10–15 T, extraordinary technical efforts are required. The necessary resources are available only at a few large magnet laboratories which have been established for this purpose.

This volume gives a comprehensive review of experiments with such strong magnetic fields. The introduction includes a survey of magnet laboratories where strong magnetic fields are available for general experimentation; most of these facilities are open to guest experiments on rather generous terms.

At the present time, most applications are in solid-state physics. In particular, very strong fields are needed to affect quantum phenomena in semiconductors. In the extreme quantum limit, when all conduction electrons are condensed into the lowest Landau level, basic scattering mechanisms can be studied and theoretically analyzed with relative ease. Research on two-dimensional semiconducting systems has recently resulted in the discovery of another macroscopic quantum effect that occurs at very high magnetic fields. This is reflected in the chapter on *magnetotransport phenomena* which includes an extensive treatment of the theoretical background and an up-to-date discussion of two-dimensional systems and the quantum Hall effect.

Research in *magnetism* has evolved from the more practical applications into a testing ground for advanced statistical physics. As the interactions in a cooperative magnetic system are well defined, these are amenable to rigorous theoretical analysis. This can be checked against experimental results from samples with special characteristics that have been synthesized for this purpose. Very high fields are needed to compete with strong exchange interactions and to detect resonances which are broadened to the extent that they cannot be observed in lower fields.

Strong magnetic fields are well suited for studying large *biomolecules* in vivo. In research on this and related topics, great progress has been made in recent years. This chapter is the first comprehensive review of this new field of research. It includes research on polymers and polymerization reactions, liquid crystals, biological membranes and artificial multilayers, the separation of macro-molecules in strong magnetic field gradients, and the influence of magnetic fields on living organisms.

At the present time, advanced electromagnets are designed mostly for plasma physics and *nuclear fusion research*. Here the general emphasis is more on large scale and special geometries than on the highest field strength. In this chapter the relative merits and trade-off considerations regarding an increase of the magnetic field strength are discussed, primarily with the example of the Alcator high-field tokamak. This gives a good view on the technical problems involved in modern high-field magnet design.

The final chapter points to the future. Techniques for experimentation with *pulsed magnetic fields* are now well developed but many researchers have been reluctant to adopt them for their experiments. Right now there is an increasing trend for switching to this method which is efficient, economical and nearly open-ended regarding the field strength. The revival of this trend initiated mainly in Japan and is now gradually spreading to Europe and to the USA. We describe methods for generating the fields as well as many experiments in the field range from 20 T to several megagauss. The megagauss ($=100$ T) is the limit above which the magnetic field violently destroys the field coil, thus requiring drastically different one-shot experimental methods.

This volume is intended to provide inspiration and guidance for making good use of the excellent magnet laboratory facilities that are now available worldwide.

Leuven, January 1985   *Fritz Herlach*

Contents

5. High Field Magnetic Confinement of Fusion Plasmas

Contributors

Dransfeld, Klaus
  Fakultät für Physik, Universität Konstanz, Postfach 5560,
  D-7750 Konstanz, Fed. Rep. of Germany

Hajdu, János
  Institut für Theoretische Physik der Universität zu Köln,
  D-5000 Köln 41, Fed. Rep. of Germany

Herlach, Fritz
  Katholieke Universiteit Leuven,
  Laboratorium voor Lage Temperaturen en Hoge-Veldenfysika,
  Celestijnenlaan 200 D, B-3030 Leuven, Belgium

Landwehr, Gottfried
  Physikalisches Institut der Universität Würzburg,
  D-8700 Würzburg, Fed. Rep. of Germany

Maret, Georg
  Max Planck Institut für Festkörperforschung, Hochfeld-Magnetlabor,
  F-38042 Grenoble-Cédex, France

Miura, Noboru
  The Institute for Solid State Physics, The University of Tokyo,
  Roppongi, Minato-Ku, Tokyo 106, Japan

Montgomery, D. Bruce
  Mass. Institute of Technology, Plasma Fusion Center, Room NW 17-288,
  Cambridge, MA 02139, USA

Motokawa, Mitsuhiro
  Osaka University, Faculty of Science, Toyonaka, Osaka 560, Japan

Roeland, Louis W.

Natuurkundig Laboratorium der Universiteit van Amsterdam,
Valckenierstraat 65, NL-1018 XE Amsterdam, The Netherlands

Schinkel, Cornelis J.

Natuurkundig Laboratorium der Universiteit van Amsterdam,
Valckenierstraat 65, NL-1018 XE Amsterdam, The Netherlands

1. Introduction

Fritz Herlach

With 7 Figures

The use of magnetic fields in experimental physics is discussed in general terms and with a view to the topics treated in this volume. The development of laboratory electromagnets is sketched from the early beginnings in 1820. Opportunities for research with very strong magnetic fields are illustrated by a review of the large magnet laboratories that are open to guest experiments. The state of the art of generating pulsed magnetic fields is briefly reviewed.

1.1 Experiments with Magnetic Fields

The general principle of most experiments in physics consists in disturbing the object under study and then observing the reaction. Combined with theoretical analysis, this can be used to obtain information on the undisturbed state and its dynamics. In practice, the magnetic field is a most suitable means of excitation for conducting a large variety of experiments.

In theory, magnetic force is the part of the force between charged particles that depends on their relative motion. As a relativistic effect, it is comparatively weak, of the order of v^2/c^2 of the electrostatic force (v: particle velocity, c: velocity of light). Yet it is the only known long-range force that can be directly experienced as such by the human observer: it is comparable to the human muscle in strength, range and size of the source. No wonder that "magnetic" has become synonymous in our folklore with everything that is mysteriously attractive, and that magnetic fields provide the muscle for most of our advanced technical apparatus. We may soon be riding in magnetically levitated trains at velocities of more than 500 km/h [1.1], and payloads may be launched magnetically into outer space or from the moon to orbiting space stations [1.2].

The reason for the apparent strength of the magnetic field is the absence of free magnetic charges – the electrostatic force may be orders of magnitude stronger but the free electric charges of opposite polarity always move into positions where they neutralize each other and thus the macroscopic effect is minimized. While an electric field of 3 kV/mm causes breakdown in air at an energy density of 40 J/m^3, a magnetic field of 1000 T has been reproducibly generated representing an energy density of 400 kJ/cm^3. Magnetic monopoles have been theoretically predicted but these are perhaps the most elusive elementary particles that have ever been speculated about: many intensive searches have been conducted but none of these has yielded any conclusive results [1.3].

A quantitative calculation of the magnetic force as a function of the relative motion of several charged particles would be fairly complicated. The magnetic field, intuitively conceived by Faraday, is a more practical description of the phenomenon. Maxwell later discovered that the magnetic field is more of a physical reality as it carries energy density. In short, it is a vector field related in a simple way to the motion of charges in the laboratory, i.e., electrical currents, and it acts on moving charges in such a way that the force is always perpendicular to the direction of motion. A freely moving particle is deflected but it does not acquire energy. Energy is transmitted only when the particle is constrained by a conducting wire or in an atomic orbit, or when it carries a permanent magnetic moment. Strong as it may appear macroscopically, the magnetic field therefore provides a most gentle excitation on the molecular level. With the exception of magnetic materials or conductors subjected to strong pulsed magnetic fields, the samples do not suffer any gross deformation or irreversible damage. This is a particular asset in the study of biological systems. On the other hand, the observable effects are usually quite weak, and some quantum effects occur only above certain threshold fields that may be quite high. This provides the incentive for continued efforts to generate stronger magnetic fields needed for advancing into interesting regions of nonlinear response and extreme quantum limits.

The magnetic field provides a fixed direction of reference in space. In this straightforward application, it is used to align atomic nuclei or molecules against the random influence of thermal motion, in order to study asymmetries in their behavior. Very strong fields are needed to align molecules that do not carry a strong magnetic moment, e.g., large biomolecules.

In elementary particle physics, magnetic fields are used to guide and focus particle beams, and to analyze the momentum of reaction products. For many years, these where the primary applications of large-scale magnets and advanced magnet technology. An entire series of conferences has been dedicated to this topic: the first International Conference on Magnet Technology was held in 1965 at the Stanford Linear Accelerator, and subsequent ones at Oxford (1967), Hamburg (1970), Rome (1975), Bratislava (1978), Karlsruhe (1981) and Grenoble (1983). At the Rome conference, there was a shift in emphasis towards large magnets for nuclear fusion devices, which now has become the leading topic. It is a sign of healthy optimism that huge superconducting coils are now designed and tested for the confinement of plasma in future thermonuclear reactors.

With particles in a solid, the magnetic field interacts in two ways: either with their permanent magnetic moment (Zeeman splitting, magnetic resonances, magnetization) or with their movement (splitting of the energy bands into Landau levels, magnetotransport phenomena, magnetooptics); between these two groups there is no sharp boundary.

Although it does not result in strong macroscopic effects, the splitting of the energy bands into Landau levels is a profound change of the band structure. The distance between the Landau levels is a function of the effective mass, thus they are not equally spaced if the effective mass is a function of energy. This nonlinear effect can be observed in very high fields. Transitions between the Landau levels are known as cyclotron resonance, and the passage of subsequent Landau levels through the Fermi energy results in oscillatory changes in the magnetization (de Haas-van Alphen effect) and in the resistivity (de Haas-Shubnikov effect) as a function of the magnetic field B, with a periodicity in $1/B$. All these effects can be used to determine the shape of the Fermi surface if they are studied as a function of crystal orientation. They can be observed only at sufficiently low temperatures and high magnetic fields, when the splitting of the Landau levels exceeds their thermal broadening. This is called the quantum limit; the "extreme quantum limit" is reached when the lowest Landau level approaches the Fermi energy, and all electrons are condensed into this level. Interesting and surprising effects are now being discovered in this regime, and there has been much speculation about the elusive phenomena of charge density waves and Wigner crystallization. In semiconductors, on which most of this research is carried out, there are additional effects due to impurity states and the phonon spectrum; these can be studied in great detail by different combinations of magnetotransport and magnetooptical measurements.

In the high-temperature region above the quantum limit, the nonresonant magnetooptical effects can still be used: the Faraday effect (rotation of the plane of polarization of light propagating parallel to the magnetic field), the Voigt effect (with light perpendicular to the magnetic field; magnetic birefringence) and the Cotton-Mouton effect which is similar but it is caused by the magnetically induced orientation of molecules.

An interesting object of study has been provided by the transistor industry: the MOSFET which contains a thin layer of electrons surrounded by insulating material. The thickness of this quasi-two-dimensional layer can be continuously varied by means of the gate voltage. This gives rise to a particle-in-a-box quantization in one dimension perpendicular to the plane of the layer. In a magnetic field, we have thus a two-dimensional quantum system with two external quantization parameters that can be varied independently. Extensive research on these and related systems has led to the discovery of another fundamental macroscopic quantum effect: the quantum Hall effect [1.4]. While superconductivity is destroyed by magnetic fields, the quantum Hall effect is brought about by very high magnetic fields. At the present time, this surprising effect is still not fully understood.

In research on superconductivity, strong magnetic fields are used to study critical behavior. Superconductors with high critical fields and related high transition temperatures are needed for the construction of magnets and other technological applications [1.5]. In the present book, we refer only to those special materials with very high critical fields up to 60 T. The presently known materials of this type are not well suited for technological applications; they are brittle and have a low critical current density.

Magnetism is the common label for all phenomena related to the interactions of elementary magnetic moments (electron spin and orbit, also nuclear spin which is ~ 1000 times smaller). Dependent on the crystal structure, a rich variety of differently ordered states and phase transitions is produced by the competition between exchange interaction, magnetic dipole interaction, thermal motion and an external magnetic field. As these interactions are well defined, a magnetic system can be a good model for sophisticated theoretical analysis and statistical physics. In recent years, there has been much progress in the synthesis of new magnetic materials both concentrated and dilute, with interesting and unusual properties, e.g., lower dimensionality in either the exchange interaction or the spatial arrangement of magnetic atoms. Much of the essential information can be derived from measurements of the bulk magnetization as a function of temperature, field and crystal orientation provided the sensitivity and the precision are high enough [1.6]. Other experimental methods are neutron diffraction, specific heat measurements, the Mössbauer effect, Faraday rotation, magnetoresistance, and the different magnetic resonances. Resonance frequency and linewidth give information on the local magnetic and electric fields and their fluctuations at the site of the magnetic dipole in resonance. High magnetic fields are needed to resolve lines that are broadened by high concentrations of impurities, high temperatures or by interactions in undiluted spin systems. The highest fields are of interest for the study of exchange interactions which are of the order megagauss.

Besides the regular conferences on semiconductors and magnetism with their recent satellites dedicated to high magnetic field research (Hakone 1980 [1.7], Grenoble 1982 [1.8] for semiconductors and Osaka 1982 [1.9] for magnetism), there has been a somewhat irregular series of conferences on the generation and applications of very high magnetic fields. A particular milestone was the first of these, held in 1961 at MIT [1.10]. Subsequent conferences were held at Oxford (1963), Grenoble (1966), Nottingham (1969), Grenoble (1974) and MIT (1978) [1.11].

The United States Academy of Sciences has recently conducted a broad survey of research opportunities in high magnetic fields. This included feasibility studies for building stronger magnets, both continuous and pulsed, and proposals for many experiments that can be carried out only in magnetic fields that are stronger (or of longer pulse duration) than those now available. This resulted in the recommendation to develop facilities for generating stronger magnetic fields [1.12].

1.2 Development of Electromagnets for Scientific Research

The history of early magnet development [1.13] originated in 1820 when Oersted discovered the magnetic forces encircling an electric current in a wire. Shortly afterwards, Ampère got the idea to wind the wire into a solenoid; he used this first electromagnet to magnetize steel needles and it inspired him to find the correct interpretation of permanent magnetism. Five years later, Sturgeon built the first horseshoe-type electromagnets with soft iron and demonstrated a greater weight-lifting capability than that of the strongest permanet magnets

Table 1.1. The development of laboratory electromagnets

Year	Field [T]	i.d. [mm]	Gap [mm]	Power [kW]	Weight [10³ kg]		
1894	3.8	3	2	5	0.27	Du Bois	ring magnet
1899	3.7	6	1	16	0.175	Du Bois	half-ring magnet
1911/13	5.5	6	1	16	1.4	Du Bois	heavy yoke, coils close to pole tips
1907/17	2.4	25	20	18	1	Weiss	
	4.6	3.6	2	18	1	Weiss	conical pole pieces
1928/30	3.9	60	40	100	100	Cotton	
	7	3	2	100	100	Cotton	largest Weiss magnet
1932	2.4	100	60	80	16	De Haas	
	6.7	1.8	1.2	80	16		
1935	4.4	60	40	140	37	Dreyfus	strongly conical pole pieces
1951	3	146	25	20	2	Bitter	fully enclosed by iron shell
	4.2	25	6	20	2		
			Axial length		Solenoid magnets without iron		
1914	5	20	20	340		Deslandres and Perot (Paris)	
1939	10	28	40	1,700		Bitter (MIT)	
1946/58	7	40		1,330		Tsai, Gaume (Bellevue)	
1960	9.5	100	80	7,500		Giauque and Lyon (Berkeley) cooling with kerosene	
1966	22.5	32	760	10,000		Montgomery (MIT)	
1969	20	110		1,000		Coles (NASA, Cleveland) pure aluminum, cryogenic	
1972	29.4	50	200	30,000		Carden (Canberra) peak field for 1 s, total pulse 14 s	
1976	25	32		5,200		MIT/Nijmegen hybrid magnet	
1981	29.5	35	192	8,700		MIT hybrid magnet	
1982	25	50		10,000		Grenoble polyhelix	

then known. Faraday used these magnets for his experiments; he discovered the Faraday effect in 1845. In 1846, Ruhmkorff gave the laboratory electromagnet the shape that it has retained essentially until today, with a yoke and two coils on opposite cylindrical pole pieces, where they are most efficient. Weiss improved the efficiency by concentrating the coil windings towards the poles, and by properly dimensioning the yoke to avoid saturation. All further optimization consisted in avoiding saturation in parts of the magnetic circuit, and in reducing the stray fields. A further improvement results from the tapering of the pole pieces inside the coils. Originally proposed by Weiss in 1898, this principle was not applied until 1928, due to the more complicated construction. It was used for the first time in the large magnet built by Cotton and others at Bellevue, the first "National Magnet Laboratory". The Bellevue magnet was not yet well optimized; weighing 100 tons it was much heavier than need be. A much lighter magnet with similar performance was built by Dreyfus in 1935 for the University of Uppsala. The improvement was mainly due to the more strongly tapered pole pieces. The final conclusion of this development is given by *Bitter's* elegant design [1.14]. This magnet with optimized all-around flux return path was manufactured by A.D. Little and is used in many laboratories. In Table 1.1, values for the peak fields of the different magnets are given. These can be divided into two categories: the record values which are achieved by reducing the field volume to the absolute minimum, and the more conservative values for convenient experimentation.

For fields in excess of 5 T, there is no advantage in using iron and it is best left out, which brings us back to the original solenoid, with the difference that several megawatts of power are dissipated! The first of these "brute force" magnets was

Fig. 1.1. Francis Bitter with his original high field magnet

▲

Fig. 1.2. Francis Bitter in the 1960s, in one of the magnet cells of the National Magnet Laboratory at MIT that was later named after him

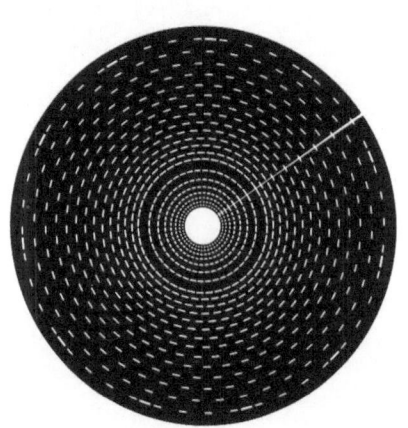

Fig. 1.3. A recent design of a copper plate used to build a Bitter stack. In place of the original round holes for passage of the cooling water, there are now slits for improved cooling

built by Deslandres and Perot around 1914. Unfortunately, this work was disrupted by World War I and not taken up again until 1936, when Bitter initiated the work that finally resulted in setting up his first magnet laboratory in 1939, with an available power of 1.7 MW (Fig. 1.1). Again this was disrupted by war, but in the postwar bonanza of physics several big magnet laboratories have been set up; the Francis Bitter National Magnet Laboratory at MIT still takes the first place among these. This is documented by the comprehensive and well-edited progress report that is published every year. Bitter's original design, a stack of interleaved metal and insulator discs with small holes to provide passages for the coolant, usually demineralized water, is still most widely used (Figs. 1.2–5). In most laboratories, two standard types have evolved with a bore of 5 cm: one type generates ~ 15 T with 5 MW and the other ~ 20 T with 10 MW. The flow rate of the cooling water is of the order of 100 litres per second.

Fig. 1.5 ▶

Fig. 1.4

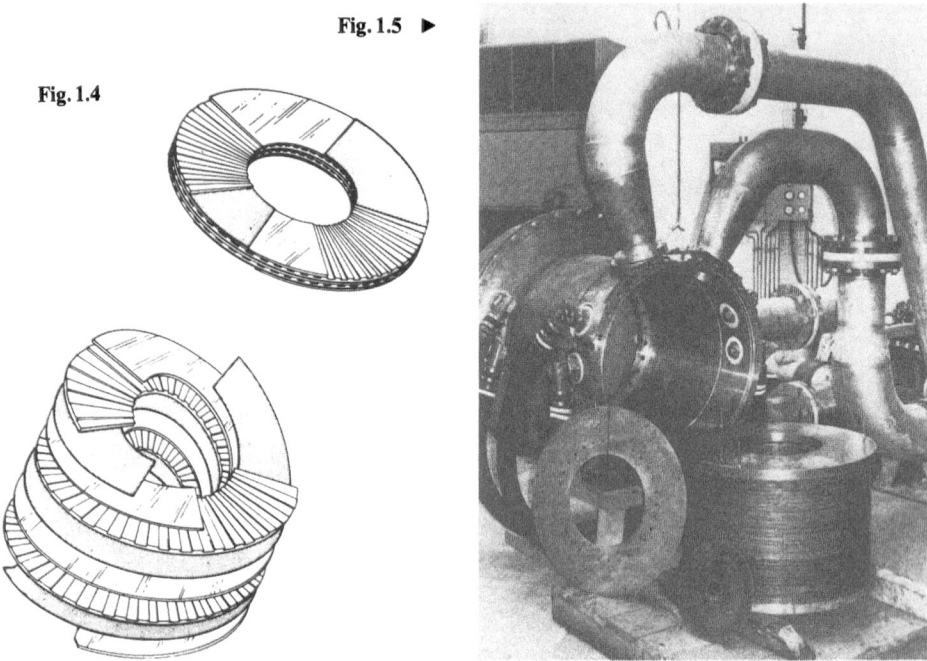

Fig. 1.4. Plate assembly for a Bitter magnet with radial flow of the cooling water (Francis Bitter National Magnet Laboratory)

Fig. 1.5. One of the Bitter magnets of the Grenoble Laboratory

◀ **Fig. 1.6**   **Fig. 1.7**

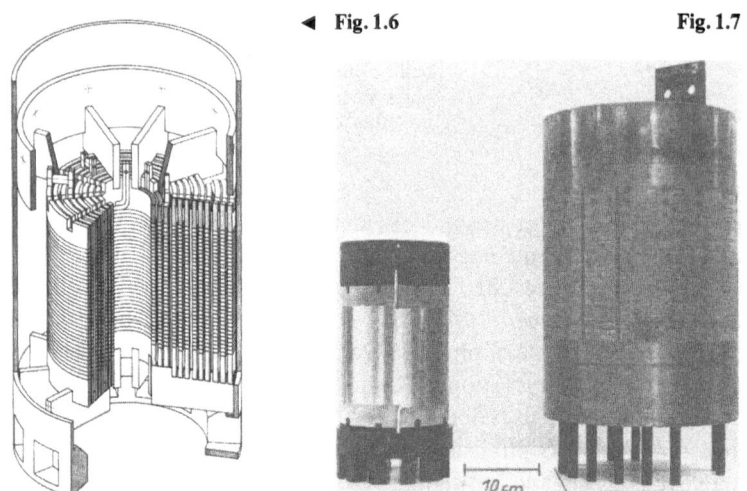

Fig. 1.6. Cutaway drawing of a polyhelix magnet (Schneider-Muntau, Grenoble)

Fig. 1.7. A polyhelix magnet (*left*) and a Bitter magnet which both generate 15 T with 5 MW in a bore of 50 mm diameter (Schneider-Muntau, Grenoble)

Many other types are available for special purposes, in particular split coils with radial access and coils with high homogeneity.

A recent development are "polyhelix" magnets which consist of a number of independent solid helices with the coolant flowing axially in the narrow gaps between the coils [1.15, 16]. These can be fully optimized with regard to current distribution and stress, and consequently they are smaller and more efficient (Figs. 1.6, 7). The development of these magnets has required large technological efforts and their manufacture is quite elaborate. Therefore they are used only where the advantage really counts, such as a high field insert for a Bitter magnet or a hybrid magnet. At Grenoble, a polyhelix magnet has now been in trouble-free operation for almost 1000 hours at field levels of ~ 23 T. This magnet is capable of generating 25 T but it is standard practice in all magnet laboratories to run the magnets at slightly lower fields than the possible maximum in order to increase their lifetime.

At present, fields up to ~ 12 T are available from superconducting laboratory magnets at moderate expense. A ~ 15 T magnet is still quite expensive but certainly competitive with a water-cooled magnet. While the field can be varied more rapidly in a water-cooled magnet, a superconducting magnet is inherently more stable and it is easier to obtain a more homogeneous magnetic field. A 20 T superconducting magnet is technically feasible but still considered excessively expensive. A cost-effective synthesis of available technologies has been made in the hybrid magnet, which consists of a large superconducting magnet generating ~ 10 T with a 20 T water-cooled insert (for the 30 T magnet at MIT [1.17], the precise numbers are 7 T and 23 T). In Table 1.2, estimates are given for the power that would be needed for extending this technology to higher fields [1.18].

Table 1.2. Estimated power requirement for water-cooled and hybrid magnets with a 3 cm bore

Field [T]	Resistive [MW]	Hybrid (10 T) [MW]	Hybrid (20 T) (from superconducting magnet) [MW]
25	18	6	
50	75	45	25
75	165	125	90

1.3 Magnet Laboratory Facilities

The design and operation of a water-cooled high field magnet is a challenging engineering problem [1.19]. In several countries, national magnet laboratories have been set up to provide adequate resources; most of these will accommodate guest experiments by qualified researchers on rather generous terms.

To give a survey on the available facilities, an inquiry among the known magnet laboratories has been conducted. The laboratories that responded to the inquiry are listed in Table 1.3, and some others in Table 1.4. (The response from the NRL was concerned only with scientific research.) The addresses and a few more details are given below.

1.3.1 Francis Bitter National Magnet Laboratory (FBNML)

MIT building NW 14
170 Albany Street
Cambridge, MA 02139, USA.

This is by far the largest and oldest of the existing magnet laboratories [1.20]. The present facility became operational in 1963 and in 1980 it had a permanent staff of 80 technical/administrative and 60 scientific collaborators at a yearly budget of $ 4,000,000. Excellent brochures are available for the information of prospective visitors, and a comprehensive progress report is published every year. The power supply consists of four motor generators with dc output. These are now considered old-fashioned but they have the advantage of inherent stability and an overload capability by a factor of 3.2 for 5 seconds (flywheel-assisted) and 1.2 for 15 minutes. The basic stability is of the order 10^{-3}; for special requirements this can be improved to 10^{-5} by means of a feedback system with a 1.5 kW power amplifier. An important feature is the sweep and modulation of the field in a variety of patterns, including a $1/B$ sweep and polarity reversal by sweeping continuously through zero, with sweep times from 30 s to 2 hours. There are 24 magnet stations to accommodate different experiments on a 15 hour schedule. Besides the standard magnets, there are magnets with smaller and larger bores (e.g., 15 T in 10 cm or 7.7 T in 25 cm with 10 MW), with high homogeneity and with lateral access (split coil). Two hybrid magnets are now in operation, one generating 29 T in a 3.3 cm bore and the other 24.5 T in a 5.4 cm bore. Plans for the future include 15 T superconducting magnets and further development of pulsed magnets (now 50 T for 10 ms, later possibly 70 T).

1.3.2 Grenoble (GREN)

Service National des Champs Magnétiques Intenses (SNCI) and
Max Planck Institut für Festkörperphysik, Hochfeld-Magnetlabor
166 X
F-38042 Grenoble, France.

This laboratory was established by the Centre National de la Recherche Scientifique (CNRS) in 1971 [1.21]. In 1972, a contract of cooperation with the Max Planck Institut für Festkörperphysik at Stuttgart provided for doubling the available power to 10 MW. The facility is now operated jointly by the two organizations. It is somewhat smaller than the FBNML, with a resident staff of 50 in total and 8 magnet stations. However, the available power is the same and the magnets are operated around the clock. Usually, two of the 5 MW magnets

Table 1.3. Magnet laboratories that responded to the inquiry (dc fields only)

Laboratory		FBNML	GREN	WROC	NIJM	BRAU	SEND	OXF	IBAR
Power	[MW]	10	10	6.8 (20)[a]	6	5.5	3.5 (8)	2	
Voltage	[V]	240	335	550 (330)	300	250	350	460	
Subunits		4	4	3 (4)	2	2			
Type[b]		mg	ds	mg (ds)	ds	ds	dm (ds)	mg	
Water-cooled magnets	(≈15 T)	19	11	2	3	3	7	3	
	(≲20 T)	5	5	1	1	(1)			
Hybrid magnets	(25/30 T)	1/1	−/1	−	1/(1)	−	(3/1)	1/−	
Superconducting magnets	(<12 T)	−	7	2			−	3	2
	(>12 T)	(2)	2	1	(1)		2	1	2

[a] Values in brackets refer to units under construction.
[b] Types of power supply (mg: motor-generator, ds: direct with semiconductor devices, dm: direct with mercury rectifier)

Table 1.4. Some other dc magnet laboratories

Laboratory	Power [MW]	Number of magnets	
		~15 T	25 T hybrid
C.E.A. Saclay (France)	3.9		
Kurchatov Institute (Moscow)	6		1
Lebedev Institute (Moscow)	9	3	
Naval Reasearch Laboratory (Washington, DC)	6	3	

are run during the daytime and the 10 MW for a single magnet are available during the night shift. The power supply is all solid state, with a possible stability of 10^{-5} to 5×10^{-6}. The specifications of the magnets are similar to those at the FBNML, with exception of the polyhelix magnets. Much effort has been invested in the design and construction of these, which has been quite successful. One 25 T magnet is in regular operation at the 23 T level, and the hybrid magnet which is now under construction will have a polyhelix insert.

1.3.3 Wroclaw (WROC)

International Laboratory of High Magnetic Fields and Low Temperatures
Prochnika 95
53-529 Wroclaw, Poland.

This laboratory with a permanent staff of 35 (technical/administrative) and 7 (scientific) was set up in 1968 with joint funding from the Academies of Science of Bulgaria, the German Democratic Republic, Poland and the USSR [1.22]. Guest experiments are welcome and an informative brochure is available. The power is provided by motor generators; stability and noise are of the order of 10^{-3}, similar to FBNML. On the whole, this survey has revealed that at the level of technical performance there is not so much difference between laboratories with similar type of equipment. There are now three magnets: 20 T in 2.5 cm, 15 T in 3.4 cm and 10 T in 4.3 cm, respectively. There is also a 70 kJ, $\gtrsim 50$ T pulsed field apparatus with 10 ms pulse duration. Plans for the future provide for an increase of the power to 20 MW with thyristor control (now under construction).

1.3.4 Nijmegen (NIJM)

Fysisch Laboratorium, Katholieke Universiteit Nijmegen
Toernooiveld
NL-6526 ED Nijmegen, The Netherlands.

This laboratory became operational in 1976 and since 1978 it contains the first one of a new generation of large hybrid magnets. [1.23]. This magnet was built under contract at the FBNML. It generates 25 T in a room temperature bore of 3.2 cm with 6 MW and with 8 T from the superconducting coil. The helium consumption is 5 l/h in standby and 10 l/h in high field operation. Power supply and cooling system have interesting design fatures: the coarse regulation is achieved by mechanically rotating the primary coils in the two transformers with diode rectification, followed by passive filters and a transistor bank for fine regulation. The stability is 10^{-4} and the noise/ripple 5×10^{-5} T. The transformers have an overload capability of 12 MW for one minute flat top, once every 15 minutes. The local power company restricts the use of high power in the daytime, according to season. As there is no river in the vicinity, cooling is provided from a reservoir with 150 tons of ice which can be frozen in 16 h and is sufficient for the removal of 18 MWh (with provisions for extension to 36 MWh). There are three conventional 15 T Bitter magnets and a 20 T Bitter magnet. A second hybrid magnet for 30 T in a 32 mm bore will be installed in

1984. The 11 T superconducting coil of this magnet will be cooled with superfluid helium at 1.8 K and atmospheric pressure.

1.3.5 Braunschweig (BRAU)

Hochfeldmagnetanlage der Physikalischen Institute, Technische Universität Braunschweig
Mendelssohnstraße 1 B
D-3300 Braunschweig, Fed. Rep. of Germany.

This is a smaller laboratory with a staff of 3 (technical/administrative) and limited accommodations for guest experiments [1.24]. The thyristor-controlled power supply has a stability of 1.5×10^{-6} which appears to be typical for this type of regulation. Cooling is provided by means of a cooling tower. There are three magnet stations with conventional 16 T Bitter magnets and an 18 T Bitter magnet is under construction.

1.3.6 Sendai (SEND)

The Research Institute for Iron, Steel and Other Metals (RIISOM)
Tohoku University
Sendai 980, Japan.

This laboratory was originally established in 1939 with a Kapitza-type magnet using a turbo-alternator as a flywheel to provide a single pulse in the shape of a half sine wave (27 T in a 10 mm bore) [1.25]. After complete destruction in 1945, the present installation described in Table 1.3 was set up in 1959. The power supply with mercury rectifiers can be overloaded by a factor of 4 for 0.1 s; this has been used to generate a 45 T pulsed field in a 9 mm bore. For some time, the generation of megagauss fields by means of explosive-driven flux compression was pursued. At present, efforts are concentrated on a new project of setting up several advanced hybrid magnets, mainly for research on high field supercon-ductors. Two high field superconducting magnets are now in operation: 16.5 T in a 47 mm bore and 13 T in a 50 mm bore.

1.3.7 Oxford (OXF)

Clarendon Laboratory
Parks Road
Oxford OX1 3PU, United Kingdom.

This laboratory is small compared to some of the others, but it has a long tradition in outstanding scientific research and magnet development. It was established in 1949 [1.26]. One of the first hybrid magnets was developed here and has been in service since 1975. Built with a polyhelix insert, it generates 16 T in a 5 cm bore. Bitter magnets are not used at all; the standard magnets are either stacks of pancakes (Tsai-type) or layer-wound helices and generate 10 T in 5 cm. The stability of the motor-generator power supply is 10^{-4}. The technique of winding polyhelix coils from a continuous length of wire has been developed recently [1.27] and is expected to increase the field in the hybrid magnet from 16.5 T to 20 T in a reduced bore of 3 cm. Other recent developments are a 15.8–16.5 T (at 4.2–3 K) portable superconducting magnet for shared use with a 25 mm cold bore [1.28] and a 10–11 T superconducting magnet with an 8 mm axial gap for radial access. The development of a 50 T pulsed magnet has been initiated.

1.3.8 Ibarakiken (Tsukuba Science City) (IBAR)

Superconducting High-Field Magnet Center, National Research Institute for Metals
1-2-1, Sengen, Sakamura, Niiharigun
Ibarakiken 305, Japan.

This is the home of the most advanced superconducting magnets. Instead of megawatt power supplies, there are helium liquefiers with a capacity of 48 l/h, and the consumption of electrical power is of the order of 100 kW. There is a 7 T split coil with a 2.5 cm gap and a 6 cm cold bore (4.2. K), a 10 T magnet with a 3 cm cold bore and a 17.5 T magnet with a 3.1 cm bore. The outer coil of this large magnet generates 13.5 T in a cold bore of 16 cm diameter. A 20 T superconducting magnet is under construction. This laboratory also has a pulsed magnet: a 125 kJ capacitor bank generates 25–35 T in a 1.2–2.7 cm bore with a pulse duration of 10–15 ms. This is useful for the study of the critical behavior of superconducting materials at higher fields.

1.3.9 Laboratories for Pulsed Magnetic Fields

Table 1.5 is a list of major laboratories where pulsed magnetic fields are now used for scientific research. Most of these will accept visitors on a basis of mutual collaboration. This is because most pulsed field generators are still designed and operated in close relation with experiment, in particular those for the highest fields. Therefore, in Chap. 6 an extensive description of the design and operation of pulsed field generators is given. As a short summary, it can be said that fields close to 50 T are now easily available in a 20 mm bore (exceptionally, 40 mm) with a pulse duration in the range 1 ms–1 s. Efforts are made to push this towards the megagauss limit; in the difficult range between 50 and 100 T the pulse duration drops to ~0.1 ms and the volume will be smaller. Above 100 T, four methods are in use for experiments. The capacitor discharge into single-turn coils has been used for experiments up to 150 T in a 5 mm bore and with ~3μs pulse duration. This technique shows promise for reaching at least 300 T for experimental applications. Single-turn coils powered by explosively driven generators have reproducibly generated ~200 T with a slightly longer pulse duration. Electromagnetically driven implosions are now in the 200 T range in volumes of a few millimetres diameter and for a fraction of a microsecond. This will soon be extended to much higher values with the heavy equipment that is now installed at the Institute for Solid State Physics (University of Tokyo).

Fields up to 1000 T are now available for experimentation from explosive-driven flux compression, but only at laboratories that are not easily accessible because they are involved in highly classified research.

Table 1.5. Major pulsed field laboratories

	Energy [kJ]	Voltage [kV]	Methods[a]
Tokyo University	285	30	⎫ Electromagnetic flux compression
(The Institute for	77*	3.3*	⎭ 280 T for about 1 μs
Solid State Physics)	1000	40	⎫ Multimegagauss electromagnetic
	4000	40	⎬ flux compression (under development)
	1500*	10	⎭ * designates capacitor banks for initial flux
	200	5/10	⎫ wire-wound coils, 45 T in 20 ms
	32	3.3	⎭
	100	40	150–250 T from a single turn coil
Osaka University	250	20	Free-standing helices made from maraging steel
	1250	26.6/40	40–70 T for less than a millisecond
Toulouse	1250	10	Wire-wound coil, 40 T for ~1 s
Leuven University	1000	5/10	Wire-wound coil, solid helix and single turn (under construction)
Moscow (Kurchatov Instit.)	190	5	Wire-wound coil (wire with superconducting core for extra strength) 50 T for 15 ms
Amsterdam	1500	0.66	8.2 MW directly from the mains, thyristor-controlled, 40 T for ~1 s

[a] The time intervals refer to the total pulse duration, e.g., a half period of a sine wave

References

1.1   K.Oshima: IEEE-Trans. **MAG-17**, 2338 (1981)
1.2   H.Kolm, K.Fine, F.Williams, P.Mongeau: Proc. 2nd IEEE Intern. Pulsed Power Conference, Lubbock, Texas (1979) p. 42
1.3   Phys. Today **35**, No. 6, 17 (June 1982)
1.4   K. von Klitzing, G.Dorda, M.Pepper: Phys. Rev. Lett. **45**, 494 (1980)
1.5   General references on superconducting magnets:
      M.N.Wilson: *Superconducting Magnets* (Clarendon, Oxford 1983), and S.Foner, B.Schwartz (eds.): *Superconducting Machines and Devices* (Plenum, New York 1974)
1.6   R.Pauthenet: In [Ref. 1.9, p. 77]
1.7   S.Chikazumi, N.Miura (eds.): *Physics in High Magnetic Fields, Springer Ser. Solid-State Sci.*, Vol. 24 (Springer, Berlin, Heidelberg 1981)
1.8   G.Landwehr (ed.): *Application of High Magnetic Fields in Semiconductor Physics*, Lecture Notes in Physics Vol., 177 (Springer, Berlin, Heidelberg 1983)
1.9   M.Date (ed.): *High Field Magnetism* (North-Holland, Amsterdam 1983)
1.10 H.Kolm, B.Lax, F.Bitter, R.Mills (eds.): *High Magnetic Fields* (MIT Press/Wiley, New York 1962)
1.11 R.L.Aggarwal, A.F.Freeman, B.B.Schwartz (eds.): J. Magn. Magn. Mater. **11**, 1–437 (1979)
1.12 *High Magnetic Field Research and Facilities* (National Academy of Sciences, Washington 1979)

Available from: Solid State Sciences Committee, National Research Council, 2101 Constitution Avenue, Washington, DC-20418, USA

1.13   D.De Klerk: *The Construction of High Field Electromagnets* (Newport Instruments 1965)

1.14   F.Bitter, F.E.Reed: Rev. Sci. Instrum. **22**, 171 (1951)

1.15   M.Wood: In Proc. Intern. Conf. The Application of High Magnetic Fields in Semiconductor Physics, ed. by J.F.Ryan, Clarendon Laboratory, Oxford (1978) p. 67

1.16   H.-J.Schneider-Muntau: IEEE Trans. **MAG-17**, 1775 (1981)

1.17   M.J.Leupold, J.R.Hale, Y.Iwasa, L.G.Rubin, R.J.Weggel: IEEE Trans. **MAG-17**, 1779 (1981)

1.18   D.B.Montgomery: In [Ref. 1.11, p. 293]

1.19   D.B.Montgomery: *Solenoid Magnet Design* (Wiley Interscience, New York 1969)

1.20   L.G.Rubin, R.J.Weggel, M.J.Leupold, J.E.C.Williams, Y.Iwasa: In [Ref. 1.9, p. 249]

1.21   H.-J.Schneider-Muntau, J.C.Picoche, P.Rub, J.C.Vallier: In [Ref. 1.8, p. 531]
       J.C.Picoche, P.Rub, J.C.Vallier, H.-J.Schneider-Muntau: In [Ref. 1.9, p. 257]

1.22   K.Trojnar, B.Drys, M.Luczak: J. Phys. **E11**, 925 (1978)

1.23   K. van Hulst, C.J.M.Aarts, A.R. de Vroomen, P.Wyder: In [Ref. 1.11, p. 317]

1.24   E.Justi, D.Schneider: Physikalische Bl. **29**, No. 2, 59 (1973)

1.25   S.Miura, A.Hoshi, Y.Nakagawa, K.Noto, K.Watanabe, Y.Muto: In [Ref. 1.9, p. 331]

1.26   M.F.Wood: In [Ref. 1.10, p. 387]
       P.A.Hudson, H.Jones, H.M.Whitworth: In [Ref. 1.29, p. Cl-35]

1.27   P.A.Hudson, H.Jones, H.M.Whitworth: In [Ref. 1.29, p. Cl-55]

1.28   P.A.Hudson, H.Jones: Cryogenics **19**, 397 (1979)

1.29   Proc. 8th International Conference on Magnet Technology, Grenoble Sept. 5–9, 1983 (Suppl. Journal de Physique Cl-1984)

Additional Reference

Recently, rather unusual and surprising effects of pulsating low-strength magnetic fields (typically a few gauss at 50 Hz square wave) have been reported. Although this is on the fringe and not directly related to the topic of this volume, readers may find it of interest. A short report "Magnetic Fields in Biology" is available from Professor José M. R. Delgado, Centro Ramón y Cajal, Madrid 34, Spain. This contains a list of further references.

2. Quantum Transport Phenomena in Semiconductors in High Magnetic Fields

János Hajdu and Gottfried Landwehr

With 56 Figures

According to quantum mechanics the transverse components of the gauge invariant momentum in the presence of a magnetic field are non-commuting quantities. This fact is responsible for a variety of equilibrium and non-equilibrium (transport) macroscopic quantum effects. Examples are: the diamagnetism of conduction electrons, the oscillations of the magnetisation as a function of the magnetic field (de Haas-Van Alphen effect), similar oscillations of the electric conductivity (Shubnikov-de Haas effect) and the recently discovered quantum Hall effect in two-dimensional semiconducting systems (v. Klitzing). Due to the quantisation by the magnetic field the concept of the gauge invariant momentum space and the momentum distribution function breaks down. Since the discovery of magnetic quantum effects about 50 years ago it has been a continuous challenge for theoreticians to set up a quantum transport theory which is able to explain all the details of the observations. The motivation for this challenge is not only the interest in fundamental research; rather, the analysis of magnetic quantum effects is the most important method by which the electronic properties of semiconductors can be determined.

In this chapter we give an introduction to the quantum transport theory and describe a great variety of experimental results. The main topics are: Shubnikov-de Haas effect, cyclotron resonance and thermomagnetic effects in 3 and 2 dimensions. Our aim is to guide the reader through the basic concepts to current research. To this end special emphasis is put on the quantum Hall effect.

2.1 Introduction

For a long time the highest stationary magnetic fields available for solid state experiments were of the order of 1 T. For pure metals at low (helium) temperatures such fields can be considered to be "high fields" since the radius of the corresponding cyclotron orbits of the conduction electrons is smaller than the mean free path. Once this condition is satisfied *quantum effects* brought about by the periodic cyclotron motion are observable. In the case of semiconductors, the regime of high fields sets in at much higher values. Therefore, extensive high field research on semiconductors began only in the late fifties, after magnetic fields of 10 T and more were available at several laboratories. It was widely recognised at that time that the study of quantum

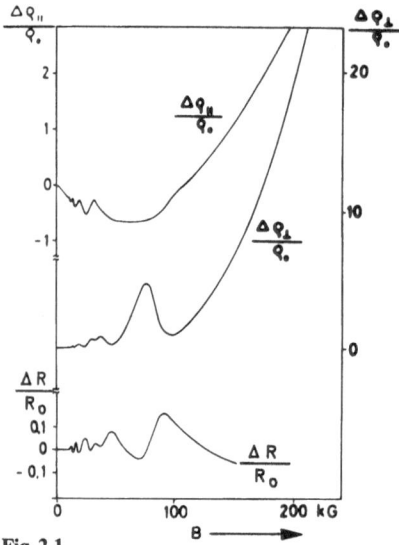

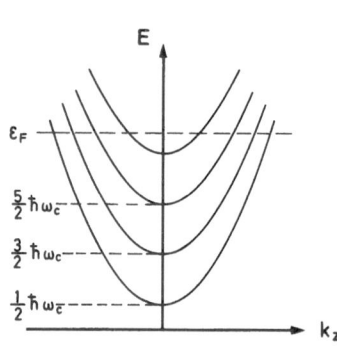

Fig. 2.1   Fig. 2.2

Fig. 2.1. Relative change of the resistivity of n-type InSb ($n = 6 \times 10^{16}$ cm^{-3}) in longitudinal (*upper curve*) and transverse magnetic fields (*middle curve*). *Lower curve*: relative change of Hall coefficient $\Delta R/R_0$ as a function of the magnetic field. All data were taken at 4.2 K [2.2]

Fig. 2.2. The energy spectrum of a free electron in a magnetic field as a function of the longitudinal wave number k_z

effects can yield detailed information about the electronic structure of semiconductors.

The first high field quantum effect was observed much earlier: in 1930 *Shubnikov* and *de Haas* investigated the properties of the semimetal bismuth at low temperatures and discovered an oscillatory change of the resistivity as a function of the magnetic field [2.1] – an effect now named after these researchers. Somewhat later, similar oscillations in the magnetic susceptibility of Bi were observed by de Haas and van Alphen. Figure 2.1 shows – as an example of more recent observations – the Shubnikov-de Haas (SdH) effect in semiconducting n-type indium-antimonide [2.2]. Both the transverse and the longitudinal resistivity as well as the Hall coefficient show oscillatory behaviour. Beyond a certain value of the magnetic field the resistivities cease to oscillate and increase monotonically.

The origin of the SdH oscillations can easily be understood. The motion of a free electron in a homogeneous magnetic field $\boldsymbol{B}$ is a superposition of a free motion parallel to the magnetic field axis and a circular (cyclotron) motion in a plane perpendicular to the field. According to quantum mechanics the energy of the free motion is continuous and, putting $\boldsymbol{B}$ parallel to the z axis, equal to $\hbar^2 k_z^2/2m$, whereas the energy of the circular motion (being equivalent to harmonic oscillations) is quantised, such that the energy spectrum consists of discrete equidistant levels given by $\hbar\omega_c(v+1/2)$, $v=0,1,2,\ldots$ (Fig. 2.2). Here $\omega_c=eB/m$ is the cyclotron frequency. Thus, the possible energies of the three-dimensional motion are [2.3]

$$\varepsilon^0_{vk_z}=\hbar\omega_c\left(v+\frac{1}{2}\right)+\frac{\hbar^2k_z^2}{2m}. \tag{2.1.1}$$

For simplicity the spin has been omitted. At sufficiently low temperatures N independent electrons populate all quantum states with energies (2.1.1) up to the Fermi energy ε_F. Obviously, a highest quantum number v_{max} exists satisfying the condition $\hbar\omega_c(v_{max}+1/2)\leq\varepsilon_F\leq\hbar\omega_c(v_{max}+3/2)$. With increasing magnetic field v_{max} drops step by step. The total depopulation of all energy eigenvalues with a certain fixed v occurs at a critical magnetic field B_v determined by $\varepsilon_F=\varepsilon^0_{v,0}=\hbar eB_v(v+1/2)/m$. Assuming that ε_F remains constant as B increases, the process of depopulation is periodic on the $1/B$ scale with the period

$$\Delta\left(\frac{1}{B}\right)=\frac{1}{B_{v+1}}-\frac{1}{B_v}=\frac{\hbar e}{m}\frac{1}{\varepsilon_F}. \tag{2.1.2}$$

This periodic depopulation of energy levels (with $k_z=0$) gives rise to the observed SdH oscillations in the resistivity and to similar phenomena in all electronic (equilibrium and transport) properties of metals and semiconductors. In fact, ε_F depends on the magnetic field in an oscillatory manner as well; this dependency is, however, much weaker than in the case of the resistivity (Sect. 2.3).

To see all this in a slightly more formal way, it is sufficient to refer to the density of quantum states on the energy scale per unit volume – a quantity which has a crucial influence on all macroscopic electronic properties. According to *Landau* [2.3], the density of states of free electrons in a magnetic field is given by

$$n(\varepsilon)=\frac{\sqrt{2m}eB}{(2\pi\hbar)^2}\sum_v\frac{1}{\sqrt{\varepsilon-\hbar\omega_c(v+1/2)}} \tag{2.1.3}$$

(Sect. 2.2). For fixed $\varepsilon=\varepsilon_F=$ const this quantity diverges periodically on the $1/B$ scale at the critical fields B_v.

The singularities of the density of states are rather "academic". In a real solid the interaction of the electrons with lattice imperfections gives rise to a finite lifetime τ which is equivalent to a level broadening $\Gamma=\hbar/\tau$ [2.4]. In most cases this lifetime is of the same order of magnitude as the average collision time appearing in the conductivity formula for vanishing magnetic field, $\sigma_0=e^2n\tau/m$, where n is the electron concentration. Assuming τ to be constant, the influence of the collisional damping on the density of states can be calculated by folding the

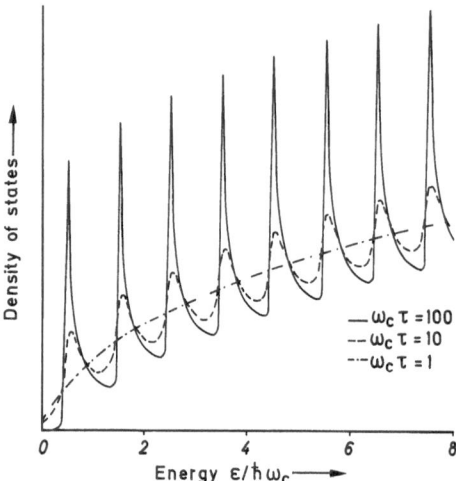

Fig. 2.3. The density of states of an electron in a magnetic field as a function of the energy (in units of $\hbar\omega_c$) for different values of the damping parameter Γ, corresponding to $\omega_c\tau = 100$ (——), 10 (- - -) and 1 (-·-) [2.5]

free electron expression (2.1.3) with the Lorentzian [2.4]

$$D(\varepsilon - \varepsilon') = \frac{1}{\pi} \frac{\Gamma/2}{(\varepsilon - \varepsilon')^2 + \Gamma^2/4}, \tag{2.1.4}$$

$$n_\Gamma(\varepsilon) = \int D(\varepsilon - \varepsilon') n(\varepsilon') d\varepsilon'. \tag{2.1.5}$$

The results are shown for several values of the parameter $\omega_c\tau$ in Fig. 2.3.

Quantisation by the magnetic field requires the cyclotron frequency to be much higher than the collision frequency $1/\tau$,

$$\omega_c\tau \gg 1. \tag{2.1.6}$$

In other words, the separation $\hbar\omega_c$ of adjacent energy levels with $k_z = 0$, must be much larger than the level broadening Γ. With the zero field resistivity $\varrho_0 = 1/\sigma_0$ and the Hall resistivity $\varrho_H = B/en$, the inequality (2.1.6) yields the macroscopic condition

$$\varrho_H \gg \varrho_0. \tag{2.1.7}$$

Only if condition (2.1.6) [or (2.1.7)] is satisfied will the Shubnikov-de Haas oscillations be readily observable. For similar reasons the level spacing must be large compared to the thermal energy $k_B T$

$$\hbar\omega_c \gg k_B T. \tag{2.1.8}$$

No further oscillations occur once the magnetic field has reached the quantum limit defined by $1/2\hbar\omega_c < \varepsilon_F < 3/2\hbar\omega_c$ (Fig. 2.1). In this regime the low temperature resistivity is a monotonically increasing function of the magnetic field and can be well approximated by a power law

$$\varrho \sim B^{\alpha}. \tag{2.1.9}$$

A typical value is $\alpha = 2$. Some model calculations assuming elastic scattering of the electrons by localised impurities were successful in explaining the experimental results. They indicated that the value of α depends both on the band structure and the range of the impurity potential (as compared to the magnetic length $l = \sqrt{\hbar/eB}$, which is the radius of the classical cyclotron orbit corresponding to the zero point energy $\varepsilon_0 = \hbar\omega_c/2$).

The application of a high magnetic field to a three-dimensional system leads to quantisation of the electron motion perpendicular to the magnetic field, whereas the motion parallel to the field is unaffected. A two-dimensional electronic system, however, can be fully quantised by a strong magnetic field resulting in very spectacular magnetotransport properties. In the last 15 years various two-dimensional systems have been realised experimentally. One of these is the Metal-Oxide-Semiconductor Field Effect Transistor, the MOSFET. By means of a surface condensor on top of a silicon substrate, an inversion layer can be induced with a thickness of the order of 100 Å or less. The high electric field at the surface results in boundary quantisation leading to electrical subbands. The application of a high magnetic field perpendicular to the surface causes additional Landau quantisation. Under favourable conditions the density of states consists of a sequence of spikes at the energy of the Landau levels with the density of states becoming negligibly small in between. This is reflected in the transverse conductivity, as can be seen in Fig. 2.4 where data obtained by *Fowler* et al. [2.6] have been reproduced. The conductivity has a pronounced peaked structure and becomes very small at certain gate voltages. The authors quoted above were the first to study the Shubnikov-de Haas effect in the silicon MOS system in 1966.

Recent investigations on (100) silicon MOSFETs with improved mobility led *von Klitzing* et al. [2.7] to the discovery of steps in the Hall resistivity ϱ_{xy} at the particular values $h/(e^2 i)$, $i = 1, 2, \ldots$ (quantum Hall effect). In the first experiment the precision of these values was already better than 10^{-5}. More recently [2.8], the experimental error has decreased to the order of 10^{-7}. It has been envisaged to maintain the resistance unit of the SI system (the Ohm) by comparison with the quantised Hall resistivity. New results on the quantum Hall effect obtained by *Englert* [2.9a] at fields up to 20 T and $T = 0.4$ K are shown in Fig. 2.5. The steps in the Hall resistivity – called plateaux – are well established and accompanied by vanishing transverse resistivity ϱ_{xx} over wide ranges of the gate voltage. Therefore, in these ranges, the Hall conductivity extracted from the observations is

$$\sigma_{xy} = -i\,\frac{e^2}{h} \tag{2.1.10}$$

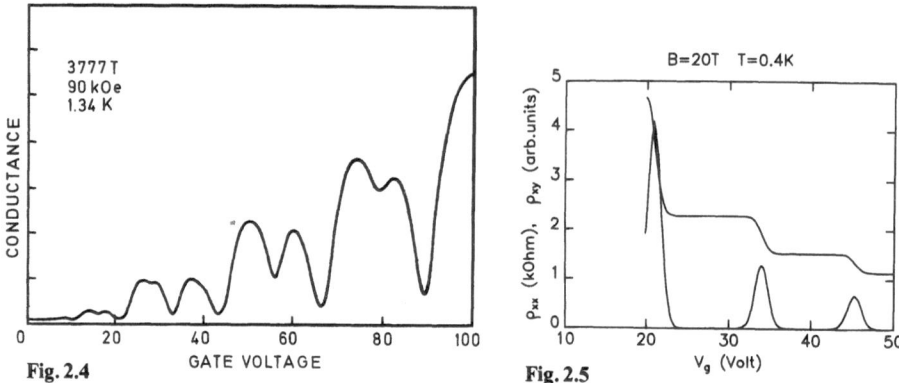

Fig. 2.4. Change of conductance of a Si(100)MOSFET at a magnetic field of 90 kOe as a function of the gate voltage [2.6]

Fig. 2.5. Resistivity ϱ_{xx} and Hall resistance ϱ_{xy} of a Si(100)MOSFET at $T = 0.4$ K in a magnetic field of 20 T as a function of the gate voltage V_g [2.9a]

and the transverse conductivity σ_{xx} vanishes. With reference to the vanishing conductivity in the plateaux regimes of σ_{xy}, the quantum Hall effect has been attributed to localisation of conduction electrons. The fundamental character of the quantum Hall effect – reflected by (2.10) – has attracted the interest of both theoreticians and experimentalists (cf. e.g. [2.10]). Although various explanations of the effect have been offered, a satisfactory predictive theory is still missing.

The previous examples demonstrate that magnetoquantum effects in transport phenomena (short quantum magnetotransport) is an exciting and challenging subject. The experimental data on quantum magnetotransport have enormously increased during the last decade. One reason for this is the widespread availability of inexpensive superconducting magnets capable of generating fields up to about 10 T. Another reason is the progress in semiconductor physics and technology and the need to obtain more precise information about the band structure of and the scattering processes in technically relevant or promising materials. Shubnikov-de Haas experiments, for instance, are now employed more or less routinely to determine the carrier concentration in two-dimensional electronic systems.

Whereas the experiments in quantum magnetotransport are fairly easy to perform, the theory is rather complicated. To achieve practically useful results it is necessary to deal with simplified models. It is usually unavoidable to make certain assumptions about the scattering processes and sometimes to neglect details of the band structure (which may be relevant in reality). In this chapter these restrictions are less significant because our emphasis is on the explanation of the fundamentals. We do not intend to review the subject of quantum magnetotransport in its full breadth. Our aim is rather to give a conceptual introduction which takes into account both important well-established matters and recent developments, showing some present limitations as well as future possibilities. Fortunately, good review articles exist to which the reader is referred. Here we only mention the article by *Roth* and *Argyres* [2.11] for 3-dimensional systems, and the most comprehensive report on 2-dimensional systems by *Ando* et al. [2.12]. Other noteworthy reviews are listed in the Selected Bibliography.

The present review is restricted to impurity scattering. The magnetophonon effect, – the resonant scattering of electrons by optical phonons – has been excluded, although this effect has given useful band structure information on several semiconductors. This is treated in Section 6.2.2 (for pulsed fields) and in the comprehensive reviews by *Harper* et al. [2.13] and *Peterson* [2.14]. Hot electron effects in quantum magnetotransport have been omitted, in spite of their increasing importance. The interested reader is referred to a detailed report by *Bauer* [2.15a] and conference proceedings [2.15b]. We shall not deal with the field of Landau emission which has recently been reviewed by *Gornik* [2.9b].

Quantum magnetotransport includes frequency-dependent conductivity and, therefore, overlaps with magneto-optics. Cyclotron resonance studies of semiconductors started in the fifties and gave a wealth of band structure information [2.16]. Although it is relatively easy to determine effective masses by this technique in pure semiconductors, complications do arise when scattering by impurities becomes relevant. Since the analysis of line shape is then by no means straightforward, we shall devote some attention to this matter.

Another area we shall touch briefly on is that of the thermomagnetic effects. The experimental study of these phenomena, especially at low temperatures, is usually more difficult and more time-consuming than resistivity measurements. The information one may obtain is, however, sometimes hard to get by other means. In this area further theoretical and experimental investigations are needed.

The experiments discussed briefly in this introduction by no means exhaust the rich variety of quantum magnetotransport phenomena. They are characteristic examples which we treat – among a few others – in the following sections in more detail. We must apologise to all researchers whose contributions to the field have not been presented in an adequate manner, or not even mentioned at all, in spite of their current importance.

2.2 Lattice Electrons in Strong Magnetic Fields

2.2.1 Electrons in an Ideal Lattice

Most electronic properties of a normal metal or semiconductor can be explained by the model of nearly independent electrons moving in a periodic lattice potential and interacting with some sort of lattice imperfections. In a perfect lattice the single-electron energy eigenstates are characterised by a wave vector

$k = (k_x, k_y, k_z)$ and a band index j. The energy spectrum shows a band structure with a certain dispersion $\varepsilon = \varepsilon_j(p)$, $p = \hbar k$ in each band. At zero temperature all allowed single-electron states below the Fermi energy ε_F are occupied. Therefore, by weak perturbation, only excitations at the Fermi surface $\varepsilon(p) = \varepsilon_F$ are possible. Assuming that the bands are well separated, we can omit all bands except the conduction band (the one in which the Fermi energy is located) and consider only conduction electrons which fill the conduction band up to the Fermi energy.

The quantum mechanical problem of a (conduction) electron moving in a periodic potential, and perhaps some other potential $U(r)$, *and* an external homogeneous magnetic field B is exceedingly complicated. According to *Peierls* this problem can be approximated by the eigenvalue problem

$$H|\alpha\rangle = \varepsilon_\alpha |\alpha\rangle, \quad H = \varepsilon(P) + U(r), \tag{2.2.1}$$

where $\varepsilon(p)$ is the conduction band energy function,

$$P = p + eA(r) \tag{2.2.2}$$

is the gauge invariant momentum and $A(r)$ is the vector potential $B = \mathrm{curl}\ A$ [2.17].

It is crucial that the components of P do not all commute but

$$P \times P = -i\hbar eB. \tag{2.2.3}$$

It is due to this relation that the concept of a classical quasi-continuous P space is limited to weak magnetic fields.

The equations of motion which correspond to the Hamiltonian H are

$$v = \dot{r} = \frac{i}{\hbar}[H, r] = \frac{\partial \varepsilon}{\partial P}, \quad \dot{p} = \frac{i}{\hbar}[H, p] = -\frac{\partial(\varepsilon + U)}{\partial r}. \tag{2.2.4}$$

They yield

$$\dot{P} = -e\frac{\partial \varepsilon}{\partial P} \times B - \frac{\partial U}{\partial r}. \tag{2.2.5}$$

In the following we put $B = (0, 0, B)$. Then the only non-vanishing commutator is

$$P_x P_y - P_y P_x = -i\hbar eB \tag{2.2.6}$$

and by (2.2.5)

$$\dot{P}_x = -eB\frac{\partial \varepsilon}{\partial P_y} - \frac{\partial U}{\partial X}; \quad \dot{P}_y = eB\frac{\partial \varepsilon}{\partial P_x} - \frac{\partial U}{\partial y}; \quad \dot{P}_z = \dot{p}_z = -\frac{\partial U}{\partial z}. \tag{2.2.7}$$

It often proves to be advantageous to introduce center coordinates (X, Y) and relative coordinates (ξ, η) [2.18] defined by

$$x = X + \xi, \quad y = Y + \eta$$
$$\xi = \frac{1}{eB} P_y, \quad \eta = -\frac{1}{eB} P_x. \tag{2.2.8}$$

They obey the commutation relations

$$[X, Y] = i l^2, \quad [\xi, \eta] = -i l^2, \tag{2.2.9}$$

where again $l^2 = \hbar/eB$. By (2.2.7)

$$\dot{X} = \frac{1}{eB} \frac{\partial U}{\partial y}, \quad \dot{Y} = -\frac{1}{eB} \frac{\partial U}{\partial x}. \tag{2.2.10}$$

Thus X, Y and p_z are constants of motion for vanishing potential U. Consequently the classical electron orbit in P space is the intersection curve of a surface of constant energy, $\varepsilon(P) = \text{const}$, and a plane of constant longitudinal momentum, $p_z = \text{const}$.

The most complicated energy function for which (2.2.1) can be solved in a closed form is an ellipsoid

$$\varepsilon(p) = \frac{p_x^2}{2m_1} + \frac{p_y^2}{2m_2} + \frac{p_z^2}{2m_3} \tag{2.2.11}$$

characterised by the effective masses m_i. Also sets of tilted ellipsoids allowed by the lattice symmetry can be handled. In the case of more complex energy functions, the semiclassical quantisation suggested by *Onsager* [2.19] can be used as a first approximation. In complete analogy to the Bohr-Sommerfeld phase integral quantum condition of the "old quantum mechanics", the semiclassical approximation of (2.2.1) (for $U=0$) takes approximately account of the commutation relation (2.2.6) by retaining only those closed orbits in P space for which

$$\oint P_x dP_y = 2\pi \hbar eB(\nu + \gamma), \tag{2.2.12}$$

with $\nu = 0, 1, 2, \ldots$ and $0 < \gamma < 1$ is satisfied, i.e. orbits which surround an area

$$A(\varepsilon, p_z) = 2\pi \hbar eB(\nu + \gamma). \tag{2.2.13}$$

This relation determines the energy spectrum $\varepsilon_\nu(p_z)$ in a semiclassical approximation. According to (2.2.13) the possible areas are quantised in units of

$$\Delta A = 2\pi \hbar eB. \tag{2.2.14}$$

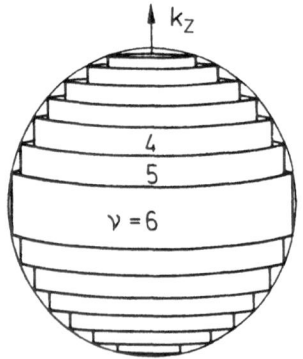

Fig. 2.6. Quantised cylinders in **k** space. States on seven cylinders are occupied

Defining the level spacing $\Delta\varepsilon = \hbar\omega_c$ for fixed p_z by $\Delta A \simeq (\partial A/\partial\varepsilon)\,\Delta\varepsilon$, we obtain $\omega_c = eB/m_c$ with the cyclotron mass m_c defined by

$$m_c = \frac{1}{2\pi}\frac{\partial A}{\partial\varepsilon}. \tag{2.2.15}$$

For the spherical energy function

$$\varepsilon = \frac{p^2}{2m} \tag{2.2.16}$$

with some effective mass m

$$A(\varepsilon, p_z) = \pi(p^2 - p_z^2) = \pi(2m\varepsilon - p_z^2) \tag{2.2.17}$$

and (2.2.13) leads to the exact Landau levels (2.1.1) for $\gamma = 1/2$. Furthermore, by (2.2.15), $m_c = m$. Whereas in the ground state for $\boldsymbol{B} = 0$ the electrons occupy all states in momentum space with $p_i = \pm(2\pi/L_i)\hbar n_i$ ($i = x,y,z$, n_i integer, $L_x L_y L_z = \mathrm{Vol}$) within a sphere with the radius corresponding to the maximum energy ε_F, for $\boldsymbol{B} \neq 0$ the occupied states lie on a series of cylinders about the p_z axis, with cross-sectional areas $(2\nu + 1)\pi e\hbar B$ (Fig. 2.6). With increasing magnetic field these areas grow and the heights of the cylinders decrease. The maximum area $A_0 = A(\varepsilon_F, p_z = 0) = \pi p_F^2 = 2\pi m\varepsilon_F$ is reached at a critical field B_ν given by

$$\frac{1}{B_\nu} = 2\pi\hbar\frac{e}{A_0}(\nu + \gamma) \tag{2.2.18}$$

with $\gamma = 1/2$. For still larger fields the νth cylinder will be depopulated.

Obviously, by (2.2.18), for a Fermi surface of arbitrary shape, we can define critical values of the magnetic field corresponding to stationary (i.e. extremum) values A_0 of the area $A(\varepsilon_F, p_z)$ and obtain for the period of depopulation on the

1/B scale

$$\Delta\left(\frac{1}{B}\right) = 2\pi\hbar\,\frac{e}{A_0}.$$ (2.2.19)

Thus, the periods of the Shubnikov-de Haas oscillations determine a stationary intersection area of the Fermi surface $\varepsilon(\boldsymbol{p}) = \varepsilon_F$ and a plane perpendicular to the magnetic field axis.

For the spherical energy function (2.2.16) the replacement Hamiltonian is formally the same as for free electrons in external fields

$$H = \frac{1}{2m}\,(\boldsymbol{p} + e\boldsymbol{A})^2 + U.$$ (2.2.20)

Choosing the asymmetric gauge $\boldsymbol{A} = (0, Bx, 0)$, we have by (2.2.2, 8)

$$p_y = -\hbar X/l^2 = -eBX,$$ (2.2.21)

and for $U = 0$ (2.2.20) takes the form

$$H_0 = \left[\frac{p_x^2}{2m} + \frac{m\omega_c^2}{2}\,(x - X)^2\right] + \frac{p_z^2}{2m}.$$ (2.2.22)

This Hamiltonian describes an harmonic oscillator in x direction with centre at X and a free motion along the z axis. Consequently, imposing the normalisation condition

$$L_y L_z \int\limits_{-\infty}^{\infty} |\psi_\alpha|^2 dx = 1$$ (2.2.23)

and the periodic boundary conditions

$$\psi_\alpha(y + L_y) = \psi_\alpha(y), \quad \psi_\alpha(z + L_z) = \psi_\alpha(z),$$ (2.2.24)

the solution of the eigenvalue problem

$$H_0\psi_\alpha = \varepsilon_\alpha^0\psi_\alpha$$ (2.2.25)

is given by the energy levels (2.1.1) and the wave function

$$\psi_\alpha(\boldsymbol{r}) = \frac{1}{\sqrt{L_y L_z}}\,\exp\,[i(k_y y + k_z z)]\,\varphi_\nu(x - X),$$ (2.2.26)

where $\alpha = (\nu, k_y, k_z)$, $k_i = \pm 2\pi n_i/L_i$ $(i = x, y)$, $n_i = 0, 1, 2, \ldots$ and $\varphi_\nu(x - X)$ denotes normalised oscillator wave functions centred around X [2.3].

For free electrons the spin gives rise to an additional term to the energy eigenvalues

$$\varepsilon_{\alpha s}^0 = \varepsilon_\alpha^0 + s g \mu_B B, \tag{2.2.27}$$

where g is the Landé factor, $\mu_B = e\hbar/2m_e$ the Bohr magneton and $s = \pm 1/2$ the spin quantum number. In the wave function the spin function $\chi_s(s') = \delta_{ss'}$ occurs as a factor. In the Peierls approximation a similar addition of the spin energy can, however, be misleading. For narrow-gap semiconductors, for instance, the crudest (2-band) approximation yields the non-parabolic energy function

$$\varepsilon(\boldsymbol{p}) = -\frac{1}{2}\,\varepsilon_g + \left(\frac{\varepsilon_g^2}{4} + \varepsilon_g\,\frac{p^2}{2m}\right)^{1/2}, \tag{2.2.28}$$

where ε_g is the gap energy [2.20]. In a magnetic field the corresponding energy levels are [2.21]

$$\varepsilon_{\alpha n} = -\frac{1}{2}\,\varepsilon_g + \left(\frac{\varepsilon_g^2}{4} + \varepsilon_g \varepsilon_{\alpha s}^0\right)^{1/2}. \tag{2.2.29}$$

Thus the spin energy appears under the square root sign and not as an additional term.

A prerequisite to apply the Peierls approximation is that the conduction band is sufficiently separated from the other bands [2.22, 23]. Besides the spin problem, this condition is also violated in narrow-gap semiconductors.

In the following, when not stated otherwise, we restrict ourselves to a spherical energy function for which (2.2.27) is correct.

As we have seen in the introduction, an important quantity in the transport theory is the density of states on the energy scale per unit volume

$$n(\varepsilon) = \frac{1}{\mathrm{Vol}}\,\frac{d}{d\varepsilon} \sum_{\varepsilon_\alpha \le \varepsilon} = \frac{1}{\mathrm{Vol}} \sum_{\alpha s} \delta(\varepsilon - \varepsilon_{\alpha s}). \tag{2.2.30}$$

As long as $\varepsilon_{\alpha s}$ is independent of k_y (degeneracy) the summation with respect to k_y must be cut off. For instance, counting only states with $0 \le X \le L_x$ (the cyclotron centre is inside the system volume), the density of Landau states (2.2.26) will be independent of the volume

$$n(\varepsilon) = \begin{cases} \dfrac{\sqrt{2m}}{(2\pi\hbar)^2}\,eB \displaystyle\sum_{v,s} \frac{1}{\sqrt{\varepsilon - \varepsilon_{vs}}} \\[4mm] 0 \quad \text{for} \quad \varepsilon < \varepsilon_{0-} \end{cases} \tag{2.2.31}$$

where $\varepsilon_{vs} = (v + 1/2)\hbar\omega_c + s g \mu_B B$ [2.3]. This expression shows square root singularities at the critical values of B.[1] In reality these singularities are suppressed by level broadening (collisional damping).

The simplest phenomenological way to take level broadening into account is to replace the singular δ function in (2.2.30) by the Lorentzian (2.1.4) with some constant damping parameter Γ. The quantity $T_D = \Gamma/\pi k_B$ is called the Dingle temperature (k_B is the Boltzmann constant)[2.4]. The corresponding density of states

$$n_\Gamma(\varepsilon) = \frac{4\sqrt{2m}}{(2\pi\hbar)^2} eB \sum_{vs} \left(\frac{1}{\Gamma} \frac{y_{vs} + \sqrt{1+y_{vs}^2}}{1+y_{vs}^2} \right)^{1/2}, \qquad (2.2.32)$$

where $y_{vs} = 4(\varepsilon - \varepsilon_{vs})/\Gamma$ still oscillates, the heights of the finite asymmetric peaks are proportional to $1/\sqrt{\Gamma}$ [2.11].

In thermal equilibrium the distribution of N independent lattice electrons is described by the Fermi function, $f_{\alpha s} = f(\varepsilon_{\alpha s})$,

$$f(\varepsilon) = [\exp (\varepsilon - \zeta)/k_B T + 1]^{-1}. \qquad (2.2.33)$$

The chemical potential ζ is determined by the condition

$$\frac{1}{\text{Vol}} \sum_\alpha f_\alpha = \int f(\varepsilon) n(\varepsilon) d\varepsilon = n. \qquad (2.2.34)$$

In the limit of vanishing temperature (2.2.34) reduces to

$$\int_{-\infty}^{\varepsilon_F} n(\varepsilon) d\varepsilon = n. \qquad (2.2.35)$$

This equation determines the Fermi energy ε_F as a function of B and the system parameters. When we use (2.2.32) for the density of states, the lower limit of the integration has to be replaced by the cut-off energy $\varepsilon_0^* = \varepsilon_{0-} - \Gamma$ in order to avoid unphysical results for the Fermi energy brought about by the artificial tail of $n_\Gamma(\varepsilon)$ for negative energies.

In the quantum limit ($\hbar\omega_c/2 < \varepsilon_F < 3\hbar\omega_c/2$), (2.2.32, 35) yield

$$\varepsilon_F(B, \Gamma) = \frac{\hbar\omega_c}{2} + \eta \left[(1+\sqrt{a\gamma})^2 - \frac{1}{4} \frac{\gamma^2}{(1+\sqrt{a\gamma})^2} \right], \qquad (2.2.36)$$

---

1 Assuming the same degeneracy as for a spherical energy function and omitting the spin, for an arbitrary Fermi surface (2.2.30) leads to

$$n(\varepsilon) = \frac{1}{(2\pi l)^2} \sum_v dk_z/d\varepsilon = -D(B) \sum_v [(\partial A/\partial \varepsilon)/\hbar(\partial A/\partial p_z)]_{p_z = p_z(\varepsilon, v)}.$$

This expression shows that the density of states diverges at stationary areas $\partial A/\partial p_z = 0$.

where $\eta=(2\varepsilon_F^0/3\hbar\omega_c)^2\varepsilon_F^0$, $\gamma=\Gamma/\eta$, $a=(\sqrt{5}/2)-1$; $\varepsilon_F^0=[(6\pi^2n)^{2/3}\hbar^2/2m]$ is the Fermi energy for $B=0$, $\Gamma=0$ [2.24]. At intermediate fields (2.2.35) has to be solved numerically or by applying the Poisson summation formula. At low fields we can replace ε_F by ε_F^0.

2.2.2 Interaction with Impurities

a) Level Broadening

At low temperatures the most effective interaction process limiting the mobility is the scattering of the conduction electrons by impurities. We shall assume that the impurities are randomly distributed and fixed at positions R_J and write for the impurity potential

$$U=\sum_{J=1}^{N_I} u(r-R_J). \tag{2.2.37}$$

The scattering of electrons by impurities gives rise to level broadening (collisional damping) which we want to investigate first. In Section 2.2.2b below the screening problem is considered.

To utilise the standard technique of many-particle theory, we introduce a number of useful quantities: the single-electron Green's functions

$$\hat{G}(z)=\frac{1}{z-H}, \quad \hat{G}^0=\frac{1}{z-H_0}, \tag{2.2.38}$$

which are related by

$$\hat{G}=\hat{G}^0+\hat{G}^0U\hat{G} \tag{2.2.39}$$

and the average of $\hat{G}$ with respect to all possible impurity distributions with equal weight, $G=\langle\hat{G}\rangle_I$,

$$\langle\ldots\rangle_I=\int\frac{d^3R_1}{\text{Vol}}\ldots\int\frac{d^3R_{N_I}}{\text{Vol}}(\ldots). \tag{2.2.40}$$

Assuming macroscopic translational and rotational invariance G will be diagonal in the Landau representation (2.2.26), $G_{\alpha\alpha'}=G_\alpha\delta_{\alpha\alpha'}$. Furthermore, we introduce the self-energy $\Sigma_\alpha(z)$ defined by

$$G_\alpha=G_\alpha^0+G_\alpha^0\Sigma_\alpha G_\alpha, \tag{2.2.41}$$

or, solving for $G_{\alpha'}$

$$G_\alpha = \frac{1}{z - \varepsilon_\alpha - \Sigma_\alpha(z)} \tag{2.2.42}$$

and the spectral function

$$D_\alpha(\varepsilon) = \frac{i}{2\pi} \left[G_\alpha(\varepsilon + i0) - G_\alpha(\varepsilon - i0) \right]. \tag{2.2.43}$$

With the decomposition

$$\Sigma_\alpha(\varepsilon \pm i0) = \Delta_\alpha(\varepsilon) \mp \frac{i}{2} \Gamma_\alpha(\varepsilon)$$

$$D_\alpha(\varepsilon) = \frac{1}{\pi} \frac{\frac{1}{2} \Gamma_\alpha(\varepsilon)}{[\varepsilon - \varepsilon_\alpha - \Delta_\alpha(\varepsilon)]^2 + \frac{1}{4} \Gamma_\alpha^2(\varepsilon)}. \tag{2.2.44}$$

Using the symbolic identity

$$\frac{1}{\varepsilon - i\eta - H} = \frac{P}{\varepsilon - H} + i\pi\delta(\varepsilon - H) \quad (0 < \eta \to 0), \tag{2.2.45}$$

we obtain for the density of states

$$n(\varepsilon) \equiv \frac{1}{\mathrm{Vol}} \mathrm{Tr} \langle \delta(\varepsilon - H) \rangle_1 = \frac{1}{\mathrm{Vol}} \sum_\alpha D_\alpha(\varepsilon). \tag{2.2.46}$$

Thus, $\Delta_\alpha(\varepsilon)$ and $\Gamma_\alpha(\varepsilon)$ describe level shift and level broadening, respectively. Equation (2.1.5) can be interpreted as an ad hoc approximation of (2.2.46) by putting $\Delta_\alpha(\varepsilon) = 0$ and $\Gamma_\alpha(\varepsilon) = \Gamma = \mathrm{const}$.

We can assume without restricting the generality that $\langle U \rangle_1 = 0$. Iterating and averaging (2.2.39) and comparing the result with the iteration of (2.2.41), we obtain in lowest order (Born approximation)

$$\Sigma_\alpha^0 = \sum_{\alpha'} \langle |U_{\alpha\alpha'}|^2 \rangle_1 G_{\alpha'}^0. \tag{2.2.47}$$

The intuitive self-consistent generalisation of this expression is the integral equation

$$\Sigma_\alpha = \sum_{\alpha'} \langle |U_{\alpha\alpha'}|^2 \rangle_1 G_{\alpha'} = \sum_{\alpha'} \frac{\langle |U_{\alpha\alpha'}|^2 \rangle_1}{\varepsilon - \varepsilon_{\alpha'} - \Sigma_{\alpha'}} \tag{2.2.48}$$

which defines the so-called generalised Born approximation (GBA). The essential feature of this approximation is that in the description of the scattering of an electron by one of the impurities the influence of all other impurities is taken into account in a self-consistent way.

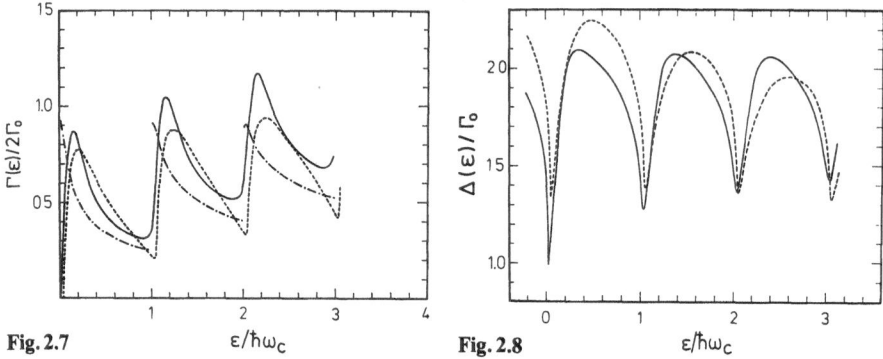

Fig. 2.7

ε/ℏω_c

Fig. 2.8

ε/ℏω_c

Fig. 2.7. Normalised damping function $\Gamma(\varepsilon)/2\Gamma_0$ as a function of energy in three approximations. (——): Generalised Born Approximation (GBA); (–·–): T-matrix Approximation (TMA); (–––): Self-consistent T-matrix Approximation (STMA). $\Gamma_0 = 2\pi n_1 \lambda_1^2 (3n/4\varepsilon_F^0)$ [2.25]

Fig. 2.8. Normalised energy shift $\Delta(\varepsilon)/\Gamma_0$ as a function of energy for point impurities in three approximations: GBA (——), TMA (–·–) and STMA (–––) [2.25]

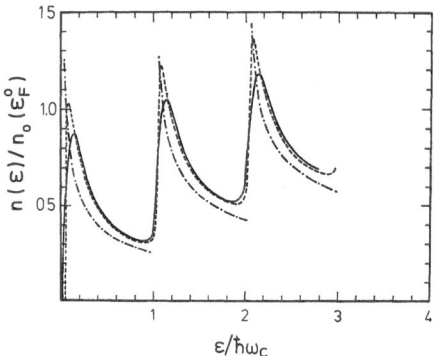

ε/ℏω_c

Fig. 2.9. Normalised density of states $n(\varepsilon)/n_0(\varepsilon_F^0)$ as a function of energy for point impurities in three approximations: GBA (——), TMA (–·–) and STMA (–––). $n_0(\varepsilon_F^0) = (3n/2\varepsilon_F^0)$ [2.25]

The solution of (2.2.48) is difficult. For point impurities,

$$u(r) = \lambda_1 \delta(r), \tag{2.2.49}$$

the self-energy turns out to be independent of the quantum numbers in GBA and (2.2.48) reduces to the algebraic equation

$$\Sigma(\varepsilon) = n_1 \lambda_1 + i\pi n_1 \lambda_1^2 \frac{\sqrt{2m}}{(2\pi l)^2} \sum_\nu \frac{1}{\sqrt{\varepsilon - \hbar\omega_c(\nu + \frac{1}{2}) - \Sigma(\varepsilon)}} \tag{2.2.50}$$

($n_1 = N_1/\text{Vol}$). The numerical solution of this equation and the resulting density of states are shown in Figs. 2.7–9 [2.26]. At the maxima $\Gamma(\varepsilon) \sim n_1^{2/3}$.

To demonstrate an analytic solution of (2.2.48) we consider the anisotropic Gaussian model potential

$$u(r) = \frac{\lambda_2}{(2\pi)^{3/2} r^2 s} \exp\left[-(x^2+y^2)/2r^2 - z^2/2s^2\right] \tag{2.2.51a}$$

with Fourier transform

$$u(q) = \lambda_2 \exp\left[-r^2(q_x^2+q_y^2)/2 - s^2 q_z^2/2\right] \tag{2.2.51b}$$

in the limit $s \to \infty$ (i.e. $s \gg 1/k_F$), when

$$\exp\left(-s^2 q_z^2/2\right) \simeq \frac{\sqrt{\pi}}{s} \delta(q_z). \tag{2.2.52}$$

(Impurities with finite range perpendicular to B and infinite range parallel to B [2.26]). Assuming $\omega_c \gg \Sigma$ [corresponding to (2.1.6)] we can simplify (2.2.48) by retaining the resonant terms with $n'=n$ only. Working out the elementary integrals we arrive at

$$\Sigma_\alpha(\varepsilon) = \frac{a_\nu}{\varepsilon - \varepsilon_\alpha - \Sigma_\alpha(\varepsilon)}. \tag{2.2.53}$$

The first two coupling parameters are

$$a_0 = \frac{a}{1+x}, \quad a_1 = \frac{a(1+x^2)}{(1+x)^3}, \tag{2.2.54}$$

where $a = (n_1/r^2 s)\lambda_2^2/(2/\sqrt{\pi})^3$ and

$$x = l^2/2r^2. \tag{2.2.55}$$

The solution of (2.2.53) is

$$\Sigma_\alpha(\varepsilon) = \tfrac{1}{2}\left[\varepsilon - \varepsilon_\alpha - \sqrt{(\varepsilon-\varepsilon_\alpha)^2 - 4a_\nu}\right]; \tag{2.2.56}$$

Im $\{\Sigma_\alpha(\varepsilon)\}$ is non-zero between $\varepsilon_\alpha - 2\sqrt{a_\nu}$ and $\varepsilon_\alpha + 2\sqrt{a_\nu}$, where it becomes

$$\Gamma_\alpha(\varepsilon) = \sqrt{4a_\nu - (\varepsilon-\varepsilon_\alpha)^2}. \tag{2.2.57}$$

Consequently, the spectral functions are non-zero over the same intervals and

$$D_\alpha(\varepsilon) = \Gamma_\alpha(\varepsilon)/2\pi a_\nu. \tag{2.2.58}$$

As a result of scattering by impurities, the free electron δ-type spectral functions, cf. (2.2.30), are changed to semi-ellipses situated symmetrically around the Landau energy, with a width $4\Gamma_v$ and central height $1/\pi\Gamma_v$, where $\Gamma_v = 2\sqrt{a_v}(\sim n_{\mathrm{i}}^{1/2})$ [2.26,27]. Occasionally it is advantageous to approximate the semi-elliptic spectral functions by Lorentzians having the same heights and covering the same area

$$D_\alpha(\varepsilon) = \frac{1}{\pi} \frac{\frac{1}{2}\Gamma_v}{(\varepsilon - \varepsilon_\alpha)^2 + \frac{1}{4}\Gamma_v^2}. \tag{2.2.59}$$

The density of states then has the same form as given by (2.2.32), except that in y_v Γ has to be replaced by Γ_v. The corresponding approximation for the Green's function is

$$G_\alpha(z) = \frac{1}{z - \varepsilon_\alpha + i\Gamma_v/2}. \tag{2.2.60}$$

b) Screening

We shall see that the model potential (2.2.51) which contains three fitting parameters (λ_2, r, s) is useful for studying some qualitative features of quantum magnetotransport phenomena. Nonetheless the real effective impurity potential might have a rather different form. In particular it will depend on the magnetic field, the temperature and the system parameters. On the other hand, it will be free of any fitting parameters. The potential of an isolated ionised impurity is the usual Coulomb potential. In a conductor this potential will more or less be effectively screened by the conduction electrons. Since the screening depends on the state of motion of the electrons, we are faced with a delicate self-consistency problem: the impurity potential determines the self-energy, the self-energy influences the screening or, more generally, the dielectric polarisation (described by the polarisation function Π), and the polarisation function determines the effective impurity potential.

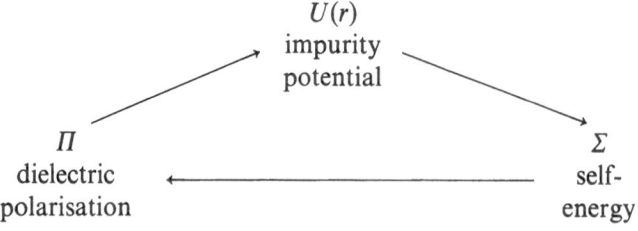

The solution of this self-consistency problem is very difficult and not yet available in a satisfactory approximation.

The effective impurity potential can be expressed by the dielectric function ε. Using Fourier transforms

$$u(q) = \frac{u_0(q)}{\varepsilon(q)}, \tag{2.2.61}$$

where $u_0(q) = -4\pi e e_I / \varkappa q^2$, e_I is the charge of the impurity and $\varkappa$ is the lattice dielectric constant. Furthermore $\varepsilon(q)$ is related to the polarisation function by

$$\varepsilon(q) = 1 + u_c \Pi(q), \tag{2.2.62}$$

where $u_c = 4\pi e^2 / \varkappa q^2$. In the simplest random phase approximation (RPA) which is sufficient for normal systems

$$\Pi(q) = \frac{1}{\pi} \int dz\, f(z)\, \mathrm{Tr}\, [e^{i\boldsymbol{q}\cdot\boldsymbol{r}}\, \mathrm{Im}\, \{F_q(z, r)\}], \tag{2.2.63}$$

where $f(z)$ is the Fermi function and

$$F_q(z; r) = \langle \hat{G}(z)\, e^{-i\boldsymbol{q}\cdot\boldsymbol{r}}\, \hat{G}(z) \rangle_I, \tag{2.2.64}$$

a so-called vertex function. Equations (2.2.39, and 61–64) form a coupled set for the Green's function and the impurity potential which has to be solved self-consistently. Severe mathematical difficulties force us to adopt essential simplifications to satisfy this requirement. The simplest approximation is to factorise the impurity average in (2.2.64)

$$\langle \hat{G}(z)\, e^{-i\boldsymbol{q}\cdot\boldsymbol{r}}\, \hat{G}(z) \rangle_I \simeq \langle \hat{G}(z) \rangle_I\, e^{-i\boldsymbol{q}\cdot\boldsymbol{r}}\, \langle \hat{G}(z) \rangle_I \tag{2.2.65}$$

(i.e. neglect vertex corrections). This approximation has to be handled with special care because it violates the compressibility sum rule

$$\Pi(q \to 0) = \frac{\partial n}{\partial \zeta} = \int \left(-\frac{\partial f}{\partial \varepsilon} \right) n(\varepsilon)\, d\varepsilon. \tag{2.2.66}$$

In the limit $q \to 0$

$$\varepsilon(q) = 1 + \frac{K^2}{q^2}, \tag{2.2.67}$$

$$K = \sqrt{4\pi e^2\, \frac{\partial n}{\partial \zeta}} \tag{2.2.68}$$

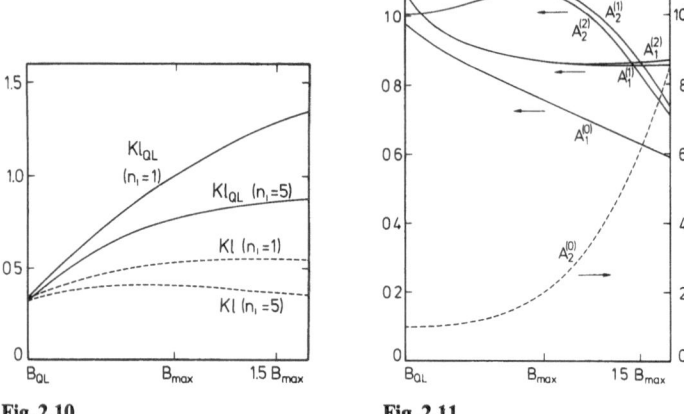

Fig. 2.10   **Fig. 2.11**

Fig. 2.10. Field dependence of the screening parameter K for n-type InSb with $n = 10^{16}$ cm^{-3}, $n_i = n_i/n$. l_{QL} is the magnetic length for the quantum limit field $B_{QL} = 17.7$ T; $B_{max} = 64.7$ T. (——): self-consistent calculation. (---): collisional damping neglected [2.24]

Fig. 2.11. Field dependence of the anisotropy parameters A_1 and A_2 for n-type InSb. $n_i = 1$; subscripts 1, 2: self-consistent calculation (different approximations), superscript 0: collisional damping neglected [2.24]

is the Thomas-Fermi screening wave number and consequently

$$u(\boldsymbol{q}) = -4\pi \frac{ee_1}{\varkappa} \frac{1}{K^2 + q^2}, \quad u(\boldsymbol{r}) = -\frac{ee_1}{\varkappa} e^{-Kr}. \tag{2.2.69}$$

For $T = 0$ $\Pi(q \to 0) = \partial n/\partial \zeta = n(\varepsilon_F)$.

Unfortunately not even the above-formulated lowest order correlation approximation (LOCA) for the self-consistent calculation of u and Σ could be carried out in full generality. Calculations for small q (Thomas-Fermi limit) yield in the quantum limit the anisotropic potential

$$u(q) = -\frac{4\pi ee_1}{\varkappa} \frac{1}{K^2 + A_1 q_\perp^2 + A_2 q_\parallel^2}, \tag{2.2.70}$$

where $q_\perp^2 = q_x^2 + q_y^2$, $q_\parallel^2 = q_z^2$; A_1, A_2 and K are functions of B and the system parameters [2.24, 28]. The results of a numerical computation are shown in Figs. 2.10, 11 [2.24]. Note that the potential (2.2.70) reflects the cylindrical symmetry brought about by the magnetic field. In the limit of vanishing level broadening ($\Gamma = 0$) $u(\boldsymbol{q})$ disappears at the critical values B_ν of the magnetic field. This unacceptable result indicates once more the importance of collisional damping.

A further example is the behaviour of the polarisation function for zero temperatures at the critical wave number $\boldsymbol{q} = \boldsymbol{q}_c = (0, 0, 2q_0)$, $q_0 = \sqrt{2m}$

$(\varepsilon_F - \hbar\omega_c/2)/\hbar^2$. Approximating the averaged Green's function by $G_\alpha = 1/(\varepsilon - \varepsilon_\alpha + i\Gamma)$ with $\Gamma = \text{const}$, one obtains (for small Γ) [2.29].

$$\Pi(q_c; \Gamma) = \frac{2m}{(2\pi\hbar l)^2} \frac{1}{q_c} \log \frac{\Gamma/2}{\hbar^2 q_c^2/2m}. \qquad (2.2.71)$$

Thus, for $\Gamma \to 0$ the polarisation function diverges logarithmically. [Equation (2.2.71) refers to quantum limit fields; in general as many logarithmic singularities occur as Landau states are occupied.] But even for realistic finite values of Γ, the value of $\Pi(q_c; \Gamma)$ is remarkably high. Consequently the long-wavelength limit of the longitudinal phonon frequency, given by

$$\omega^2(q) = \frac{\omega_{pl}^2}{\varepsilon(q)}, \qquad (2.2.72)$$

where ω_{pl} is the plasma frequency, $\omega_{pl}^2 = 4\pi e^2 n/m$, will drop to a small value. This phenomenon, known as the Kohn anomaly, is considerably stronger in high magnetic fields than in zero field. The reason for this phenomenon is again the breakdown of the p space into a series of concentric cylinders, i.e. the quantisation due to the magnetic field.

2.3 Galvanomagnetic Effects

2.3.1 Fundamentals

In the presence of a homogeneous magnetic field the conductivity tensor of a solid, defined by Ohm's law

$$j_\mu = \sum_\nu \sigma_{\mu\nu}(\boldsymbol{B}) E_\nu, \qquad (2.3.1)$$

depends on the magnetic field strength $\boldsymbol{B}$. In (2.3.1) j is the electric current density and $\boldsymbol{E}$ is the electric field strength; $\mu, \nu = x, y, z$. The resistivity – which is in most cases the directly measurable quantity – is $\varrho_{\mu\nu} = (\sigma^{-1})_{\mu\nu}$. The ratio between the projection of the electric field strength in the direction of the current and the magnitude of the electric current density is often referred to as "measurable resistivity"

$$\varrho = \frac{\boldsymbol{E} \cdot \boldsymbol{j}}{|\boldsymbol{j}|^2}. \qquad (2.3.2)$$

For an isotropic solid and a magnetic field parallel to the z axis, the non-vanishing conductivity and resistivity components are

$\sigma_{xx} = \sigma_{yy}$, σ_{zz} (in general different from σ_{xx}), $\sigma_{xy} = -\sigma_{yx}$,

$\varrho_{xx} = \sigma_{xx}/(\sigma_{xx}^2 + \sigma_{xy}^2) \equiv \varrho_{\perp}$ (transverse resistivity),

$\varrho_{xy} = \sigma_{xy}/(\sigma_{xx}^2 + \sigma_{xy}^2) \equiv \varrho_H \equiv RB$ (Hall resistivity),

$\varrho_{zz} = 1/\sigma_{zz} \equiv \varrho_{\parallel}$ (longitudinal resistivity).

The relative change of the measurable resistivity with the magnetic field, $\Delta\varrho/\varrho_0$, is called magnetoresistance.

The most simple phenomenological friction model for the conductivity based on the classical equation of motion

$$m\dot{v} = -ev \times B - eE - \frac{m}{\tau} v \qquad (2.3.3)$$

and the current formula

$$j = -env \qquad (2.3.4)$$

(Drude-Lorentz model) lead under stationary conditions (E, B constant in time, $\dot{v} = 0$) to

$$\sigma_{xx} = \frac{\sigma_0}{1 + \omega_c^2 \tau^2}, \quad \sigma_{xy} = -\frac{\omega_c \tau \sigma_0}{1 + \omega_c^2 \tau^2}, \quad \sigma_{zz} = \sigma_0, \qquad (2.3.5)$$

where $\sigma_0 = e^2 n\tau/m$ is the conductivity for $B = 0$. Consequently $\varrho_{\perp} = \varrho_{\parallel} = \varrho_0 = 1/\sigma_0$ and $\varrho_H = -B/en$. This result indicates that as long as all carriers have the same velocity (degenerate electron gas) magnetoresistance is a quantum effect. In fact it can be shown that the monotonic magnetoresistance observed at temperatures satisfying $k_B T \gtrsim \hbar\omega_c$ is caused by anisotropy of the Fermi surface. (The "classical" two-band model [2.30, 31] must also be ascribed to this case). This monotonic magnetoresistance can be fully described by a semiclassical transport theory based on the Boltzmann-Bloch equation for the electron distribution function in P space, and will not be considered here (for a comprehensive presentation cf. [2.32]). In contrast to this, to explain (oscillatory and quantum limit) magnetoresistance at low temperatures and high fields ($k_B T \ll \hbar\omega_c$) requires a full quantum description. Indeed, under these conditions the commutation relation (2.2.3) becomes crucial and the semiclassical distribution function $f(P_x, P_y, P_z)$ becomes meaningless.

The exciting history of quantum magnetotransport theory began in 1935 with the ingenious work by *Titeica* [2.33]. Following his ideas is still the best approach for understanding quantum magnetotransport phenomena; it is also an excellent introduction to more elaborate theories.

Let us start with the *transverse conductivity* $\sigma_{xx}(B)$. It is obvious that regular cyclotron motion does not contribute to the net charge transfer through a plane perpendicular to the applied electric field. The current is, rather, due to the migration of the cyclotron centre caused by the impurities *and* the electric field. Putting $E=(E,0,0)$ and inserting the total Hamiltonian of an electron $H_T = H + eEx$ in (2.2.4) we obtain

$$\dot{X} = \frac{1}{eB} \frac{\partial U}{\partial y}, \qquad \dot{Y} = -\frac{1}{eB} \frac{\partial U}{\partial x} - \frac{E}{B}. \tag{2.3.6}$$

$\dot{X}$ is independent of E and vanishes for $U=0$. But the unperturbed $(U=0)$ energy levels, i.e. the eigenvalues of $H_0^F = H_0 + eEx$,

$$\varepsilon_\alpha^E = \varepsilon_\alpha + eEX\left(+\frac{1}{2}\frac{e^2E^2}{m\omega_c^2}\right) \tag{2.3.7}$$

depend on E, indicating the breaking of the right-left symmetry by the electric field. Consequently, for $U \neq 0$, the transition probabilities for transitions $X_1 \to X_2$ and $X_2 \to X_1$ are different. In other words, a net current occurs parallel to the electric field. Denoting the transition probability averaged with respect to the impurity distribution by $\langle w_{\alpha\alpha'}^E \rangle_I$ and counting the charge transfer through a plane perpendicular to E at $x=0$, one obtains for the current

$$j_x = -\frac{eN}{L_y L_z} \sum_{vv'} \sum_{k_z k_z'} \sum_{k_y}' \sum_{k_y'}' [\langle w_{\alpha\alpha'}^E \rangle_I f_\alpha^E(1 - f_{\alpha'}^E) - \langle w_{\alpha'\alpha}^E \rangle_I f_{\alpha'}^E(1 - f_\alpha^E)]. \tag{2.3.8}$$

The primes indicate that the summations with respect to $k_y = -X/l^2$ and $k_y' = -X'/l^2$ are restricted:

$$-\frac{L_x}{2} \leq X \leq 0, \qquad 0 \leq X' \leq \frac{L_x}{2}.$$

f_α^E is the electron distribution function in the presence of the electric field and the impurities obeying the condition of stationarity

$$\frac{\partial f^E}{\partial t} = \sum_{\alpha'} [\langle w_{\alpha\alpha'}^E \rangle_I f_\alpha^E(1 - f_{\alpha'}^E) - \langle w_{\alpha'\alpha}^E \rangle_I f_{\alpha'}^E(1 - f_\alpha^E)] = 0. \tag{2.3.9}$$

The factors $(1 - f)$ account for the Pauli principle. The current given by (2.3.8) vanishes for $E=0$. Equation (2.3.8) further demonstrates that a finite current is a consequence of the simultaneous action of the impurities and the electric field during collisions. The current expression can be rewritten in a handier form:

$$j_x = -en \sum_{\alpha\alpha'} \langle w_{\alpha\alpha'}^E \rangle_I (X - X') f_\alpha^E(1 - f_{\alpha'}^E)$$

$$\equiv -en \sum_{\alpha\alpha'} \frac{(X - X')}{\tau_{\alpha\alpha'}^E} f_\alpha^E(1 - f_{\alpha'}^E) \equiv -en\langle \dot{X} \rangle. \tag{2.3.10}$$

Note that this expression for the current is not restricted to weak electric fields and can be used to study non-linear transport phenomena. Here we consider Ohmic conductivities only and, therefore, linearise (2.3.10) with respect to E. Assuming elastic scattering, $w_{\alpha\alpha'}^E$ contains a factor $\delta(\varepsilon_\alpha^E - \varepsilon_{\alpha'}^E)$ expressing energy conservation. To lowest order in E

$$\delta(\varepsilon_\alpha^E - \varepsilon_{\alpha'}^E) = \delta(\varepsilon_\alpha - \varepsilon_{\alpha'}) - eE(X - X')\delta'(\varepsilon_\alpha - \varepsilon_{\alpha'}) \tag{2.3.11}$$

and, by (2.3.9), $f_\alpha^E = f_\alpha + O(E^2)$. Consequently

$$\sigma_{xx} = e^2 n \sum_{\alpha\alpha'} \langle w_{\alpha\alpha'} \rangle_1 \frac{1}{2}(X - X')^2 \left(-\frac{\partial f}{\partial \varepsilon_\alpha}\right), \tag{2.3.12}$$

where $w_{\alpha\alpha'}$ is the transition probability for $E=0$. Equation (2.3.12) is often referred to as Titeica's formula. With the random walk expression for the diffusion constant

$$D = n\langle \tfrac{1}{2}(\varDelta X)^2/\tau \rangle, \tag{2.3.13}$$

(2.3.12) takes the form of the Einstein relation

$$\sigma_{xx} = e^2 \frac{\partial D}{\partial \zeta}. \tag{2.3.14}$$

[For Boltzmann statistics $(-\partial f/\partial \varepsilon_\alpha) = f_\alpha/k_B T$ and $\sigma_{xx} = e^2 D/k_B T$.] This indicates once more that the electric current in the direction of the applied electric field is of diffusive character.

Unfortunately, in lowest-order Born approximation (LOBA),

$$w_{\alpha\alpha'} = \frac{2\pi}{\hbar} |U_{\alpha\alpha'}|^2 \delta(\varepsilon_\alpha - \varepsilon_{\alpha'}), \tag{2.3.15}$$

Titeica's conductivity expression (2.3.12) diverges (logarithmically) for any value of the temperature and the magnetic field[2]. This unphysical result is due to the singularities of the Landau density of states (2.1.3) and clearly indicates that in quantum magnetotransport theory one has to go beyond the LOBA by taking into account multiple-scattering [2.36–38] and collisional damping processes [2.37]. In doped semiconductors collisional damping is essential, whereas the multiple-scattering processes become significant in systems of high purity.

---

2 For degenerate electrons at $T=0$ this divergency problem can be eliminated by interchanging the order of the limiting process $T \to 0$ and the integration with respect to the energy [2.34]. This manipulation, however, cannot avoid the conductivity diverging at the critical values B_v of the magnetic field. Moreover, it leads to incorrect asymptotic behaviour of the conductivity in the quantum limit [2.35].

A simple phenomenological way to incorporate collisional damping is to write in LOBA Titeica's formula in a more symmetric form by using $\partial f/\partial \varepsilon_\alpha = \int d\varepsilon \, \delta(\varepsilon - \varepsilon_\alpha) \partial f/\partial \varepsilon$ and replacing the δ functions by Lorentzians. Equation (2.1.4) with some constant damping parameter becomes [2.2.39, 40]

$$\sigma_{xx} = e^2 n \sum_{\alpha\alpha'} \frac{2\pi}{\hbar} \langle |U_{\alpha\alpha'}|^2 \rangle_{\mathrm{I}} \frac{1}{2} (X - X')^2$$

$$\cdot \int d\varepsilon \left(-\frac{\partial f}{\partial \varepsilon} \right) D(\varepsilon - \varepsilon_\alpha) D(\varepsilon - \varepsilon_{\alpha'}). \tag{2.3.16}$$

Unfortunately, Titeica's intuitive treatment is not appropriate for deriving a formula for the *Hall conductivity* σ_{xy} similar to (2.3.12). In the limit $U \to 0$ ($\omega_c \tau \gg 1$), however, we can use (2.3.6) to calculate the current: $j_y = (en/B)E$. This yields the classical result

$$\sigma_{xy} = -\frac{en}{B} \tag{2.3.17}$$

and (since $\omega_c \tau \gg 1$)

$$\varrho_\perp \simeq \left(\frac{B}{en} \right)^2 \sigma_{xx}, \quad \varrho_H \simeq -\frac{B}{en}. \tag{2.3.18}$$

We now turn to the *longitudinal conductivity* σ_{zz}. Since the motion parallel to the magnetic field is not affected by the field, the velocity is $v_z = p_z/m$. Consequently, the longitudinal current density can be written as

$$j_z = -en \sum_\alpha \frac{\hbar k_z}{m} f_\alpha, \tag{2.3.19}$$

where f_α is the electron distribution function. In the usual kinetic approximation f_α obeys a Boltzmann-Bloch type transport equation. Assuming elastic scattering and linearising with respect to E, one obtains

$$\sigma_{zz} = \frac{e^2 n}{m} \sum_\alpha \left(-\frac{\partial f}{\partial \varepsilon_\alpha} \right) \hbar^2 k_z^2 \tau_\alpha, \tag{2.3.20}$$

τ_α being the solution of

$$\sum_{\alpha'} \langle w_{\alpha\alpha'} \rangle_{\mathrm{I}} \left(\delta_{\alpha\alpha'} - \frac{k_z'}{k_z} \right) \tau_{\alpha'} = 1. \tag{2.3.21}$$

For example,

$$\tau_\alpha = \left[\sum_{\alpha'} \left(1 - \frac{k_z'}{k_z} \right) \langle w_{\alpha\alpha'} \rangle_{\mathrm{I}} \right]^{-1} \tag{2.3.22}$$

supposing the right-hand side depends only on energy, i.e. $\tau_\alpha = \tau(\varepsilon_\alpha)$. For point impurities in LOBA [2.41]

$$\frac{1}{\tau_\alpha} = \frac{2\pi}{\hbar} n_1 \lambda_1^2 n(\varepsilon_\alpha). \tag{2.3.23}$$

We see that in LOBA, for $T=0$, σ_{zz} vanishes at the critical values B_ν of the magnetic field. To avoid this, higher order scattering processes have to be taken into account [2.36–38]. In the phenomenological damping approximation [2.2, 40]

$$\sigma_{zz} = \frac{e^2 n}{m} \sum_\alpha \hbar^2 k_z^2 \tau_\alpha \int d\varepsilon \left(-\frac{\partial f}{\partial \varepsilon} \right) D(\varepsilon - \varepsilon_\alpha) \tag{2.3.24}$$

and

$$\langle w_{\alpha\alpha'} \rangle_{\mathrm{I}} = \frac{2\pi}{\hbar} |U_{\alpha\alpha'}|^2 D(\varepsilon - \varepsilon_{\alpha'}); \tag{2.3.25}$$

furthermore, in (2.3.23) $n(\varepsilon)$ has to be replaced by $n_\Gamma(\varepsilon)$, (2.1.5).

The simple phenomenological damping approximation discussed so far is useful in the qualitative analysis of experiments. Nonetheless, from a fundamental point of view it is completely unsatisfactory. First of all, the damping, brought about by the impurity potential, should be calculated rather than described by a constant fitting parameter which requires an additional ad hoc cut-off procedure. Indeed, as we have seen in the previous section, the amount of damping depends on the energy, the magnetic field and the system parameters and ought to be calculated together with the impurity potential in a self-consistent way. Secondly, since such a calculation involves higher order perturbation theory, the conductivity expression must also be evaluated in a correspondingly higher order self-consistent perturbation approximation.

The appropriate starting point for a systematic microscopic theory of the (dynamic) conductivity is the Kubo formula [2.42]

$$\sigma_{\mu\nu}(\omega) = \int_0^\infty \int_0^\beta \langle j_\mu(t+i\lambda)j_\nu(0) \rangle \, e^{i\omega t} \, d\lambda \, dt. \tag{2.3.26}$$

Here $\langle \ldots \rangle$ denotes equilibrium average and $j(t)$ is the volume average of the current density at time t.

Mori invented a very effective method for evaluating the conductivity for a given system [2.43]. A considerably simplified (but less general) formulation was developed by *Götze* and *Wölfle* [2.44]. In the appendix we give an outline of the Mori formalism and calculate the dynamic conductivity in lowest-order correlation approximation (LOCA) [2.26, 9c]. The results for the static conductivities meet our expectations: they coincide with the expressions obtained above, if the constant damping parameter Γ is replaced by the damping function $\Gamma_\alpha(\varepsilon) = 2 \operatorname{Im} \{\Sigma_\alpha(\varepsilon)\}$ calculated in GBA. The validity of (2.3.16) is then confirmed for $\omega_c \tau \gg 1$.

2.3.2 Magnetoresistance

To extract information from measurements of high field magnetoresistance, a practical and sufficiently general theory is needed. This theory should be valid in a wide range of the magnetic field for arbitrary band structure, and it should be free of any fitting parameter. Unfortunately, all readily applicable theories formulated so far do not meet these requirements. Sophisticated self-consistent transport theories have been worked out only for spherical energy functions and simple model impurity potentials [2.25, 38, 45]. Moreover, in most cases tedious numerical computation is required.

Nevertheless, a fundamental understanding of magnetoresistance has been achieved which in many cases allows the interpretation of experimental data. Since both the theoretical and the experimental situations depend on the strength of the magnetic field, we consider the oscillatory Shubnikov-de Haas (SdH) regime and the quantum limit (QL) separately.

a) Shubnikov-de Haas Regime

The theoretical tools currently available are the semi-classical formula (2.2.19) for the periods of the oscillations (which allows for an arbitrary Fermi surface but disregards the spin) and the constant damping approximation for the conductivities, normally worked out for a spherical energy surface and some short- or long-range model potential. The point, as well as the Gaussian model have the advantage that most parts of the calculations can be carried out analytically. Also numerical calculations based on the Kubo formula and self-consistent perturbation theory have been performed [2.25]. As long as $\hbar\omega_c \ll \varepsilon_F$, the Poisson summation formula

$$\sum_{v=0}^{\infty} f(v + \tfrac{1}{2}) = \sum_{-\infty}^{\infty} (-1)^r \int_0^{\infty} f(x) \exp(2\pi i r x) dx \qquad (2.3.27)$$

can be applied. For point impurities and low temperatures, $k_B T \ll \hbar\omega_c$, one obtains (neglecting less important terms)

$$\varrho = \varrho_0 \left[1 + a \sum_{r=1}^{\infty} b_r \cos \left(\frac{2\pi\varepsilon_F^0}{\hbar\omega_c} r - \frac{\pi}{4} \right) \right] \qquad (2.3.28)$$

$$b_r = \frac{(-1)^r}{r^{1/2}} \left(\frac{\hbar\omega_c}{2\varepsilon_F^0} \right)^{1/2} \frac{2\pi^2 r k_B T / \hbar\omega_c}{\mathrm{sh}\,(2\pi^2 r k_B T / \hbar\omega_c)}$$

$$\cdot \cos(\pi M r) \exp(-2\pi^2 r k_B T_D / \hbar\omega_c), \qquad (2.3.29)$$

where $a = 5/2$ and 1 for $\varrho_\perp$ and $\varrho_\parallel$, respectively; $M = g\mu_B B/\hbar\omega_c = (m_c/m_e)g/2$, $T_D = \Gamma/\pi k_B$; $\varepsilon_F^0 = (6\pi^2 n)^{2/3}\hbar^2/2m$ is the Fermi energy for $B=0$ and $\Gamma=0$ [2.11]. According to these formulas the behaviour of the resistivities can be described by a superposition of harmonic oscillations in $1/B$, each modulated by the spin splitting factor and damped by temperature and level broadening. The similarity of $\varrho_\perp$ and $\varrho_\parallel$, as predicted by (2.3.28) for $k_B T \ll \hbar\omega_c \ll \varepsilon_F$, has by no means been confirmed fully by experiment (Fig. 2.1). We shall comment on this subsequently.

In the following we shall show through typical examples how information on the electronic structure can be extracted from SdH oscillations. We shall not try to cover all the numerous experimental data collected during the past two decades.

As we have pointed out, to observe SdH oscillations in the resistivity requires a degenerate electron gas with a well-defined Fermi energy $\varepsilon_F^0 \gg k_B T$. The number of occupied energy levels ($\sim \varepsilon_F^0/\hbar\omega_c$) should not be too large, however, since, as (2.3.29) shows, a large ratio $\varepsilon_F^0/\hbar\omega_c$ reduces the amplitude of the SdH oscillations substantially. This is why the SdH effect is only of limited usefulness in metal physics. For metals it is usually more advantageous to investigate the oscillations of the diamagnetic susceptibility (de Haas-van Alphen effect) which yields in principle the same information as the resistivity oscillations. Furthermore, to obtain quantisation by the magnetic field, the condition $\omega_c\tau > 1$ must be satisfield, which is equivalent to $\varrho_H > \varrho_0$, or, expressed by the mobility μ of the electrons, to $\mu B > 1$. The two conditions $\varepsilon_F > \hbar\omega_c$ and $\mu B > 1$ cannot always be met simultaneously. To obtain a sufficiently large Fermi energy the carrier concentration n must generally be above 10^{17} cm^{-3}. At such high doping levels the carrier mobility of semiconductors at helium temperature is sometimes not sufficiently high to satisfy $\mu B > 1$ in a magnetic field of 10 T. Whether the latter condition can be met depends mainly on the effective mass m (and also on the value of the dielectric constant). For semiconductors with m^* of the order 0.1 m_e, it is usually possible to observe SdH oscillations, whereas for $m^* \simeq m_e$ it is difficult.

For systems with an isotropic Fermi surface (such as n-InSb), (2.3.16, 20) worked out for the parabolic (2.2.27) or non-parabolic (2.2.29) energy spectrum and some model impurity potential can directly be compared with experiment. Assuming point impurities, two parameters M and Γ (or T_D) remain, which can easily be determined by best fit given that the agreement between observation

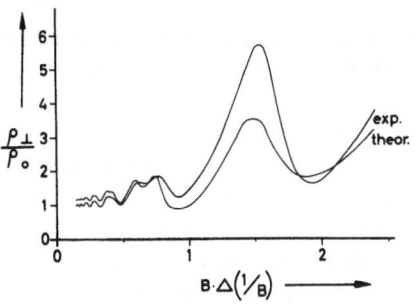
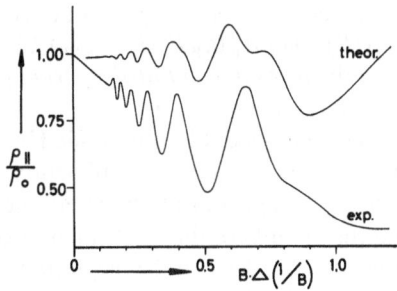

Fig. 2.12. Observed (exp) and calculated (theor) transverse resistivity vs $B\Delta(1/B)$ for n-type InSb with $n = 6 \times 10^{16} \, \text{cm}^{-3}$ [2.40]

Fig. 2.13. Observed (exp) and calculated (theor) longitudinal resistivity vs $B\Delta(1/B)$ for n-type InSb with $n = 6 \times 10^{16} \, \text{cm}^{-3}$ [2.40]

and calculation is sufficiently good. Figures 2.12, 13 show that this is the case for $\varrho_\perp$, but not for $\varrho_\parallel$ [2.40]. The large maximum to be seen in $\varrho_\perp$ corresponds to the (depopulation) of the 0^+ state ($v = 0$, $s = +1/2$). For $v = 1$ and $v = 2$, the spin splitting is well resolved and agrees with the theory. In contrast to this, in $\varrho_\parallel$, the 0^+ peak is missing (cf. also Fig. 2.1) and no spin splitting is observed.

The absence of the 0^+ maximum in $\varrho_\parallel$ can be explained very simply: it is the consequence of an accidental cancellation of singular factors occurring in both the nominator and denominator of the conductivity expression (2.3.20) with (2.3.23). It has been pointed out, however, that spin flip scattering by magnetic impurities [2.46] or spin-orbit interaction [2.47a] may well give rise to an observable 0^+ peak. In contrast, the missing spin splitting of the $v = 1$ maximum has not yet been fully explained [2.48]. Both band structure effects, together with collision damping [2.49] and the spin-orbit interaction [2.50], have been claimed to be responsible for this phenomenon.

Another peculiarity of the longitudinal resistivity is that its relative change $\Delta\varrho_\parallel/\varrho_0$ often becomes negative below QL fields (Figs. 2.1, 13). It was pointed out by *Dubinskaya* [2.51] that this effect might be due to scattering by ionized impurities. To see how this comes about, we have only to recall that for scattering by the Coulomb potential at $B = 0$ the cross section $\sigma(\vartheta)$ behaves as $\sin^{-4}(\vartheta/2)$. Thus the largest contribution to the resistivity

$$\varrho_0 \sim \frac{1}{\tau} = n_I v_F 2\pi \int_0^\pi (1 - \cos\vartheta)\sigma(\vartheta) \sin\vartheta \, d\vartheta \tag{2.3.30}$$

comes from angles around $\vartheta = 0$ (forward scattering). For $B \neq 0$ this region is excluded by the energy conservation law for elastic collisions. A simple model calculation of $\varrho_\parallel$ for degenerate electrons and screened Coulomb impurities confirms this picture [2,52] and qualitatively agrees with experiment [2.53].

To determine the period of the oscillations from experiment, it is useful to plot the positions of the extrema on a $1/B$ scale vs integers in a range where the spin splitting is not observed. For a spherical Fermi surface $B\Delta(1/B) = \hbar\omega_c/\varepsilon_F^0 \, (k_B T \ll \hbar\omega_c \ll \varepsilon_F^0)$. Since this ratio is independent of the effective mass, the *measurement of the period* provides us only with information on the electron concentration n.

The analysis of Shubnikov-de Haas oscillations is considerably simplified if the harmonic content is sufficiently small. This can always be achieved by choosing an appropriate magnetic field range. In the regime of not too small quantum numbers the fundamental oscillation dominates, so that the higher terms in (2.3.28) can be neglected. This allows the cyclotron mass to be determined rather conveniently from the temperature dependence of the amplitude of the oscillations at constant magnetic field:

$$b_1 = \frac{2\pi^2 m_c k_B T/\hbar eB}{\mathrm{sh}\,(2\pi^2 m_c k_B T/\hbar eB)}. \tag{2.3.31}$$

If a computer fit of experimental data shows that the amplitude of the quantum oscillations obeys the functional dependence of (2.3.31), one can expect that the determined effective mass is reasonably accurate. It is certainly advisable to measure the temperature dependence of the amplitude at various magnetic fields. It turns out that the masses deduced from careful SdH experiments agree well with the values obtained from cyclotron resonance data.

The Dingle temperature can be extracted from measurements of the field dependence of the amplitude b_1 at constant T; comparison of the amplitude at two different B values yields

$$\frac{b_1(B_1)}{b_2(B_2)} \simeq \sqrt{\frac{B_1}{B_2}} \, \exp\left[-2\pi^2 k_B T_D \frac{m_c}{\hbar e} \left(\frac{1}{B_1} - \frac{1}{B_2} \right) \right]. \tag{2.3.32}$$

It is advisable, however, to check (2.3.32) at several values of the magnetic field. Care should be taken to employ homogeneous samples, because inhomogeneity in the electron and impurity concentration causes anomalous dependence of the amplitude on the magnetic field, which leads to errors in the determination of the Dingle temperature.

The magnitude of the g factor can be deduced by measuring the *spin splitting*. Denoting the magnetic fields at the maxima, corresponding to $s = \pm 1/2$ for fixed v, by B^+ and B^-, respectively, and putting $\varepsilon_F \simeq \varepsilon_F^0$, (2.2.31) yields

$$\frac{B_v^+}{B_v^-} = \frac{2v+1-M}{2v+1+M}, \tag{2.3.33}$$

where again $M = g m_c/2m_e$. Normally the spin splitting of the peaks is resolved for $v = 1\text{--}5$ only. This means that the field dependence of ε_F must be taken into account. Thus, the g factor calculated by using (2.3.33) may deviate considerably from its true value.

The analysis outlined above has been carried out for many n-type III-V compounds which have their Fermi surface at the center of the Brillouin zone. For a typical n-InSb sample with $n \simeq n_\mathrm{I} \simeq 6 \times 10^{16}$ cm^{-3}, one obtains [2.2]

$$m \simeq 0.013 \, m_\mathrm{e}, \quad |g| \simeq 35, \quad T_\mathrm{D} \simeq 4 \, \mathrm{K}.$$

A slight warping of a spherical Fermi surface can give rise to beating effects. A good example of this is n-HgSe, where the beating is due to the superposition of two oscillations with similar periods. Analysis of the experimental data allows one to determine the parameters of the inverted α-Sn band structure [2.54].

A straightforward analysis of the SdH oscillations is still possible if the Fermi surface consists of a number of ellipsoids. (The classical example of this case is bismuth.) For instance, in a cubic ellipsoidal six-valley structure 3 oscillations with different periods are superimposed which reduce to 2 or 1 at points of high symmetry. One obtains information on the location of the energy minima in k space and the effective masses by *measuring the* SdH *period* at various angles between the crystal axis and the magnetic field. The reason for this is that the cyclotron mass m_c is different from the "density of state mass" m_n defined by $\varepsilon_\mathrm{F}^0 = (6\pi^2 n)^{2/3} \hbar^2 / 2 m_n$. E.g., for a single ellipsoid (2.2.11) with z axis parallel to B

$$m_\mathrm{c} = \sqrt{m_1 m_2} \quad \text{and} \quad m_n = \sqrt[3]{m_1 m_2 m_3}. \quad \text{Therefore}$$

$$\Delta\left(\frac{1}{B}\right) \sim \frac{\sqrt[3]{m_1 m_2 m_3}}{\sqrt{m_1 m_2}}. \tag{2.3.34}$$

For z valleys the value of m_n is enhanced by a factor $z^{3/2}$. To determine the Fermi energy one must know, in addition, the electron concentration n. This can be obtained by *weak field Hall effect* measurements. One must check, however, that the number of electrons which contribute to the SdH oscillations is the same as the number of electrons involved in the Hall effect. Otherwise the determination of the number of valleys would be ambiguous.

Band parameters have been determined for a number of substances by carrying out the analysis outlined above. An example is $(\mathrm{Bi}_{1-x}\mathrm{Sb}_x)_2\mathrm{Te}_3$ which has 6 degenerate ellipsoidal valleys [2.55].

b) Quantum Limit

Quantum limit means $\hbar\omega_\mathrm{c}/2 < \varepsilon_\mathrm{F} \geq 3\hbar\omega_\mathrm{c}/2$ in general and $\hbar\omega_\mathrm{c}/2 \geq \varepsilon_\mathrm{F}^0$ in a narrower sense. Sometimes the term is used for $\hbar\omega_\mathrm{c}/2 = \varepsilon_\mathrm{F}^0$. Figure 2.1 shows that in the quantum limit both resistivities increase monotonically and obey the (approximate) power law (2.1.9). Typical values of the exponent are $\alpha_\perp = 1-3, 3$ and $\alpha_\parallel = 1-2$.

The monotonic increase of the resistivity can be easily explained. In the quantum limit only the $v=0$ Landau cylinder is occupied. With increasing field the diameter of this cylinder grows and its height decreases. Thus, k_z is reduced more and more, and consequently the probability of multiple scattering of an electron by the same impurity is considerably enhanced.

Theoretical studies have shown that the value of α depends sensitively on the type of scattering and the band structure, and, unfortunately, on the approximations as well. For the isotropic Gaussian impurity model [$s=r$ in (2.2.51)] the GBA yields

$$\left.\begin{array}{l} \alpha_\perp = 1 \\ \alpha_\parallel = 0 \end{array}\right\} (x \ll 1) \qquad \left.\begin{array}{l} 3 \\ 1 \end{array}\right\} (x \gg 1)$$

($x = l^2/2r^2$, r being the impurity range). Non-parabolicity adds 0.5 to $\alpha_\perp$ and 0.25 to $\alpha_\parallel$ [2.56]. In the transverse case these values are confirmed by experiment [2.57]. (For an estimate of the actual impurity range in semiconductors, cf. [2.58].) The simplified Gaussian model (2.2.52) yields $\alpha_\perp = \alpha_\parallel = 1.5$ [2.26]. For screened Coulomb impurities $\alpha_\parallel = 2.2$ was obtained [2.52], in agreement with experimental data [2.2].

These results indicate that in principle one should be able to deduce information on the main scattering mechanism (and perhaps on other properties of the system) from magnetoresistance measurements in the quantum limit. However, the number of substances is limited in which clear-cut quantum limit conditions can be realized in a sufficiently wide range of the magnetic field. The main reason for this is that the electron concentrations for which $k_B T \ll \varepsilon_F^0 \le \hbar\omega_c/2$ is satisfied (about 10^{15}–10^{16} cm^{-3}) are effectively reduced in high magnetic fields since the impurity binding energy increases with the field. Due to this magnetic freeze-out effect the condition of degeneracy $k_B T \ll \varepsilon_F$ will be violated in many cases. As a characteristic example Fig. 2.14 shows the influence of a longitudinal magnetic field on the resistance of an InSb sample. For small values of B SdH oscillations can be recognised which are similar to those shown in Fig. 2.1. In Fig. 2.15 the magnetic field scale has been compressed and simultaneously the free electron concentration as determined from the Hall effect has been plotted. It is evident that beyond 8 T the electron concentration rapidly decreases, leading to an accelerated rise in the resistivity. The field range of true quantum limit behaviour extends only from about 4 to 7 T. By increasing n, the freeze out can be shifted towards higher fields. But for large n the quantum limit condition $\varepsilon_F^0 \le \hbar\omega_c/2$ usually requires magnetic fields which are beyond the experimentally accessible range. Thus, the experimental criteria for the observability of the magnetoresistance under true quantum limit conditions are rather limited. Moreover, the amount of information we can extract from these observations is restricted.

There is, however, a number of narrow gap semiconductors, for which magnetic freeze out does not seem to be significant, because the free carriers are derived from intrinsic lattice defects, e. g. $Hg_{1-x}Cd_xTe$, PbTe and $Pb_{1-x}Sn_xTe$.

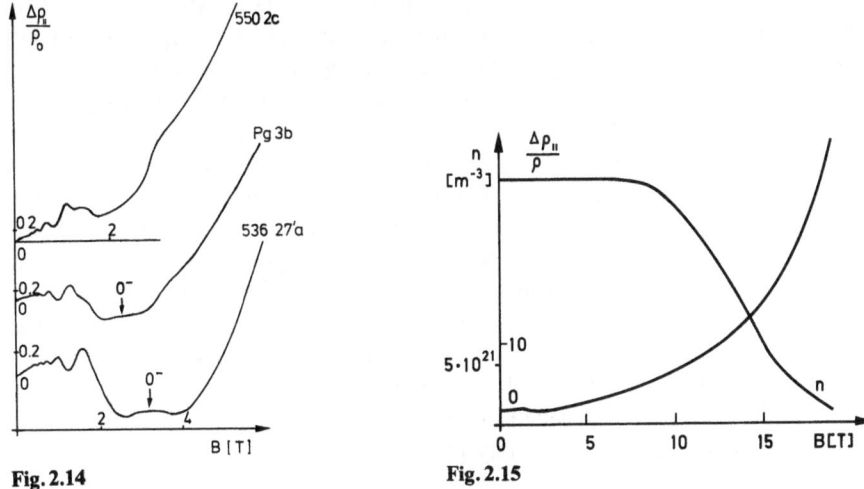

Fig. 2.14   **Fig. 2.15**

Fig. 2.14. Longitudinal magnetoresistance of InSb in the quantum limit [2.47b]. For the correct attribution of the 0^- peak cf. [2.227]

Fig. 2.15. Longitudinal magnetoresistance for sample Pg3b of Fig. 2.14 on a compressed scale, and electron concentration n as a function of the magnetic field [2.47b]

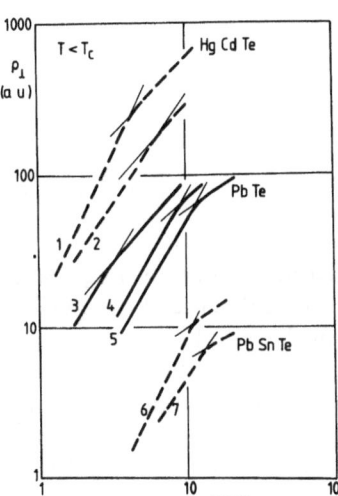

Fig. 2.16. Transverse resistivity for different samples of HgCdTe, PbTe and PbSnTe as a function of the magnetic field [2.59]. For further developments regarding electron correlation effects in HgCdTe, *see* Editor's Appendix, p. 351

Nimtz and co-workers [2.59] studied the quantum limit behaviour of these materials and found at 4.2 K a monotonous increase of the magnetoresistance as described before. When the temperature was lowered to about 1.5 K, the slope of the $\log \varrho$ vs $\log B$ curves increased significantly in the lower field range and decreased at high B values. Typical data for the transverse magnetoresistance of different samples of $Hg_{1-x}Cd_xTe$, PbTe and $Pb_{1-x}Sn_xTe$ are shown in Fig. 2.16. The magnitude of the longitudinal magnetoresistance begins to approach that of the transverse one at the highest fields and the lowest temperatures.

The authors pointed out that this anomaly of the resistivity might indicate condensation of the electron gas into a regular lattice, a phase transition predicted by *Wigner* in 1934 [2.60]. Referring to a calculation of the ground-state energy by *Kleppmann* and *Elliott* [2.61] who showed that an external homogeneous magnetic field supports the Wigner condensation, *Nimtz* and co-workers estimated the transition temperature of $Hg_{1-x}Cd_xTe$ to be about 0.6 K at $B_c = 8$ T. Refined calculations by *Gerhardts* [2.62] yielded for $Hg_{0.8}Cd_{0.2}Te$ with an electron concentration $n = 6 \times 10^{14}$ cm^{-3} a maximum critical temperature of 0.9 K which is roughly half of the experimentally observed value of about 2 K. Whereas the explanation of the observed anomaly is still controversial, *Nimtz* and co-workers succeeded in constructing a highly sensitive microwave bolometer by utilising this anomaly [2.63].

2.3.3 Hall Effect

In the limit $B \to \infty$ the Hall coefficient has the classical value $R = R_0 = 1/en$. To get the lowest order correction to this result, we must expand both σ_{xx} and σ_{xy} in terms of $1/\omega_c\tau$. In leading order $\sigma_{xx} \simeq \sigma_{xx}^{(1)}$ and

$$\sigma_{xy} \simeq \sigma_{xy}^{(0)} + \sigma_{xy}^{(2)} \quad (\sigma^{(n)} \sim (1/\omega_c\tau)^n, \ \sigma_{xy}^{(2n+1)} \equiv 0)$$

due to time reversal invariance. With $\sigma_{xy}^{(0)} = en/B$ the definition of $R(B)$ (p. 38) yields for the lowest order correction

$$\frac{\Delta R}{R_0} = \frac{\sigma_{xy}^{(2)}}{|\sigma_{xy}^{(0)}|} - \left(\frac{\sigma_{xx}^{(1)}}{\sigma_{xy}^{(0)}}\right)^2 . \tag{2.3.35}$$

For this quantity to be positive, $\sigma_{xy}^{(2)}$ must be positive and sufficiently large. Since in this case the correction for the Hall constant is the difference of two positive terms of the same order of magnitude, these terms have to be calculated very carefully. The often-used approximation $\sigma_{xy}^{(2)} \simeq 0$ always leads to a negative correction, in contradiction to observation (Fig. 2.1). (This indicates that the observed SdH oscillations in R are not entirely due to those appearing in ϱ_{xx}.) The complementary approximation, in which the second term on the right-hand side of (2.3.35) is neglected [2.64], is inconsistent as well.

The calculation of $\sigma_{xy}^{(2)}$ is difficult. The lowest order Born approximation, improved by introducing a constant damping parameter (which was useful in calculating σ_{xx}), fails completely when applied to σ_{xy}, because in this case principal value terms (Δ_α) are essential as well. A satisfactory treatment of the Hall effect requires a high-order self-consistent perturbation theory. A first step in this direction was taken by working out the Kubo formula (2.3.26), with $\mu = x$ and $\nu = y$, for point impurities in GBA [2.65], Fig. 2.17.

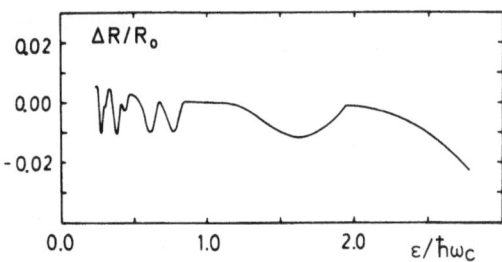

Fig. 2.17. Calculated deviation of the Hall coefficient from its free electron value as a function of the magnetic field [2.65]

In the GBA, $\Delta R/R_0$ is non-analytic in the impurity concentration n_1. An alternative approximation based on an expansion in powers of n_1 [2.66] is restricted to very pure systems.

In the Mori formalism outlined in the appendix the general expression for the Hall coefficient reads $R = R_0[1 - M'(0)/\omega_c]$, where $M'(0)$ is the real part of the relaxation kernel at zero frequency. Since in LOCA $M'(0) = 0$ [2.26], this approximation yields $R = R_0$ for all values of B. To avoid this shortcoming the approximation has to be improved, e.g. by setting up a more flexible Mori ansatz [2.9c] (cf. also the Appendix).

The discrepancy between theory and experiment and the observed phase shift between the SdH oscillations in R and ϱ_{xx} [2.48] motivate further experimental investigation of the Hall effect.

2.3.4 Cyclotron Resonance

We now turn to the dynamic magnetoconductivity defined by the quasi-stationary linear current response to a periodic electric field $E = E(\omega) \exp(i\omega t)$

$$j_\mu(\omega) = \sum_\nu \sigma_{\mu\nu}(\omega, B) E_\nu(\omega). \tag{2.3.36}$$

Cyclotron resonance is a phenomenon occurring in the transverse conductivity $\sigma(z) = \sigma_{xx}(z) + i\sigma_{xy}(z)$. Since $\sigma_{\mu\nu}(z) = -\sigma_{\nu\mu}(-z)$ and, assuming isotropy, $\sigma_{xy}(z) = -\sigma_{yx}(z)$,

$$\sigma_{xx}(z) = \tfrac{1}{2}[\sigma(z) - \sigma(-z)], \quad \sigma_{xy} = \tfrac{1}{2i}[\sigma(z) + \sigma(-z)]. \tag{2.3.37}$$

In the Drude-Lorentz model (2.3.3, 4)

$$\sigma(\omega \pm i0) = \frac{e^2 n}{m} \frac{i(\omega - \omega_c) \pm 1/\tau}{(\omega - \omega_c)^2 + (1/\tau)^2}. \tag{2.3.38}$$

The resonance occurs at $\omega = \omega_c$; the resonance line has a Lorentzian shape with line width $\Gamma_c = 2/\tau$ and height $\sigma_0 = e^2 n\tau/m$.

In a microscopic theoretical treatment it is advantageous to introduce the relaxation kernel $M(z)$ associated with $\sigma(z)$ (appendix):

$$\sigma(\omega \pm i0) = \frac{e^2 n}{m} \frac{i[\omega - \omega_c + M'(\omega)] \pm M''(\omega)}{[\omega - \omega_c + M'(\omega)]^2 + [M''(\omega)]^2}, \tag{2.3.39}$$

$[M(\omega \pm i0) = M'(\omega) \pm iM''(\omega)]$. The general formula (2.3.39) indicates that for a real system considerable deviations from the Drude-Lorentz behaviour can be expected.

The task of the theory is to calculate $M''(\omega)$. Once $M''(\omega)$ is known, $M'(\omega)$ can be obtained by using the Kramers-Kronig relation. For an electron-impurity system the LOCA yields ([2.26], Appendix)

$$M''(\omega) = \pi \frac{m\omega_c^2}{\hbar^2} \sum_{\alpha\alpha'} \langle |U_{\alpha\alpha'}|^2 \rangle_1 \frac{1}{2} (X - X')^2$$

$$\cdot \int d\varepsilon \frac{f(\varepsilon) - f(\varepsilon - \hbar\omega)}{\hbar\omega} D_\alpha(\varepsilon) D_{\alpha'}(\varepsilon - \hbar\omega). \tag{2.3.40}$$

Equations (2.3.39, 40) are a plausible generalisation of the Titeica-type formula (2.3.16). They allow us to calculate the transverse dynamic conductivity for all values of ω and B, for any impurity potential for which the spectral functions $D_\alpha(\varepsilon)$ are available in GBA. In the following however, we shall be concerned with the cyclotron resonance only.

Denoting the solution of the equation $\omega = \omega_c - M(\omega)$ by ω_c^* and assuming that $M'(\omega)$ and $M''(\omega)$ vary smoothly around ω_c^*

$$M''(\omega) \simeq M''(\omega_c^*), \quad M'(\omega) \simeq M'(\omega_c^*) + \frac{\partial M'(\omega_c^*)}{\partial \omega} (\omega - \omega_c^*) \tag{2.3.41}$$

holds in linear approximation and, from (2.3.39),

$$\sigma(\omega \pm i0) = \frac{e^2 n}{m} Z \frac{i(\omega - \omega_c^*) \pm i/\tau_c}{(\omega - \omega_c^*)^2 + (1/\tau_c)^2}, \tag{2.3.42}$$

where

$$Z = \left(1 + \frac{\partial M'(\omega_c^*)}{\partial \omega} \right)^{-1} \tag{2.3.43}$$

and

$$\frac{1}{\tau_c} = ZM''(\omega_c^*). \tag{2.3.44}$$

In this approximation $\sigma(\omega)$ shows a Lorentzian line shape. One should be aware, however, that the linear approximation is meaningless off resonance.

Even the crudest approximation $\omega_c^* \simeq \omega_c$, $Z = 1$, offers a possibility of obtaining information on the band structure by measuring the frequency

$$\omega \simeq \frac{eB}{m_c} = 2\pi eB(\partial A/\partial\varepsilon)^{-1} \tag{2.3.45}$$

at which the resonance occurs for different directions of the magnetic field relative to the crystal axis and for different electron concentrations. In particular, cyclotron resonance measurements have been a valuable tool for investigating the band structure of semiconductors [2.67a]. Because the subject has been reviewed recently in a comprehensive fashion by *Pidgeon* [2.67b], we shall give only a few examples.

A necessary condition for the observation of cyclotron resonance in semiconductors is $\omega_c\tau_c \geq 1$. Consequently, the use of submillimeter waves allows studying semiconductors with a mobility too low to observe resonance absorption with microwaves in the cm or mm wavelength range (which were employed in the early experiments). The development of molecular lasers and the application of very high (megagauss) magnetic fields has resulted in substantial progress in the field of cyclotron resonance [2.67b, 68, 69a].

Typical megagauss data for n-type Ge obtained with three different frequencies are shown in Fig. 2.18. The non-parabolicity of the conduction band, derived from the data, can be described by a two-band model.

Multivalley band structure will generally result in a multiline spectrum. Complicated spectra are observed for p-type material of group IV and III-V semiconductors [2.67a]. These so-called quantum effects arise from the degeneracy of the upper two valence bands and the third split-off band. A similar situation holds for p-type Te when the magnetic field is oriented perpendicular to the trigonal axis. Explanation of the data requires a detailed calculation of the energy levels [2.70].

However, a multiline cyclotron resonance absorption spectrum does not necessarily result from a complicated band structure. The doublet structure observed in the magnetotransmission of InSb by *McCombe* et al. [2.71] (Fig. 2.19) is caused by free carrier cyclotron resonance in an isotropic conduction band and by an electric dipole transition between an impurity ground state and an excited state corresponding to the first Landau level. A convenient way to distinguish between impurity and cyclotron resonance transitions is to study the temperature dependence of the transmission. This can be seen in Fig. 2.20, showing data for p-type Te obtained by *von Ortenberg* [2.72]. Because the magnetic field is parallel to the trigonal axis, a single cyclotron resonance line is expected. Of the 4 minima observed at 2.6 K only one survives if the temperature is raised to 13 K, indicating the cyclotron resonance transition. At sufficiently

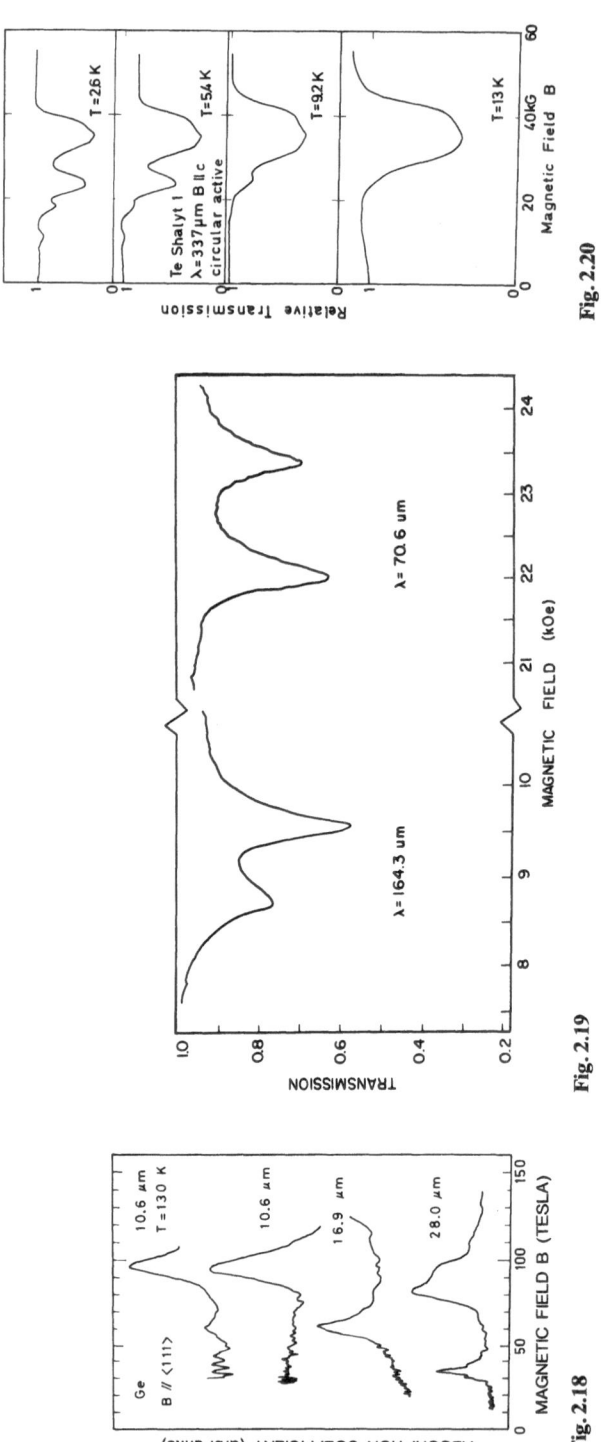

Fig. 2.18

Fig. 2.19. Cyclotron resonance in n-type Ge at various infrared frequencies at a temperature of 130 K [2.69b]

Fig. 2.19

Fig. 2.20

Fig. 2.19. Far infrared transmission in InSb at two different frequencies in the Faraday configuration. The lower field component of each doublet is an impurity transition, the upper one is cyclotron resonance [2.71]

Fig. 2.20. Magnetotransmission in Te parallel to the trigonal axis at various temperatures. The impurity transitions at lower B values vanish at higher T [2.72]

high temperature the impurities become ionised such that no optical transitions can occur.

At hole concentrations of the order 10^{14} cm^{-3}, the cyclotron resonance line shape can be substantially changed by interference [2.72]. To avoid these effects, it is advisable to use wedge-shaped samples or specimens with curved surfaces.

Transmission-type cyclotron resonance experiments in the Faraday configuration are restricted to materials with low carrier concentration because in semiconductors with high carrier density the penetration depth of the radiation is small. In this case a stripline technique can be used which has been employed in metal physics before. A review of the method has been given by *von Ortenberg* [2.73]. The analysis of the data is usually more complicated than in a straightforward transmission experiment, because the attenuation in the stripline of the transmitted radiation has to be calculated by solving Maxwell's equations under the appropriate boundary conditions. If the propagation vector of the radiation is parallel to the external magnetic field, no cyclotron resonance can be observed in a semiconductor with spherical energy surfaces like HgSe [2.73]. In a multivalley semiconductor like PbSe the coupling between radiation and electron motion occurs via tilted orbits. An example is given in Fig. 2.21. The relative transmission has been plotted for various angles between the magnetic field and the propagation vector in the (100) plane, together with results of model calculations. For B parallel to [100] and [110] directions only a single resonance line is observed due to the (111) orientation of the valence-band maxima. For a tilt angle of 22.5°, e. g. two lines can be resolved. The fitting procedure allows one to determine the components of the effective mass tensor rather accurately. Similar data have been obtained for PbTe and $Pb_{1-x}Sn_xTe$ [2.75].

Another topic of interest is the field dependence of the cyclotron resonance linewidth. In principle its analysis should give information about scattering processes. Unfortunately, the present situation is not satisfactory, both theoretically and experimentally. This can immediately be recognised from Fig. 2.22, where for InSb the linewidth (in frequency units) vs magnetic field curves obtained by various researchers have been plotted. The three dashed curves represent experimental results by *Tyssen* et al. [2.76], *McCombe* et al. [2.71], and *Apel* et al. [2.77]. The three solid curves are theoretical results with comparable input parameters. The upper data were obtained by *Heuser* and *Hajdu* [2.78] in GBA using an isotropic Gaussian potential $(s=r)$. For $x=l^2/2r^2>1$ the difference between lifetime and collision (relaxation) time (the so-called vertex correction) can be neglected. In this limit the dynamic magnetoconductivity $\sigma(\omega)$ takes a form similar to that derived by *Shin* et al. [2.79] for all values of x. The cyclotron linewidth corresponding to this formula shows saturation for $x < 1$. In this regime, however, the vertex corrections become important and cause decreasing linewidth as one expects for $x \gg 1$. In the opposite limit $x \ll 1$ (which means $B > 8$ T for typical InSb parameters) the linewidth from [2.78] asymptotically vanishes like $1/B$ in agreement with the results of *Kawabata* [2.80], and *Lodder* and *Fujita* [2.81]. Applying the LOCA formula (2.3.40) to the anisotropic Gauss potential (2.2.51, 52) the somewhat steeper asymptotic law $1/B^{3/2}$ was

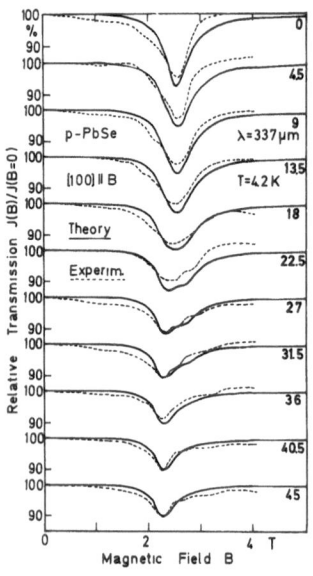

Fig. 2.21. Comparison of theoretical and experimental stripline transmission curves of PbSe for the magnetic field direction within the (100) plane. The parameter indicates the angle ((001), *B*) [2.74]

Fig. 2.22. Cyclotron resonance linewidth as a function of magnetic field. (——): different theoretical calculations; (---): experimental data [2.69c]

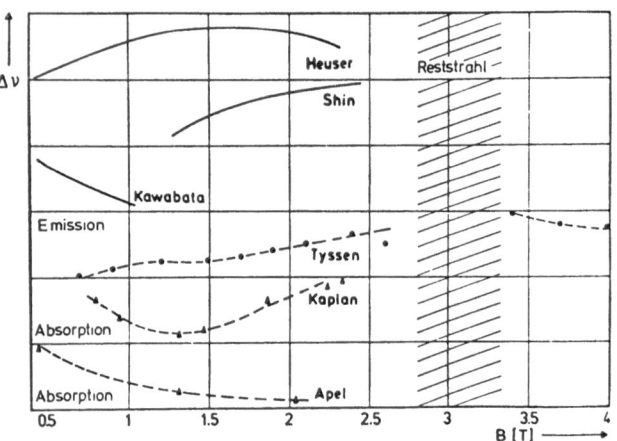

obtained in [2.9c]. The linewidth varies with the impurity concentration like $\sqrt{n_1}$.

It is possible to fit both the experimental data of *Apel* et al. [2.77], and *Tyssen* et al. [2.76] with appropriately chosen potential parameters: $r = s = 226$ Å, $\lambda_2/2\pi r^2 = 5.7$ meV, and 73 Å, 16 meV, respectively. In the first case, the part of the linewidth curve beyond the predicted maximum was matched, in the second case the increasing section. The rather large difference in the fitting parameters is surprising because the impurity concentrations of the specimens differed only by a factor of about 2.

A minimum in the linewidth was observed by *McCombe* et al. [2.71] for various samples with different carrier concentration. This behaviour cannot be explained by any of the theories cited, which take only the electron-impurity scattering into account. Furthermore, the observation of these authors that the position of the linewidth minima is independent of the carrier concentration is hard to understand, because differences in the screening should affect the range of the potential.

In recent calculations the combined effect of long-range impurity scattering (decreasing linewidth with increasing B) and electron-phonon interaction (increasing linewidth) was found to be responsible for the observed minimum [2.82, 83].

In a pure system, higher order collision processes which were not included in the previous calculations become important [2.84].

From the fact that InSb samples of not too different impurity concentration show large differences both in magnitude and field dependence of the cyclotron resonance linewidth, one must draw the conclusion that more experimental and theoretical work on this subject is needed.

2.4 Thermomagnetic Effects

Ohm's law (2.3.1) is a special case of the general phenomenological linear relation between currents and forces

$$
\begin{aligned}
\boldsymbol{j} &= L^{11}\left[\frac{1}{e}\left(\nabla\zeta - \frac{\zeta}{T}\,\nabla T\right) + \boldsymbol{E}\right] + L^{12}\left(-\frac{\nabla T}{T}\right) \\
\boldsymbol{q} &= L^{21}\left[\frac{1}{e}\left(\nabla\zeta - \frac{\zeta}{T}\,\nabla T\right) + \boldsymbol{E}\right] + L^{22}\left(-\frac{\nabla T}{T}\right).
\end{aligned}
\tag{2.4.1}
$$

Here $\boldsymbol{q}$ is the electronic energy current density. The transport coefficients L^{pq} defined by (2.4.1) are tensorial quantities and obey the Onsager relations

$$
L^{12} = L^{21}
\tag{2.4.2}
$$

(electrothermal symmetry) and

$$
L^{pq}_{\mu\nu}(\boldsymbol{B}) = L^{qp}_{\nu\mu}(-\boldsymbol{B}).
\tag{2.4.3}
$$

Note that in (2.4.1) $\nabla\zeta$ and $\boldsymbol{E}$ appear in the combination

$$
\frac{1}{e}\,\nabla\zeta + \boldsymbol{E} = \frac{1}{e}\,\nabla\zeta - \nabla\varphi = \nabla\left(\frac{\zeta}{e} - \varphi\right) = \nabla\zeta^* = \boldsymbol{E}^*,
\tag{2.4.4}
$$

where ζ^* is the electrochemical potential. Thus, the electric conductivity tensor and the diffusion tensor are identical, as required by the Einstein relation.

It has been pointed out that the transport coefficients L^{pq} consist of a regular part S^{pq} and a non-dissipative local equilibrium part D^{pq}

$$L^{pq} = S^{pq} + D^{pq};$$ (2.4.5)

$D^{11} = 0$, $D^{pq}_{\mu\nu} = -D^{pq}_{\nu\mu}$ [2.85–88]. The general Kubo formulae for the regular transport coefficients

$$S^{pq}_{\mu\nu} = \int_0^\infty \int_0^\beta \langle j^p_\mu(t+i\lambda) j^q_\nu(0) \rangle \, d\lambda \, dt$$ (2.4.6)

satisfy the Onsager relations as well [2.89].

The experimental results are usually expressed in terms of the thermoelectric power α or $\beta = \sigma\alpha$ and the heat conductivity $\varkappa$ (all tensorial) defined by

$$j = \sigma(E + \alpha\nabla T) = \sigma E + \beta\nabla T$$
$$q = \gamma E - \varkappa\nabla T.$$ (2.4.7)

They are related to L^{pq} by

$$\alpha = \frac{1}{T}\left[L^{12}(L^{11})^{-1} + \frac{\zeta}{e} \right]$$ (2.4.8)

and

$$\varkappa = \frac{1}{T}[L^{22} - L^{21}(L^{11})^{-1}L^{12}].$$ (2.4.9)

Owing to the Onsager relations (2.4.2, 3) $\gamma_{\mu\nu}(\boldsymbol{B}) = T\beta_{\nu\mu}(-\boldsymbol{B})$.

For a system consisting of a degenerate electron gas and fixed impurities, $\varkappa$ and α are connected with the conductivity σ by the Wiedemann-Franz law

$$\varkappa = LT\sigma$$ (2.4.10)

and the Mott rule

$$\alpha = (-e)LT\sigma^{-1}\frac{d\sigma}{d\zeta},$$ (2.4.11)

where $L = (\pi k_B/e)^{2/3}$ is the Lorentz number, for arbitrary values of the magnetic field [2.90].

Putting $\boldsymbol{B} = (0, 0, B)$ the leading order contributions $1/\omega_c \tau$ to the transverse transport coefficients are

$$\sigma_{xy} = -\frac{en}{B}, \tag{2.4.12}$$

$$\beta_{xy} = -\frac{ns_1}{B} \tag{2.4.13}$$

[2.85], and

$$\varkappa_{xy} = \frac{T}{eB} \int_{-\infty}^{\zeta} \left(\frac{\partial s_1}{\partial T}\right)_{\zeta'} d\zeta', \tag{2.4.14}$$

where s_1 is the entropy of the system per electron [2.86], and

$$\sigma_{xx} = S_{xx}^{11}, \quad \beta_{xx} = -\frac{1}{T} S_{xx}^{12}, \quad \varkappa_{xx} = \frac{1}{T} S_{xx}^{22}, \tag{2.4.15}$$

with [2.91]

$$S_{xx}^{pq} = e^{4-p-q} n \sum_{\alpha\alpha'} \frac{2\pi}{\hbar} \langle |U_{\alpha\alpha'}|^2 \rangle_I \frac{1}{2} (X - X')^2 \varepsilon_\alpha^{p+q-2}$$

$$\cdot \int d\varepsilon \left(-\frac{\partial f}{\partial \varepsilon}\right) D_\alpha(\varepsilon) D_{\alpha'}(\varepsilon) \tag{2.4.16}$$

Equation (2.4.16) is an obvious generalisation of the conductivity formula (2.3.16). According to (2.4.12, 13) the so-called transverse magneto-Seebeck coefficient $\alpha_{xy} = \beta_{xy}/\sigma_{xy}$ is

$$\alpha_{xy} = \frac{s_1}{e}. \tag{2.4.17}$$

In the constant damping approximation

$$s_1 = \frac{1}{\text{Vol}} k_B T \frac{\partial}{\partial T} \sum_\alpha \int D(\varepsilon - \varepsilon_\alpha) \log (1 + \exp [(\zeta - \varepsilon)/k_B T]) d\varepsilon. \tag{2.4.18}$$

Evaluating this expression for $\hbar\omega_c \ll \zeta$ by using the Poisson summation formula (2.3.27) we obtain

$$\alpha_{xy} = \alpha_{xy}^0 \left[1 - \frac{3}{\pi^2} \sum_{r=1}^{\infty} c_r \cos \left(\frac{2\pi\zeta}{\hbar\omega_c} r - \frac{\pi}{4}\right)\right] \tag{2.4.19}$$

$$c_r = \frac{(-1)^r}{r^{3/2}} \left(\frac{\hbar\omega_c}{2\zeta}\right)^{3/2} A_r(k_B T/\hbar\omega_c) \cos\,(\pi M r) \exp\left(-\frac{2\pi\Gamma}{\hbar\omega_c}\,r\right) \qquad (2.4.20)$$

$$A_r(x) = \frac{1}{x}\left[(1-2\pi^2 rx)\,\frac{\text{cth}\,(2\pi^2 rx)}{\text{sh}\,(2\pi^2 rx)}\right], \qquad (2.4.21)$$

where

$$\alpha_{xy}^0 = (\pi^2/2)\,(k_B/e)\,(k_B T/\zeta). \qquad (2.4.22)$$

Equation (2.4.19) is similar to (2.3.28). Nonetheless α_{xy} is analogous to ϱ_{xy} rather than to ϱ_{xx} since, first, it is of 0th order in $1/\omega_c\tau$ and the first correction is of order $(1/\omega_c\tau)^2$, and secondly, no factor $(-\partial f/\partial\varepsilon)$ occurs. For degenerate systems this factor indicates that the transport is restricted to the vicinity of the Fermi energy. For non-degenerate statistics $-\partial f/\partial\varepsilon = f/k_B T$ becomes small at high temperatures. Measuring the transverse magneto-Seebeck coefficient α_{xy} is in some cases preferable to measuring the resistivity, although it is more difficult, because the quantum oscillations are observable even under non-degenerate conditions (up to ~ 100 K).

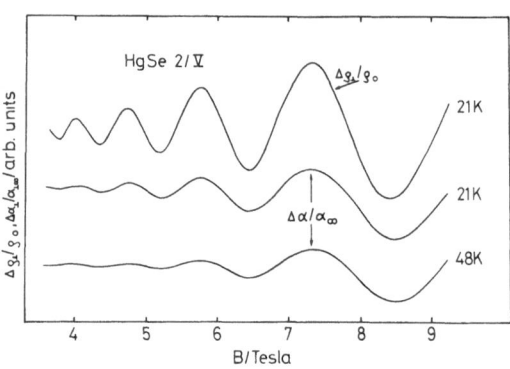

Fig. 2.23. Quantum oscillations of the transverse magnetoresistance and the transverse magneto-Seebeck coefficient in HgSe [2.92]

Figure 2.23 shows the relative change of α_{xy} and ϱ_{xx} observed in HgSe with an electron concentration of 8.5×10^{17} cm^{-3}. Comparison with (2.4.19) yields $m_c = 0.042\, m_e$ and $T_D \simeq 5$ K [2.92] in agreement with SdH data [2.93]. The longitudinal transport coefficients are given by

$$L_{zz}^{pq} = \frac{e^{4-p-q}n}{m} \sum_\alpha \hbar^2 k_z^2 \tau_\alpha \varepsilon_\alpha^{p+q-2} \int\left(-\frac{\partial f}{\partial\varepsilon}\right) D(\varepsilon-\varepsilon_\alpha)\,d\varepsilon, \qquad (2.4.23)$$

where τ_α is the solution of (2.3.22 with 25) [2.91a].

2.5 Two-Dimensional Systems

2.5.1 Electric Subbands in Inversion Layers and Similar Systems

Most of the research in magnetotransport has been performed on bulk material. During the last decade, however, several two-dimensional electronic systems have been produced and investigated. The term "two-dimensional" implies that the carriers are free to move parallel to a plane, but not perpendicular to it. One possibility of achieving these conditions is to generate a high electric field at the surface of a semiconductor so that an inversion or accumulation layer is formed. A practical example of such a system is the silicon MOSFET.

An n-channel MOSFET, shown in Fig. 2.24, consists of a p-type substrate into which n-type source and drain contacts have been implanted. The surface is covered with silicon dioxide about 100 nm thick. On top of the dielectric is a metal film (gate). Application of a positive gate voltage of several volts induces an n-type inversion layer. An important advantage of the MOSFET is that the electron concentration in the surface channel may be varied continuously between 10^{11} and 10^{13} cm^{-2} by changing the gate voltage.

To obtain a two-dimensional system, it is necessary to produce an almost atomically flat Si-SiO$_2$ interface and simultaneously to have a low impurity concentration in both surface channel and oxide. The low temperature mobility of four n-channel (100) devices, investigated by *Dorda* and co-workers [2.94a], is shown in Fig. 2.25 as a function of the surface electron concentration n. The highest mobility is measured for the device with a threshold voltage V_{th} of 4.4 V and amounts to about 19,000 cm^2/Vs. The lower mobility in the other samples with smaller V_{th} is caused by a larger amount of ionised impurity scattering. The increase of the mobility in the lower n range is caused by localisation and screening effects, and the decrease at high electron concentrations by surface roughness scattering. It is obvious that the mobilities are high enough to meet the condition $\omega_c \tau > 1$ in magnetic fields of a few tesla.

The silicon MOSFET is a rather singular device for two reasons: amorphous SiO$_2$ is a high quality dielectric and the interface Si-SiO$_2$ can be produced with a very low density of trap states. Various attempts to make MOSFETs of other semiconductors have had only very limited success.

Fortunately, it has been possible in the last decade to generate high quality 2d systems in a different fashion. The molecular beam epitaxy technique (MBE) has allowed the production of heterostructures made of III-V semiconductors and their alloys, in which carriers are confined in a narrow 2d channel caused by the interface potential of two lattice-matched materials and by an impurity space charge. The best known example is the GaAs-Ga$_{1-x}$Al$_x$As heterostructure. Very high electron mobilities at helium temperatures ($>10^6$ cm^2/Vs) have been achieved by the modulation doping technique [2.10a]. The essential point is that only the Ga$_{1-x}$Al$_x$As – having a larger gap than the GaAs – is doped and that free carriers spill over into the GaAs. The reason for this high mobility is that the spatial separation between impurities and carriers reduces the Coulomb

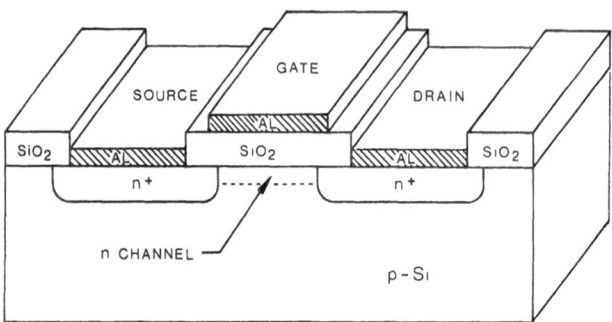

Fig. 2.24. An n-channel MOSFET

scattering substantially. Especially high mobilities are achieved if an undoped $Ga_{1-x}Al_xAs$ layer is inserted between the doped sheet and the GaAs layer [2.94b]. The energy band structure of a heterojunction is shown schematically in Fig. 2.26.

According to quantum mechanics the motion in a one-dimensional potential well gives rise to discrete energy levels which, in the present context, are called electric subbands. This quantisation is essential if the width of the well is comparable to the de Broglie wavelength of the electrons. For electric fields of the order of 10^5 V/cm subband splittings of about 10 meV or more are expected. Thus, quantisation effects in inversion layers and similar systems can even be important at room temperature. The electrons confined to a layer behave as a two-dimensional electron gas as long as only states within the lowest electric subband are occupied. In the following, with a few exceptions, we shall restrict ourselves to this important case.

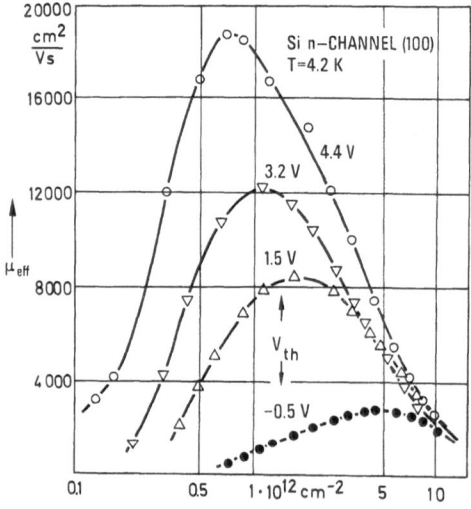

Fig. 2.25. Effective mobility μ_{eff} of n-channel MOSFETs as a function of the the electron concentration n with the threshold voltage V_{th} as a parameter [2.94a]

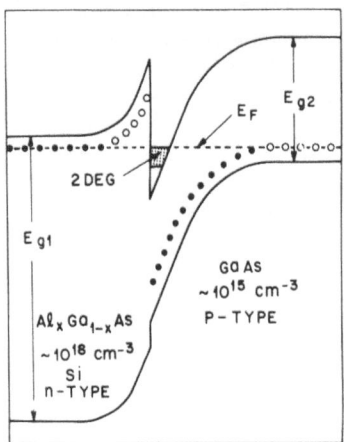

Fig. 2.26. Energy level diagram of a modulation doped GaAs-(GaAl)As heterojunction (2DEG: 2d electron gas) [2.95]

Quasi-two-dimensional electron gas systems are not only useful for studying quantum transport phenomena. They are appropriate tools to investigate many-body effects (brought about by the Coulomb interaction between the electrons), in particular correlations possibly leading to ordering (charge density waves, Wigner crystallisation), and different disorder effects (especially localisation).

Following the pioneer work of *Stern* and *Howard* [2.96] we consider now an electron gas confined by a potential $V(z)$ to a thin layer parallel to the (x,y) plane. The energy eigenfunctions and eigenvalues of an electron in the single-particle ellipsoidal effective mass approximation, representative for the (100) orientation in n-type silicon, are

$$\psi(r) = N \exp\left[i(k_x x + k_y y)\right] \varphi_l(z) \tag{2.5.1}$$

$$\varepsilon_l(k_x, k_y) = \varepsilon_l + \frac{\hbar^2}{2m_\parallel}(k_x^2 + k_y^2) \tag{2.5.2}$$

$(m_\parallel = m_1 = m_2,\ m_\perp = m_3)$. The terms $\varphi_l(z)$ and ε_l are determined by the one-dimensional Schrödinger equation

$$\frac{d^2\varphi_l}{dz^2} + \frac{2m_\perp}{\hbar^2}\left[\varepsilon_l - V(z)\right]\varphi_l = 0; \tag{2.5.3}$$

$l = 0, 1, 2, \ldots$, denumerates the electrical subbands. The layer potential $V(z)$ has to be determined self-consistently by solving (2.5.3) and the one-dimensional Poisson equation

$$\frac{d^2 V}{dz^2} = \frac{e}{\varkappa_0 \varkappa} \varrho(z) \tag{2.5.4}$$

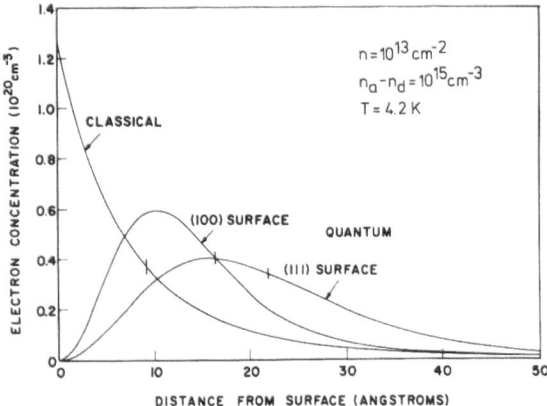

Fig. 2.27. Electron density as a function of the distance from the surface [2.97]

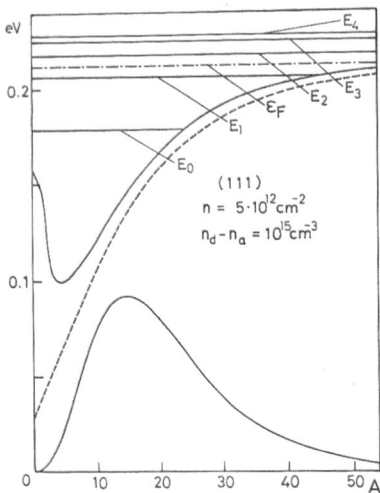

Fig. 2.28. The energy of holes in a p-type (111) Si inversion layer calculated in the Hartree approximation taking into account the image potential. E_0, E_1, ... are the electric subband energies [2.98]

simultaneously, where

$$\varrho(z) = e(n_{\mathrm{d}} - n_{\mathrm{a}}) - e \sum_l n_l |\varphi_l(z)|^2 \tag{2.5.5}$$

is the charge density in the layer. In (2.5.5) n_{d} and n_{a} denote the density of donors and acceptors in the layer, respectively, and n_l denotes the concentration of electrons in the lth subband; $\varkappa$ is the static dielectric constant.

Self-consistent solutions of (2.5.3, 4) show that the charge distribution close to the surface is very different for the classical and the quantum case. In Fig. 2.27 the electron concentration has been plotted as a function of the distance from the surface [2.97]. In the classical case the electron concentration has its highest value at the surface and decreases steadily with increasing distance. In contrast to this, the highest electron concentration in the quantum case is found between 1–2 nm, depending on the orientation of the surface. The calculated self-consistent potential for a p-type (111) silicon surface (see below) for the indicated hole density and impurity concentration is shown in Fig. 2.28 as a dotted line. The bending over of the potential at distances above 20 Å is due to the depletion layer. The solid curve is the result of a refined calculation by *Ando* [2.103c], taking into account the image potential and many-body effects. The influence of the image potential is significant, because the difference in the polarisability of Si and SiO_2 is substantial and increases the subband energies. On the other hand, many-body effects such as exchange and correlation lower the energy levels relative to the values obtained in a Hartree calculation. One can recognise that in Fig. 2.28 the energy difference between the lowest subband ε_0 and the next higher one ε_1 is about 25 meV, that is about k_BT at room temperature. Consequently, one expects that surface quantisation effects play a role not only at helium temperatures.

The density of states (2.5.1) on the energy scale per unit area is constant:

$$n_l(\varepsilon) = \frac{2}{L_x L_y} \sum_{k_x k_y} \delta(\varepsilon - \varepsilon_l(k_x, k_y)) = \frac{m}{\pi \hbar^2}, \tag{2.5.6}$$

where

$$m = \sqrt{m_1 m_2}. \tag{2.5.7}$$

If, in the (k_x, k_y) plane, the lines of constant energy consist of g_v identical ellipses (valley degeneracy), and the spin degeneracy g_s is not specified to be 2, an additional factor $g_v g_s/2$ appears in the density of states formula (2.5.6). For a n-type Si-SiO_2 interface in the (111) plane, for instance, the single-electron approximation yields $g_v = 6$ and $g_s = 2$.

If only states in the lowest ($l = 0$) electrical subband are occupied the Fermi energy corresponding to (2.5.6) for arbitrary valley and spin degeneracy, and electron density $n = N/L_x L_y$ is

$$\varepsilon_F = \frac{2\pi \hbar^2}{g_s g_v m} n. \tag{2.5.8}$$

For p-type silicon inversion layers the situation is not fundamentally different. However, due to the valence band structure the self-consistent calculations are more complicated than in the n-type case. It is necessary to take all three valence bands into account because the energy difference between the degenerate light and heavy hole band ($k = 0$) and the third valence band is only 44 meV. Within the framework of the effective mass approximation the actual valence band structure can be taken into account by employing the well-known

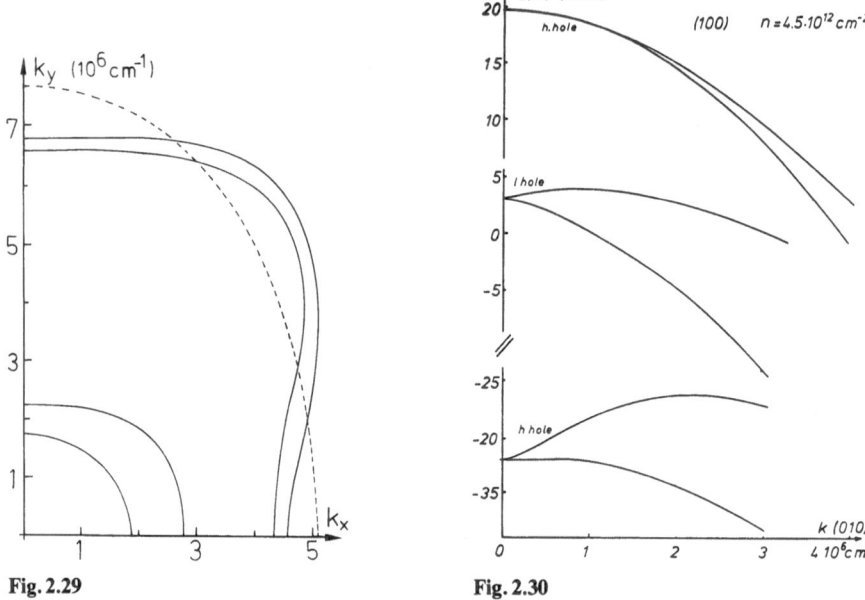

Fig. 2.29

Fig. 2.30

Fig. 2.29. Two-dimensional Fermi contours of a p-type (110)-Si inversion layer in the E_0 and E_0' subbands (———) and the bulk analogy of the E_0 contour in the (k_x, k_y) plane [2.101]

Fig. 2.30. Spin splitting of the three highest subbands of a p-type (100)-Si surface for k_x in [010] direction and $k_y = 0$ [2.103a]

6×6 *Luttinger-Kohn* Hamiltonian [2.99]. The calculations were independently performed by *Bangert* et al. [2.100] (cf. also [2.101]), and by *Ohkawa* and *Uemura* [2.102]. Electric subbands, hole distributions and constant energy contours were calculated. It turned out that for the heavy and light hole subbands the spin degeneracy is lifted for finite wave vectors due to spin-orbit interaction. There are two heavy holes and two light holes with different masses. The splitting of the constant energy contours can be visualised in Fig. 2.29 for the (110) orientation. The constant energy contour indicated by the dashed line was derived from the bulk $\varepsilon(k)$ relation. It was constructed so that the enclosed area at $k_x = k_y = 0$ was the same as for the ε_0 electric subband. It is obvious that the surface electric field has significantly changed the shape of the two-dimensional $\varepsilon(k)$ surfaces. The calculations have also been performed for (100) and (111) surfaces [2.101, 102]. In both cases the spin degeneracy is lifted and the anisotropy of the two-dimensional $\varepsilon(k)$ surfaces differs from the corresponding ones in the bulk.

The $\varepsilon(k)$ dispersion relation for the three highest subbands for a (100) surface for $k_x = [010]$ and $k_y = 0$ is shown in Fig. 2.30. The splitting of the highest heavy hole subband is rather small. It should be pointed out, however, that a slightly different set of band structure parameters gives a larger splitting [2.102, 103a]. In silicon the lifting of the spin degeneracy by the surface electric field has only recently been observed experimentally [2.224]. Recent data by *Störmer* et al. [2.104] on the Shubnikov-de Haas effect in a p-type GaAs-GaAlAs heterostructure with high hole mobility have given evidence of a splitting of the heavy hole band. Subsequent self-consistent calculations in the Hartree approximation showed that for GaAs the subband splitting should be substantial although no quantitative agreement with the experimental data was achieved [2.105a]; possible reasons for the discrepancy will be discussed in the subsequent chapter on 2d-cyclotron resonance.

2.5.2 2D Electron Gas in Strong Magnetic Fields

A strong magnetic field parallel to the z axis splits the energy spectrum (2.5.2) into discrete Landau levels $\varepsilon_\alpha = \varepsilon_l + \varepsilon_{vs}$

$$\varepsilon_{vs} = \hbar\omega_c \left(v + \frac{1}{2}\right) + sg\mu_B B, \quad \alpha = (l, v, s). \tag{2.5.9}$$

The corresponding density of states consists of a series of δ functions at the energies (2.5.9)

$$n(\varepsilon, B) = \frac{eB}{h} \sum_{v,s} \delta(\varepsilon - \varepsilon_{vs}) \tag{2.5.10}$$

for each electrical subband, and is shown schematically in Fig. 2.33 ($h = 2\pi\hbar$).

The degeneracy factor $D = eB/h$ is the number of states per unit area within a single Landau level;

$$\eta = n/D = n\frac{h}{eB} \tag{2.5.11}$$

is called the filling factor. Since $\phi = BL_x L_y$ is the magnetic flux through the system and $\phi_0 = h/e$ is the elementary flux quantum

$$\eta = \frac{\text{Number of electrons}}{\text{Number of flux quanta}}. \tag{2.5.12}$$

The Fermi energy as a function of the electron concentration and the magnetic field is depicted in Figs. 2.31, 32, respectively.

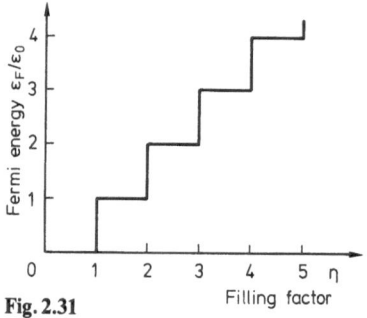

Fig. 2.31

Filling factor

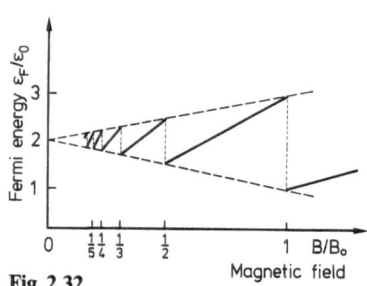

Fig. 2.32

Magnetic field

Fig. 2.31. Fermi energy of a 2d free electron gas as a function of the filling factor at a constant magnetic field

Fig. 2.32. Fermi energy of a 2d free electron gas as a function of the magnetic field at constant electron concentration

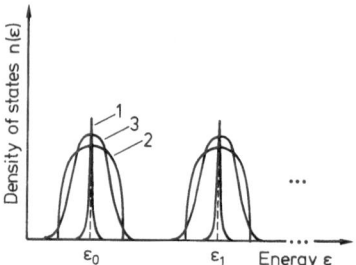

Fig. 2.33. Density of states of a 2d electron-impurity system in different approximations. (1) vanishing impurity potential (δ functions); (2) GBA (semi-ellipses); (3) higher order perturbation theory or cumulant approximation (Gaussians)

Scattering by impurities gives rise to level broadening. For randomly distributed Gaussian impurities the GBA leads to the semi-elliptical density of states

$$n(\varepsilon) = \frac{eB}{hc} \sum_{v,s} \frac{2}{\pi \Gamma_v} \left[1 - \left(\frac{\varepsilon - \varepsilon_{vs}}{\Gamma_v} \right)^2 \right]^{1/2} \tag{2.5.13a}$$

derived in Sect. 2.2 and shown in Fig. 2.33. The coefficient appearing in (2.2.54) is now $a = (n_i/r^2) \lambda_2^2/4\pi$. In the limit of point impurities

$$\Gamma_v^2 = \Gamma^2 = \frac{2}{\pi} \hbar \omega_c \hbar/\tau, \tag{2.5.14}$$

where τ is the relaxation time for zero magnetic field. Both (2.5.13a, 14) were first obtained by *Ando* and *Uemura* [2.106].

According to (2.5.13a) the spectral functions (partial density of state) vanish at $\varepsilon = \varepsilon_{vs} \pm \Gamma_v$ with vertical tangent. Higher order approximations (which include multiple-scattering processes) lead to short analytic band tails [2.107]. Different non-perturbation approaches yield the Gaussian density of states [2.107–109]

$$n(\varepsilon) = \frac{eB}{h} \sum_{v,s} \frac{\sqrt{2/\pi}}{\Gamma_v} \exp\left[-2\left(\frac{\varepsilon - \varepsilon_{vs}}{\Gamma_v}\right)^2\right]. \tag{2.5.13b}$$

The assumption of Gaussian impurities simplifies the calculations considerably. In a more ambitious theory the impurity potential has to be calculated in a self-consistent way, as described in Sect. 2.2. Such a calculation has been carried out for the quantum limit by *Das Sarma* [2.110] (who uses the factorisation approximation (2.2.65), i.e. he neglects the important vertex corrections) and in [2.9d].

The randomly distributed impurities are not only responsible for electron scattering and level broadening but also for individual bound states, satellite impurity bands and different types of localisation (cf. the following sections).

The Coulomb interaction between the electrons may become important in two dimensions. At low electron concentration (about $\eta < 1/7$) and sufficiently low temperatures ($k_B T \ll \varepsilon_F$), the average Coulomb energy per electron will exceed the average kinetic energy (ε_F), allowing in principle a highly correlated many-electron state (Wigner crystal or similar "charge density waves"). In the presence of impurities such a state may be suppressed or pinned.

These few remarks indicate once more the fundamental importance of 2d systems (inversion layers with variable electron concentration and high carrier mobility heterojunctions) in high magnetic fields (leading to complete quantisation, i.e. separate degenerate energy levels) to prove or disprove sophisticated theoretical concepts and speculations.

2.5.3 Magnetoresistance

An approximate expression for the static conductivity of a 2d system can be obtained by using (2.3.39) for $\omega = 0$ and inserting in (2.3.40) the GBA spectral functions corresponding to (2.5.13a) [2.111] [$M'(\omega = 0) = 0$ in LOCA]. The resulting expression for high magnetic fields, $\omega_c \tau \gg 1$,

$$\sigma_{xx} = \frac{e^2}{2\pi^2 \hbar} \sum_s \int \left(-\frac{\partial f}{\partial \varepsilon}\right) \left(\frac{\Gamma_v^{xx}}{\Gamma_v}\right)^2 \left[1 - \left(\frac{\varepsilon - \varepsilon_{vs}}{\Gamma_v}\right)^2\right] d\varepsilon \tag{2.5.15}$$

was first derived by *Ando* and *Uemura* [2.106, 112] and later by *Gerhardts* [2.108, 103b] – in both cases by direct approximate evaluation of the Kubo formula (2.3.26) for $\omega = 0$. In (2.5.15) the chemical potential is assumed to lie in the vth Landau band. The factor $(\Gamma_v^{xx}/\Gamma_v)^2$ can easily be calculated for Gaussian impurities. One obtains

$$(\Gamma_0^{xx}/\Gamma_0)^2 = \frac{x}{1+x} \tag{2.5.16}$$

and similar rational expressions in $x = l^2/2r^2$ for $v = 1, 2, \ldots$. [The 2d Gauss potential and its Fourier transform is obtained from (2.2.51) by putting $z = 0$, $s = 1/\sqrt{2\pi}$ and $q_z = 0$.] In the case of point impurities ($r \to 0$)

$$(\Gamma_v^{xx}/\Gamma_0)^2 = 2v + 1 \tag{2.5.17}$$

and, for $T = 0$, the peak values of the conductivity depend only on elementary constants and the quantum number v [2.107, 108, 103b], i.e.

$$(\sigma_{xx})_{\text{peak}} = \frac{e^2}{\pi^2 \hbar} (v + \tfrac{1}{2}). \tag{2.5.18}$$

The analysis of the SdH oscillations is similar to the 3d case, with some exceptions, however. Since the average effective mass (2.5.7) appears in the cyclotron frequency $\omega_c = eB/m$, the period of oscillations on the $1/B$ scale

$$\varDelta \left(\frac{1}{B} \right) = \frac{\hbar e}{m} \frac{1}{\varepsilon_F} = \frac{g_s g_v e}{h} \frac{1}{n} \tag{2.5.19}$$

turns out to be independent of m. If n is known (e.g. from measurements of the Hall coefficient), the product $g_s g_v$ can be deduced. In the case of a MOSFET n can also be determined by using the condensor equation

$$n = -\frac{\varkappa_0 \varkappa}{ed} (V_g - V_{\text{th}}), \tag{2.5.20}$$

where d is the layer thickness, V_g is the variable gate voltage and V_{th} the threshold voltage (which can be determined, e.g., from the extrapolated period of SdH oscillations). If several electric subbands are occupied it is possible to deduce the electron density n_l in each subband from the period of the corresponding SdH oscillations.

The expression for the amplitude b_r is similar to the one derived for bulk (3d) systems, (2.3.29): only the factor $(\hbar\omega_c/2\varepsilon_F)^{1/2}$ is missing in the 2d case [2.113]. Again, by studying the variation of the $r = 0$ amplitude with the temperature and the magnetic field strength, the effective mass and the Dingle temperature can be determined.

The most intensely investigated 2d system is the silicon MOSFET. The main orientations (100), (110) and (111) have been studied both for n- and p-type inversion. Most information is available for the (100) n-type inversion, mainly because the electron mobility at 4.2 K is the highest for this orientation. The scattering length is considerably smaller than the magnetic length and, therefore, the model of point scatterers is applicable. Shubnikov-de Haas oscillations in an n-channel (100) MOSFET have been shown already in Fig. 2.4. In that experiment, the gate voltage V_g was varied at a constant magnetic field. Data

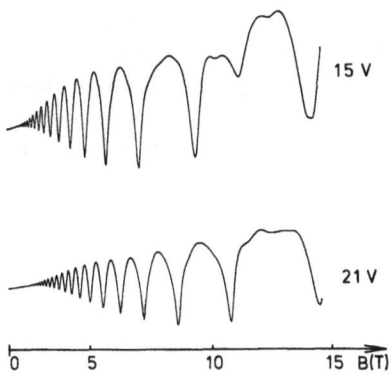

Fig. 2.34. SdH oscillations for a (100) Si MOSFET at different values of the gate voltage [2.114]

obtained at two different constant gate voltages (i.e. electron concentrations) as a function of a transverse magnetic field [2.114] have been plotted in Fig. 2.34. One can recognise that the period decreases with increasing gate voltage. In the last maximum of the lower curve a structure appears which is resolved in the upper trace ($V_g = 15$ V). The two broad maxima are due to spin splitting, whereas the additional structure is caused by the so-called *valley splitting*. For silicon (100) layers only the twofold degenerate subband is populated and the second, fourfold degenerate subband system is so much higher in energy that it is empty at 4.2 K. The data indicate that the twofold degeneracy of the lowest electrical subband, predicted by the calculations of *Stern* and *Howard* [2.96], is lifted.

Various explanations of the observed valley splitting have been offered. *Ohkawa* and *Uemura* [2.115] have pointed out that "k-space tunnelling" may cause overlapping of the wave functions (corresponding to the two occupied valleys) at the Γ point of the Brillouin zone. This mechanism obviously lifts the twofold degeneracy. *Sham* and *Nakayama* [2.116a] discuss the role of the boundary conditions at the Si-SiO$_2$ interface and use more realistic wave functions. Their surface scattering mechanism leads to a valley splitting larger than the k-space tunnelling. *Kümmel* [2.117] has suggested that the valley splitting might be the manifestation of the symmetry breaking by the surface electric field in conjunction with spin orbit interaction. Since the valley splitting predicted by all these single-electron theories is considerably smaller than the observed one, many-body effects have been invoked to explain the experimental data [2.12].

The effect of uniaxial stress on silicon inversion layers has been extensively investigated by *Dorda* and co-workers [2.118]. Uniaxial pressure applied to the [010] direction of a (100) surface of a Si MOSFET lowers the energy (ε_0') of two valleys of the fourfold degenerate subband and raises the energy (ε_0) of the two valleys in the twofold degenerate subband. For carrier concentrations of a few times 10^{12} cm^{-2} the electrons can be transferred completely into the ε_0' subband and an effective mass $m = 0.53 \, m_c$ has been extracted from the temperature dependence of the SdH oscillations. At intermediate pressures, however (when both the ε_0 and the ε_0' subbands are partially occupied), only one SdH period could be observed. On the other hand, cyclotron resonance experiments clearly show two separate lines [2.119]. The reason for this discrepancy has not yet been understood.

For the (110) and (111) orientations $g_v = 2$ has been extracted from the experimental data, in contradiction to the values 4 and 6, respectively, predicted by the single-electron theory [2.120, 121]. It has been pointed out by *Kelly* and *Falicov* [2.122] and others [2.123] that phonon-mediated electron intervalley interaction leading to a charge density wave ground state may be responsible for lifting the valley degeneracy in the (110) and (111) surfaces. Objections to this model have been raised, however, since the required coupling constants turn out to be much too large. This problem has not yet been settled. For samples treated especially in order to relieve strain at the Si-SiO$_2$ interface a valley degeneracy factor $g_v = 6$ has been observed [2.124, 116b]. In these samples, however, the electron mobility is considerably lower than in those handled in the usual way.

It should be pointed out that accurate experimental determination of valley splitting is rather difficult because of the problems in finding the actual position of the Fermi energy.

An advantage of the MOSFET is that SdH oscillations can be produced at fixed magnetic field strength by varying the gate voltage (which is, by (2.5.20, 8), proportional to the zero-field Fermi energy). Typical data for the transverse resistivity ϱ_{xx} and transverse conductivity σ_{xx} of a (100) MOSFET with Hall geometry at $B = 14.2$ T is shown in Fig. 2.35 (*Englert* and *v. Klitzing* [2.116b]). The sample is of high quality: at 4.2 K the maximum mobility exceeds 15,000 cm^2/Vs. In the vicinity of $V_g = 7$ V the resistance drops to a value at least 10^5 times smaller than at zero magnetic field (ϱ_0). Comparison of high field transverse resistivity and conductivity data show that the minima of ϱ_{xx} and σ_{xx} appear at the same values of the gate voltage. This confirms the predicted proportionality of these quantities in leading order of $(1/\omega_c\tau)$ (i.e., at Hall angles near 90°, Sect. 2.3). The conductivity data have been calculated by making use of the Hall resistivity ϱ_{xy} shown in Fig. 2.36. It should be noted that there is a plateau in ϱ_{xy} at about 6.5 kΩ in a gate voltage range where both ϱ_{xx} and σ_{xx} vanish. This is an early indication of the quantum Hall effect which is discussed subsequently.

Inspecting Fig. 2.36 one might be tempted to determine the electronic g factor from the observed *spin splittings* at low quantum numbers v. This is, however, not possible in the usual way, because at high magnetic fields the Fermi energy depends on the gate voltage (i.e. the electron concentration) in an oscillatory way (Fig. 2.32). In spite of this it has been possible to determine the g factor by applying a more sophisticated method which makes use of the 2d nature of the electron gas in a MOSFET. If the magnetic field is tilted by an angle relative to the normal direction of the surface, the cyclotron energy $\hbar\omega_c$ is decreased by a factor $\cos\varphi$, whereas the spin energy remains unchanged. The splitting of the Landau levels $\hbar\omega_c\cos\varphi$ can be decreased until it is twice the spin splitting. This method was first employed by *Fang* and *Stiles* [2.125] to measure

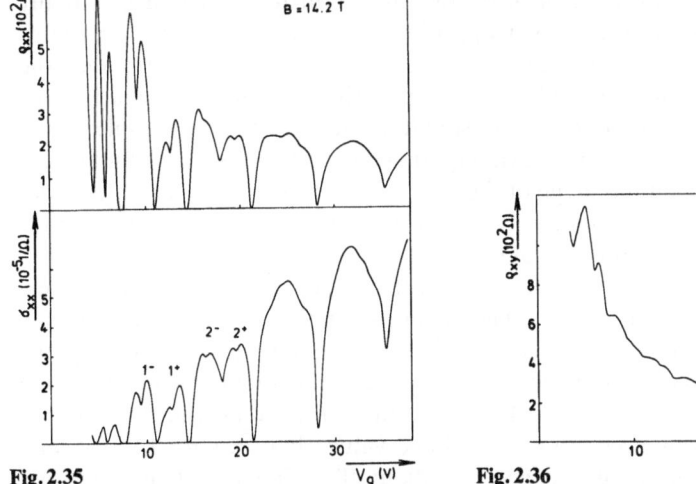

Fig. 2.35

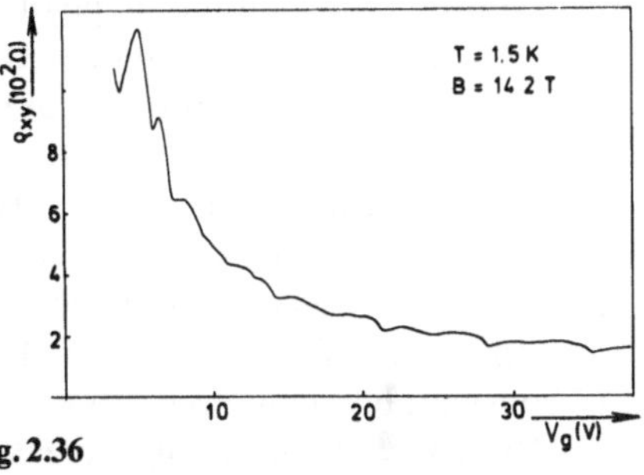

Fig. 2.36

Fig. 2.35. Quantum oscillations of ϱ_{xx} and σ_{xx} of an n-channel (100) Si MOSFET as a function of the gate voltage at $B=14.2\,\mathrm{T}$ and $T=1.5\,\mathrm{K}$ [2.116b]

Fig. 2.36. Hall resistivity of the same sample as in Fig. 2.35 [2.116b]

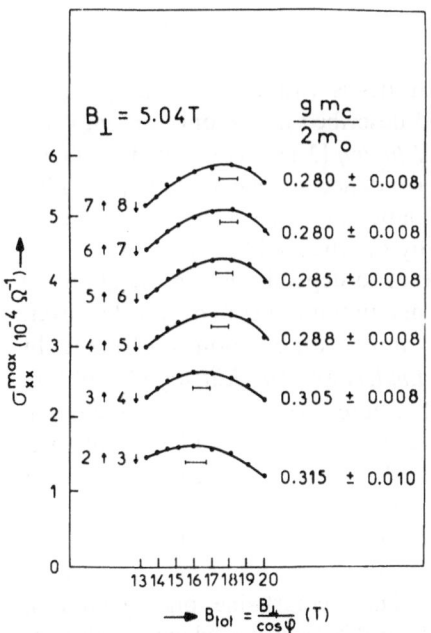

Fig. 2.37. Amplitude of SdH oscillations in a tilted field experiment. The normal component of B is kept constant at $B=5.04\,\mathrm{T}$. The maximum of the amplitude is obtained if two adjacent Landau levels (indicated on the left) coincide [2.225]

the ratio $\Delta = mg/2m_e$ as a function of the electron concentration in (100) silicon MOSFETs. Since the effective mass can be measured independently, it is possible to deduce the electronic g factor. If the tilt angle is further increased, eventually two adjacent Landau levels with an opposite spin occupation will merge. In this case a maximum of σ_{xx}, (2.5.15), is expected. Alternatively, one may keep the Landau splitting constant and increase the spin splitting by increasing the magnetic field by a factor $\cos\varphi$ and, simultaneously, tilting the sample by the angle φ. This procedure has the advantage that the oscillations do not shift on the gate voltage scale and the shape of the spectral function (density of states) does not play a role. A new equidistant level distribution is achieved when the tilt angle satisfies the condition $\Delta\cos\varphi = 3/2$.

An example of a tilted field experiment on (100) silicon inversion layers [2.225] is shown in Fig. 2.37. The magnetoconductivity σ_{xx} was directly measured in Corbino devices keeping the normal component of the magnetic field constant. Inspection of the figure shows that the first coincidence occurs at a tilt angle of 74°. A plot of Δ as a function of n reveals that the g factor is considerably enhanced above the bulk value. Between $n = 1 \times 10^{12}$ and $3 \times 10^{12}\,\mathrm{cm}^{-2}$ the g factor decreases by about 10%. Similar enhancement has been found for the (110) and the (111) orientations. *Ando* and *Uemura* [2.126] have pointed out that the g factor is renormalised by the exchange interaction between the electrons,

$$g^*\mu_B B = g\mu_B B + (\Sigma_\downarrow - \Sigma_\uparrow), \tag{2.5.21}$$

where $\Sigma_\downarrow$ and $\Sigma_\uparrow$ are the corresponding self-energies of electrons with spin-down and spin-up (relative to the direction of the magnetic field). The effective g factor g^* which appears in all macroscopic quantities is a function of the spin state occupation. With the coincidence method described above only the maximal value of g is determined. *Englert* and *von Klitzing* [2.116b] extracted a g factor from measurements of the temperature dependence of the conductivity. Within the experimental error – including a systematic one – they found indications of the oscillations of the g factor predicted by the theory [2.126].

The *peak values* of the conductivity seem to depend only on the quantum number v [2.127], Fig. 2.38. The measured maximum values of σ_{xx} are, however, smaller than predicted by (2.5.18). A similar observation was made by *Wakabayashi* and *Kawaji* [2.128] and by *Englert* and *von Klitzing* [2.116b] for long samples. However, on specimens of different origin the agreement was satisfactory [2.129]. The reason for this discrepancy has not yet been explained. The peak values of the conductivity obtained from Corbino disc geometry agree reasonably well with the prediction for $v = 1$ and 2.

Already the first experiments by *Fowler* et al. [2.6] clearly showed that the conductivity σ_{xx} vanishes for several finite ranges of the electron concentration at magnetic fields of 14.3 and 16.5 T. This remarkable phenomenon is accompanied by plateaux in σ_{xy} (quantum Hall effect), discussed in the following section. More recent data [2.9a] are shown in Fig. 2.5.

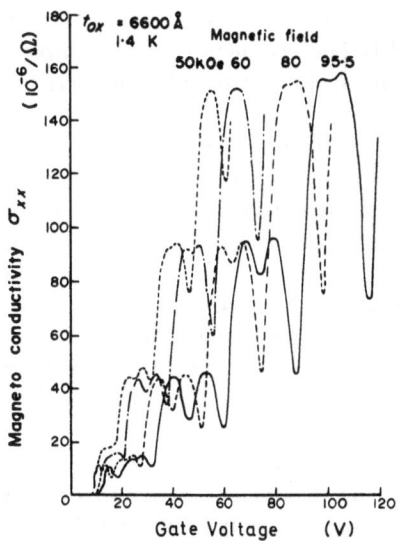

Fig. 2.38. Magnetoconductivity of a sample with Corbino geometry as a function of the gate voltage at four different transverse magnetic fields [2.127]

The range-wise vanishing conductivity σ_{xx} is an indication of some sort of "localisation". Possible mechanisms are the Anderson localisation and many-electron correlation effects brought about by the Coulomb interaction.

The *Anderson* localisation [2.130] (cf. also [2.131, 132], for recent developments [2.133–135]) occurs in a rapidly varying strong random (impurity) potential for sufficiently small values of the density of states; it is – in some respects – similar to bound states (with exponentially vanishing wave functions). The ranges of localisation (mobility gaps) are separated by finite mobility regimes which terminate in mobility edges, Fig. 2.39 [2.136]. It is generally believed that the density of states is not affected by localisation [2.109]. In the model of a single symmetric band particle-hole symmetry requires equal distance of the mobility edges from the band centre. Unfortunately, no satisfactory theory (which is free of ad hoc assumptions) of the Anderson localisation in the presence of a high magnetic field is currently available. The investigations of *Tsukada* [2.137], *Aoki* and *Kamimura* [2.138], and *Aoki* [2.139] clearly demonstrate, however, that Anderson localisation in high magnetic fields may well be responsible for the observed vanishing of the conductivity. According to

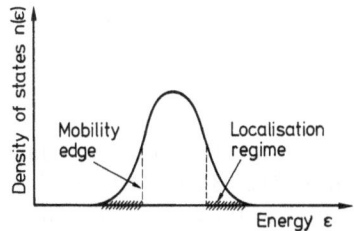

Fig. 2.39. Density of states band with mobility edges and localisation regimes

a recent elaborate theoretical investigation, at least one state at the centre of each Landau band is always delocalised [2.140]. This prediction is in complete agreement with observations, numerical results [2.133a, 226a, 141] and percolation arguments (see below).

The (strong) Anderson localisation should be distinguished from the "weak" (Anderson) localisation which is essentially a $\Delta k = 2 k_F$ scattering interference effect appearing in the limit of weak disorder. According to *Abrahams* et al. [2.143] (cf. also [2.135]) an infinite 2d system with weak disorder in the limit $T = 0$ and $B = 0$ is always in a localised state (insulator). A weak magnetic field does not destroy the weak localisation [2.142]. The observed negative magneto-resistance indicates, however, that the tendency to localisation is reduced considerably, *Kawaji* and *Kawaguchi* [2.9e]. At finite temperatures only a small (logarithmic) correction to the conductivity survives. This area of research is rapidly developing [2.133–135].

As mentioned in Sect. 2.3, a strong magnetic field supports Wigner condensation of an ideal electron gas [2.61, 62]. The ground state structure of a 2d system in high magnetic fields has been studied extensively [2.9f, 10, 144–148]. The fractional quantum Hall effect (Sect. 2.5.4h) seems to indicate that at sufficiently small filling ($\eta < 1/5$) the competition between Anderson localisation and ordering by electron correlation is won by the latter.

The quantum limit analysis is similar to the 3d case. We leave it to the interested reader to determine the asymptotic resistivity exponent.

2.5.4 Hall Effect

a) Perturbation Theory

According to (2.5.6) the Hall conductivity of an ideal 2d electron gas is

$$\sigma_{xy} = -\frac{en}{B} = -\eta \frac{e^2}{h}, \tag{2.5.22}$$

where n is the electron (surface) concentration and η the filling factor given by (2.5.11). If the first i Landau levels are fully occupied and the others empty, i.e. the Fermi energy lies betwen the $(i-1)$th and the ith Landau level, corresponding to $\eta = i$, then the "classical" Hall conductivity (2.5.22) takes the special "quantised" values

$$\sigma_{xy} = -i \frac{e^2}{h}. \tag{2.5.23}$$

In other words, for an integer filling factor $\eta = i$ the free electron Hall conductivity measured in atomic conductivity units $e^2/h = 3.873 \times 10^{-5} \, \Omega^{-1}$ is equal to this integer i. ($\eta = 1$ defines the natural unit of the electron density.)

As pointed out in Sect. 2.3, in the Mori formalism one has to go beyond the simple LOCA and take into account vertex corrections to obtain a non-trivial Hall conductivity. For high magnetic fields ($\omega_c \tau \gg 1$), the lowest order approximation [2.149] coincides with the result obtained by *Ando, Matsumoto* and *Uemura* (AMU) [2.150] by direct evaluation of the Kubo formula in GBA,

$$\sigma_{xy} = -\frac{en}{B} + \Delta\sigma_{xy} \tag{2.5.24}$$

$$\Delta\sigma_{xy} = \frac{e^2}{\pi^2\hbar} \int \left(-\frac{\partial f}{\partial\varepsilon}\right) \frac{\Gamma_v}{\hbar\omega_c} \left(\frac{\Gamma_v^{xy}}{\Gamma_v}\right)^4 \left[1 - \left(\frac{\varepsilon - \varepsilon_v}{\Gamma_v}\right)^2\right]^{3/2} d\varepsilon, \tag{2.5.25}$$

with

$$\Gamma_0(\Gamma_0^{xy}/\Gamma_0)^4 = \frac{\Gamma}{2} \frac{x^2}{(1+x)^2} \tag{2.5.26}$$

and similar expressions for $v = 1, 2, \ldots$ for Gaussian impurities and

$$\Gamma_v(\Gamma_v^{xy}/\Gamma_v)^4 = \Gamma(v + \tfrac{1}{2}) \tag{2.5.27}$$

for point impurities. The term Γ is given by (2.5.14). The chemical potential is assumed to lie in the vth Landau band (for simplicity the spin is ignored). Examining (2.5.24, 25) we immediately recognise that in GBA, for zero temperature and integer filling factor, the value of the Hall conductivity coincides with the free electron special values given by (2.5.23). As shown by AMU, this result is not restricted to the GBA but holds also in the self-consistent t-matrix approximation (STMA) which is the most general single center approximation, even when two center contributions are taken into account. The STMA is of particular interest because the t matrix describes both multiple scattering and bound states brought about by a single impurity. At finite impurity concentration the bound states are broadened to impurity bands associated with each "main" Landau band. To underline the importance of the conclusion drawn by AMU we quote it in full:

Let us consider the case that concentrations of scatterers are sufficiently small and impurity bands are separated from the vth Landau level. At the energy where the spectrum has a gap between two impurity bands or between an impurity band and the main Landau level, $1/G_v(\varepsilon)$ becomes a negative real number in the case of attractive scatterers and a positive one in the case of repulsive scatterers. When the Fermi level lies in those spectral gaps at zero temperature, therefore, the Hall conductivity becomes $-ve^2/h$ in the case of attractive scatterers and $-(v+1)e^2/h$ in the case of repulsive scatterers. This means that electrons which fully occupy impurity bands do not contribute to the Hall current, while those which occupy the main Landau level give rise to the

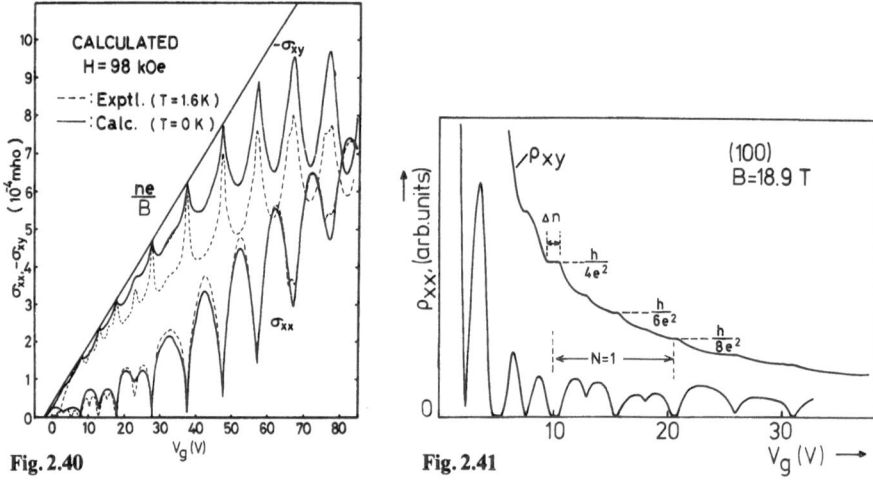

Fig. 2.40   **Fig. 2.41**

Fig. 2.40. Calculated and observed transverse conductivity σ_{xx} and Hall conductivity σ_{xy} vs gate voltage for an n-channel (100) Si MOSFET at $B=9.8$ T (——). (–––) show experimental data obtained at 1.6 K [2.151]

Fig. 2.41. ϱ_{xx} and ϱ_{xy} of a (100) Si MOSFET as a function of the gate voltage [2.7b]

same Hall current as that obtained when all, i.e. $1/2\pi l^2$, electrons of the Landau level move freely [2.150].

Surprisingly, in spite of the existence of bound states which do not contribute to the current (Sect. 2.5.4c) at integer filling, the value of the Hall conductivity remains exactly the same as in the case of free electrons. The significance of this result has been fully recognised only more than five years after publication. The bound state problem will concern us later in this chapter; for the time being it is sufficient to keep in mind that for integer filling the impurity scattering does not give rise to any deviation of the Hall conductivity from its free electron value (2.5.23), at least in high-order perturbation theory. This prediction has been verified by *Kawaji* et al. [2.151] at magnetic fields below 10 T, as can be recognised in Fig. 2.40. For any other filling the Hall conductivity is decreased below the free electron value, in agreement with the AMU theory [2.150]. To compare theory with experiment, the spin and valley splitting were incorporated. The agreement can be improved by replacing in (2.5.25) the semi-elliptical density of state factors by the Gaussian (2.5.13b) and by taking into account the first correction in $1/\omega_c\tau$ [2.151].

b) The Quantum Hall Effect (Integer Quantisation)

When measurements of the Hall resistivity were extended to fields significantly higher than 10 T, plateaux were observed by *Englert* and *v. Klitzing* [2.116b] (cf.

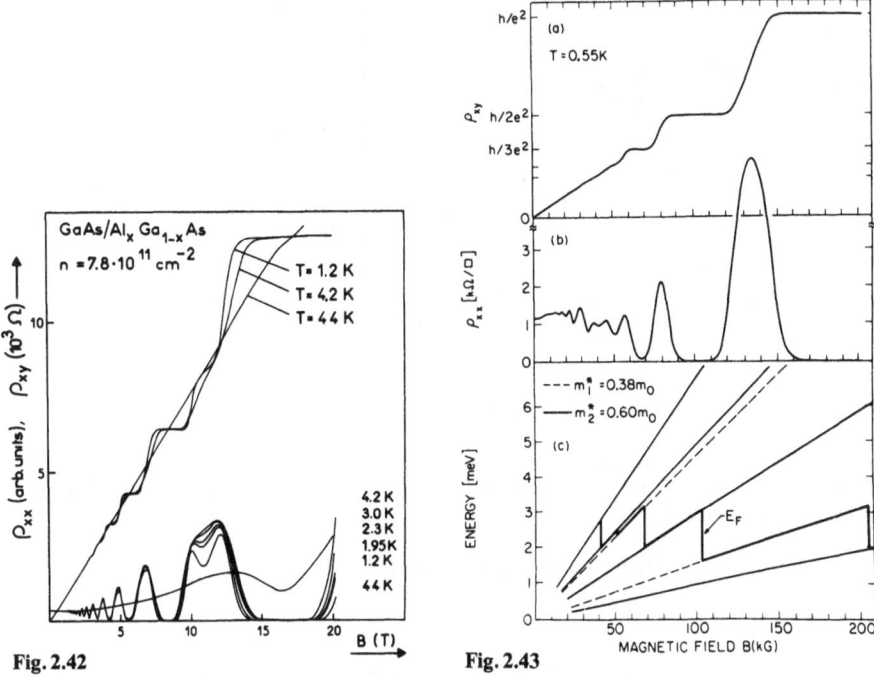

Fig. 2.42. ϱ_{xx} and ϱ_{xy} of a Ga(GaAl)As heterojunction as a function of the magnetic field at various temperatures [2.226b]

Fig. 2.43. Transverse resistivity ϱ_{xx} and Hall resistivity ϱ_{xy} of a modulation doped GaAs-(GaAl)As heterojunction as a function of the magnetic field at $T = 0.55$ K. The *lower part* shows the energy levels and the Fermi energy using masses determined by cyclotron resonance [2.156]

Fig. 2.36), and *Wakabayashi* and *Kawaji*, in [2.152a] (which correspond to plateaux in σ_{xy}), in contrast to the garland type structure which can be seen in Fig. 2.40. An extension of the magnetic field range to 19 T in conjunction with high precision measurements of ϱ_{xx} and ϱ_{xy} on high quality (100)Si MOSFETs (Fig. 2.41) led to the discovery by *von Klitzing* et al. [2.7a] that σ_{xy} is quantised in units of e^2/h (quantum Hall effect, QHE). There is no satisfactory theory of predictive power for this highly interesting effect presently available. There are, however, several remarkable theoretical suggestions and observations – possible building blocks for a future theory – worth being described in some detail. This will be done in the remainder of this section after reporting a few more recent experiments.

Soon after its discovery in Si MOSFETs, the QHE was found in n-type GaAs-(GaAl)As heterostructures by *Tsui* and *Gossard* [2.153]. Due to the small electron mass in GaAs in conjunction with high mobility at helium temperatures, pronounced plateaux were observed even at "moderately" high magnetic fields (< 10 T). Extension of the experiments in the mK range yielded sharp steps

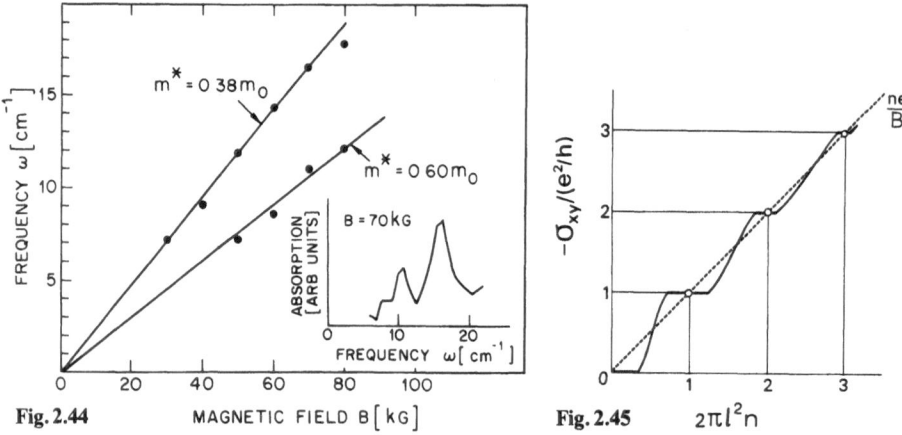

Fig. 2.44.   MAGNETIC FIELD B [kG]   **Fig. 2.45.** $2\pi l^2 n$

Fig. 2.44. Hole cyclotron resonance frequencies as a function of the magnetic field for a GaAs-(GaAl)As heterojunction. The insert shows an absorption spectrum at a magnetic field of 70 kG as a function of frequency [2.156]

Fig. 2.45. The quantum Hall effect (schematic) [2.157]

with wide plateaux [2.154, 155]. High resolution measurements [2.155] showed that the plateaux corresponding to $i = 1$ did not change by more than 6×10^{-8} if the magnetic field was changed by as much as 1 T.

The development of plateaux in ϱ_{xy} for a GaAs-(GaAl)As heterojunction with $n = 7.8 \times 10^{11}$ cm^{-2} as a function of temperature can be seen in Fig. 2.42 [2.226b]. One can recognise that the straight line describing the free electron Hall resistivity intersects the plateaux near the centre. At 1.2 K the width of the last plateau is about 5 T.

Recently the QHE has been demonstrated for holes in a modulation doped GaAs-(GaAl)As heterojunction by *Störmer* et al. [2.156]. It was shown that steps in the Hall resistivity occur at values $h/e^2 i$ within the experimental accuracy of about 10^{-4}. This proves once again that the QHE is independent of the details of the band structure. The data are shown in Fig. 2.43. The SdH oscillations in the middle part of the figure indicate beating effects caused by two kinds of holes. For the first time lifting of the spin degeneracy, predicted by the subband calculations for p-type silicon [2.103a], was experimentally verified. The existence of two types of heavy holes was corroborated by cyclotron resonance measurements, yielding effective masses of $0.38\,m_e$ and $0.60\,m_e$. The data are shown in Fig. 2.44. They agree, however, only partially with recent calculations of Landau-levels in GaAs-(GaAl)As heterojunctions [2.208a], cf. Sect. 2.5.4i.

The essential features of the QHE (at sufficiently low temperatures) are depicted schematically in Fig. 2.45. Notice that for integer filling the actual values of σ_{xy} (marked by small circles) coincide exactly with the corresponding values (2.5.23) derived for free electrons. This result is universal in the sense that it is entirely independent of the individual properties of the system. It is crucial to keep in mind that the plateaux of σ_{xy} are always accompanied by vanishing σ_{xx}. Thus, σ_{xy} and σ_{xx} are essentially related. A theory can be regarded as satisfactory only if it explains both phenomena (the plateaux of σ_{xy} and the range-wise vanishing of σ_{xx}) in a unified way.

c) The Role of Localisation

It is intuitive to assume that the vanishing transverse conductivity σ_{xx} – and therefore also the plateaux of the Hall conductivity σ_{xy} – reflect localisation in the random impurity potential. Localised states are similar to quantum mechanical bound states since neither carries electric current. Let $|a\rangle$ be any bound energy eigenstate of the system. The expectation value of the average current density in this state is

$$\langle a|j|a\rangle = -\frac{e}{L_x L_y} \langle a|v|a\rangle \qquad (2.5.28)$$

$$\langle a|v|a\rangle = \frac{i}{\hbar} \langle a|[H,r]|a\rangle = \frac{i}{\hbar} (\varepsilon_a - \varepsilon_a) \langle a|r|a\rangle = 0, \qquad (2.5.29)$$

since for a bound state $\langle a|r|a\rangle$ is bounded. The total current density, being an additive quantity, can be calculated by taking the contribution of the energy eigenstates one by one. If, at $T=0$, all occupied states are bound states there will be no net current. The current starts to flow once the Fermi energy is increased to enter the regime of mobile (delocalised) states, Fig. 2.46. Let us assume the Fermi energy to lie in one of the regimes of localised states. In this case j_x, the current parallel to the applied electric field, vanishes (since only states at the Fermi energy contribute) and the Hall current j_y takes some finite value. Varying now

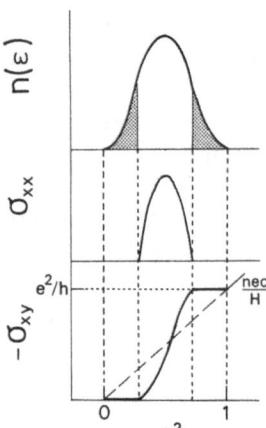

Fig. 2.46. Interrelation between QHE and localisation (schematic) [2.157]

the Fermi energy within the regime of localised states, both currents remain constant, since, again, the contributions of the energy eigenstates can be counted one by one. This argument is good for explaining the range-wise vanishing of σ_{xx} and the plateaux in σ_{xy}. To avoid σ_{xy} vanishing identically, the existence of mobile states must be postulated. What are the plateaux values of σ_{xy}? If there is one Landau band gap in each regime of localised states, by referring to the AMU result, we conclude that the plateaux values coincide with the special values (2.5.23) of the free electron system, in complete agreement with the observed QHE. Clearly, there must be at least one mobile (delocalised) state in each Landau band, otherwise the plateaux could continue to overlap, which is prohibited since the Hall conductivity must be a unique single-valued function of the Fermi energy [2.140, 157].

However appealing, the above argument cannot be considered to be a theory of the QHE, since for instance we did not explain (nor even describe) the (statistical mechanical) state of the system consistently.

The explanation of the observed increase of the Hall conductivity above the free electron value by localisation appears rather paradoxical. To understand how this happens *Prange* [2.158] studied the model of free electrons interacting with a single point impurity (δ function) which can be treated exactly. He concluded that at integer occupation ($\eta = 1$) the Hall current is exactly the same as for free electrons because the loss of current caused by the reduction of the number of mobile states by one, accompanied by the appearance of a single bound state (an exact property of the model), is exactly compensated by an appropriate increase of the Hall current carried by the remaining mobile states. This result holds for any finite number of point impurities [2.159], as (implicitly) established by AMU, and even for any finite range scattering regime [2.160]. These investigations may serve as a guide to a unified explanation of the plateaux and vanishing conductivity, which is still an open question.

d) Various Investigations

Whereas the AMU approach is based on perturbation theory – it becomes, however, exact in the limit of vanishing impurity concentration and thus covers the models considered in [2.158, 159] – *Thouless* [2.161] has demonstrated using non-perturbative theory that for integer filling, the Hall conductivity (as given by the Kubo formula) is invariant against weak local variation of the impurity potential.

A useful rearrangement of the Kubo formula (2.3.26) valid for additive systems (no Coulomb interaction) was invented by *Středa* [2.162]:

$$\sigma_{\mu\nu}(T) = \int \left(-\frac{\partial f}{\partial \varepsilon} \right) \sigma_{\mu\nu}(T=0, \varepsilon_F = \varepsilon)\, d\varepsilon, \tag{2.5.30}$$

where $f(\varepsilon)$ is the Fermi function and

$$\sigma_{\mu\mu}(\varepsilon) = \pi\hbar e^2 \left\langle \mathrm{Tr}\left\{ v_\mu \delta(\varepsilon - H) v_\mu \delta(\varepsilon - H) \right\} \right\rangle_I \tag{2.5.31}$$

$$\sigma_{xy} = \sigma_{xy}^I + \sigma_{xy}^{II} \tag{2.5.32}$$

$$\sigma_{xy}^I(\varepsilon) = \frac{e^2}{2} \, i\hbar \, \langle \text{Tr} \, \{v_x \hat{G}^+(\varepsilon) v_y \delta(\varepsilon - H) - v_x \delta(\varepsilon - H) v_y G^-(\varepsilon)\} \rangle \tag{2.5.33}$$

$$\sigma_{xy}^{II}(\varepsilon) = -e \, \frac{\partial N(\varepsilon)}{\partial B} \tag{2.5.34}$$

$$\hat{G}^{\pm}(\varepsilon) = \hat{G}(z), \quad z = \varepsilon \pm i\sigma$$

$$N(\varepsilon) = \int_{-\infty}^{\varepsilon} n(\varepsilon') d\varepsilon' \tag{2.5.35}$$

is the integrated density (or total number) of states with energy $\leq \varepsilon$. If, at $T = 0$, the Fermi energy falls in a gap of the energy spectrum of the system, $\sigma_{\mu\mu}$ and σ_{xy}^I vanish. If that gap is the ith Landau band gap and the degeneracy of the Landau bands is the same as that of the free electron Landau levels (a very plausible assumption [2.109, 140]), then $N(\varepsilon_F) = iD = ieB/h$ which once again yields (2.5.23) [2.162]. The Středa formula (2.5.30–35) has two shortcomings which should be kept in mind: it is applicable only if the Coulomb interaction between the electrons is ignored, and the splitting (2.5.32) distinguishes gaps in the density of states (e.g. band gaps), whereas in reality the mobility gaps play a distinct role.

If, as generally assumed (cf. however [2.163]), the QHE is due to localisation, the universality of the effect indicates that only general features of localisation, i.e. percolation properties, are significant [2.133b, 164–166]. In the framework of a simple classical percolation model, *Iordansky* [2.164] found that at the percolation threshold U_c, there are unbounded trajectories which give rise to a net Hall current. Estimation of this current, in the limit of vanishing diffusion, yields

$$\sigma_{xy} = -\frac{en}{B} f(U_c)/N, \tag{2.5.36}$$

where $f(U_c)/N$ is the average occupation at U_c. This result does not explain the QHE but it indicates the possibility of a finite Hall current in spite of localisation (vanishing diffusion). In the quantum limit ($v = 0$) and arbitrary values of ε_F within the $v = 0$ Landau band [2.133b] obtained for $T = 0$

$$\sigma_{xy} = -\frac{e^2}{h} \theta(\varepsilon_F - U_c), \tag{2.5.37}$$

where $\theta(x)$ is the unit step function. The essential assumption leading to (2.5.37) is that for a sufficiently large system, the random impurity potential at the boundaries perpendicular to the applied electric field can be replaced exactly by

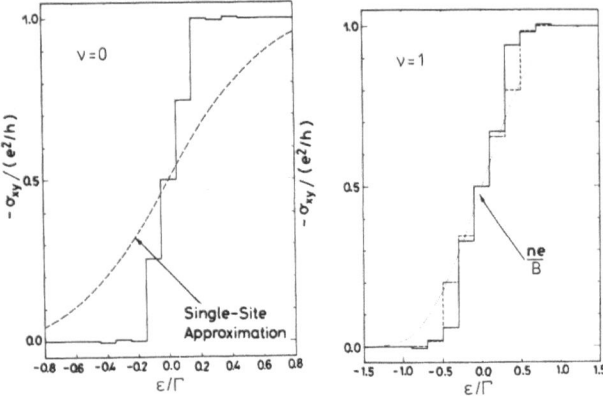

Fig. 2.47. Numerical calculation of the Hall conductivity as a function of energy. $\phi/\phi_0 = L^2 eB/h = 144$; on the rhs $(---)$ corresponds to $\phi/\phi_0 = 64$ [Ref. 2.133, p. 176]

the unique percolation energy U_c at which an equipotential line percolates from infinity to infinity. Although intuitive, an exact proof of this assumption is still missing. According to (2.5.37) the transition from the zero level to the first plateau (at $-e^2/h$) does not necessarily happen at the centre of the $v=0$ Landau band. It has not yet been demonstrated how the existence of a single percolation threshold can give rise to a multistep structure.

Models with one and two Landau bands have been investigated numerically by *Ando* [2.133a, 226a] and *Aoki* [2.141]. The results look very promising, Fig. 2.47, for they indicate the possibility of explaining the QHE within the framework of a simple electron-impurity model. The appropriate boundary conditions [2.167] and the coupling of a single band to the electric field are, however, delicate problems which make further investigations desirable. (The electron wave function for finite electric fields cannot be expanded in terms of the zero-field Landau states corresponding to a single Landau band.) Moreover, no analogous numerical calculations of σ_{xx} are available so far.

e) The Gauge Argument

The very general character of the QHE suggests that it could be explained by some kind of fundamental principle instead of going through a detailed and certainly very sophisticated quantum magnetotransport theory in the presence of localisation. *Laughlin's* "gauge argument" [2.168, 226c] (cf. also [2.2h, 169, 170, 172]) is a prominent example of this attitude.

In order to present Laughlin's approach a few preliminary remarks are necessary. The (quasi) 2d samples investigated in QHE experiments are, of course, finite systems of rectangular (or equivalent) shape. At the boundaries of such systems the electron wave functions vanish. This "vanishing boundary

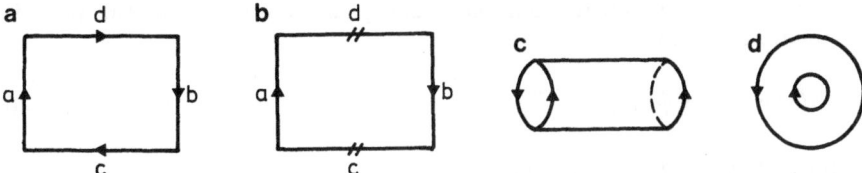

Fig. 2.48a–d. Systems with different boundary conditions and geometries. (a) real finite system with circular edge currents (single connected), (b) periodic boundary conditions identifying the edges c and d, (c) cylinder geometry, (d) annular disc geometry. The double-connected models b, c and d are topologically equivalent

condition" is equivalent to an infinitely high potential step at the boundaries of the system. Unfortunately the Schrödinger equation of an electron in a magnetic field with vanishing boundary conditions cannot be solved analytically. The reason for this is that due to the potentials at the boundaries none of the components of the linear momentum is a conserved quantity and, therefore, the 2d boundary value problem does not separate into two 1d problems. The usual way to circumvent this difficulty is to replace, in the direction of the vector potential, i.e. in the y direction when choosing $A = (0, Bx, 0)$, the vanishing boundary conditions by periodic ones. In this case p_y is conserved and the problem separates. Since $\psi(x, y + L_y) = \psi(x, y)$ means identifying the edges c and d, Fig. 2.48, the topology of the system has changed: it is now double-connected and thus equivalent to a cylinder and an annular disc. This has physical consequences: associated with the Landau diamagnetism is a finite edge current (with density $j = \text{curl } M$) which is reponsible for the jump of the magnetisation at the boundaries of the system. In a real system this current is circular, as indicated in Fig. 2.48. Imposing periodic boundary conditions, the currents along the corresponding edges (c and d) must vanish. Consequently the remaining edge currents (along a and b) become disconnected. The direction of these currents is opposite from and their magnitude in thermal equilibrium is equal to each other. A small difference of the electrostatic potentials at the edges a and b gives rise to a net Hall current. Since the edge currents are confined to a layer of thickness ~ 1 (see below), the difference in topology appears to be insignificant for the Hall conductivity, provided the system is sufficiently large, $1 \ll L_x, L_y$.

As long as $j_y \neq 0$ every gauge transformation

$$A_y \to A_y' = A_y + \delta A_y = Bx + \frac{\partial g}{\partial y}$$

$$\psi_\alpha \to \psi_\alpha' = \exp\left(-i\frac{eg}{\hbar}\right)\psi_\alpha \tag{2.5.38}$$

is associated with the work

$$\delta W = -\int j_y \delta A_y d^2 r. \tag{2.5.39}$$

Assuming $g = -Bay$ with some constant gauge parameter a, the periodic boundary condition

$$\psi(x, y + L_y) = \psi(x, y) \tag{2.5.40}$$

imposed both on ψ_α and $\psi_{\alpha'}$ requires

$$a \frac{eB}{\hbar} = \frac{2\pi}{L_y} \quad (0, \pm \text{integer}). \tag{2.5.41}$$

Thus, the smallest non-vanishing allowed value of a is $\Delta a = 2\pi\hbar/eBL_y = \phi_0/BL_y$. According to (2.3.7) the change of the electrostatic energy of the electrons associated with the gauge translation $A_y = Bx \rightarrow A_y' = B(x - a)$ is

$$\delta U = eNE_x a. \tag{2.5.42}$$

The value of the current j_y is determined by the energy conservation law $\delta W = \delta U$ which yields, as expected for free electrons, the well-established formula (2.5.22). Thus, we have not learned anything new so far.

Note, however, that our considerations offer a new interpretation of the special values (2.5.23). For, writing for the average current

$$j_y = -\frac{1}{L_x L_y} \frac{\delta U}{\delta A_y} \tag{2.5.43}$$

we arrive at (2.5.23) by *assuming* that the change of A_y by a gauge quantum $A_y = -B\Delta a = -\phi_0/L_y$ is associated with the transfer of i electrons from edge a to edge b,

$$j_y = -\frac{1}{L_x L_y} \frac{ieE_x L_x}{(-\phi_0/L_y)} = i \frac{e}{\phi_0} E_x = i \frac{e^2}{h} E_x. \tag{2.5.44}$$

As long as we are dealing with free electrons at $T = 0$, this particular current excitation process is possible only if the Fermi energy lies in the gap between the ith and the $(i+1)$th Landau level ($\eta = i$, $n = iD$). The assertion of Laughlin's gauge argument is now that in a real disordered electron-impurity system at $T = 0$, this particular process (with some appropriate integer i) is the *only possible* excitation of current if the Fermi energy lies in *any* gap of the energy spectrum or, respectively, in any mobility gap of the system.

f) Edge Currents

A few words have to be said about the case of finite L_x. The oscillator functions $\varphi_\alpha(x) = \varphi_\nu(x - X)$, as the x-dependent factor in the Landau wave functions, cf. (2.2.26), and the degenerate Landau levels $\varepsilon_\alpha = \varepsilon_\nu$ have been obtained by

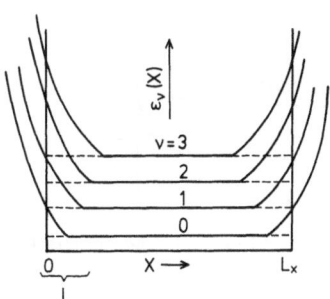

Fig. 2.49. Energy spectrum of a finite system in x direction [2.88, 9g]

assuming $L_x \to \infty$ and the normalisation condition for $\varphi_\alpha(x)$, cf. (2.2.23), and periodic boundary conditions in the y direction. Imposing, for finite L_x, vanishing boundary conditions at $x=0$ and $x=L_x$, $\varphi_\alpha(0)=\varphi_\alpha(L_x)=0$, the problem still separates but the character of the solution is changed considerably [2.88, 170]. The eigenfunctions are the so-called Weber functions $w_\nu(x, X)$ which do not depend on the difference $x - X$ only. The degeneracy of the energy eigenvalues is lifted, $\varepsilon_\alpha = \varepsilon_\nu(X)$ and cannot be given in a closed analytical form, Fig. 2.49. The deviation of the energy spectrum from the Landau levels is significant only at the edges (and beyond); in the bulk, $1 \lesssim X \lesssim L_x - 1$, the Landau structure is essentially (i.e. up to exponential corrections) maintained. The lift of the degeneracy is connected with a finite average current density at the edges a and b (at $x=0, L_x$) confined to ranges of thickness ~ 1 [2.88, 171]. In the state ψ_α the velocity is

$$v_{y\alpha\alpha} = \left(\frac{\partial H}{\partial p_y}\right)_{\alpha\alpha} = \frac{1}{\hbar}\frac{\partial \varepsilon_\alpha}{\partial k_y} = -\frac{l^2}{\hbar}\frac{\partial \varepsilon_\nu(X)}{\partial X} = -\frac{1}{eB}\frac{\partial \varepsilon_\nu(X)}{\partial X}. \tag{2.5.45}$$

Since the spectrum is symmetric with respect to $X=0$ there is no net charge transport in thermal equilibrium. If, however, the chemical potentials at the edges a and b are different, a net current will result. Assuming, e.g., $\zeta(X) \simeq \zeta + \zeta'X$ and a stationary state described by the distribution function

$$f_\alpha(X) = \{\exp[\varepsilon_\alpha - \zeta(X)]/k_B T + 1\}^{-1} \tag{2.5.46}$$

$(\hat{X} = -l^2 p_y/\hbar)$ is a conserved quantity)

$$j_y = -\frac{e}{L_x L_y}\,\mathrm{Tr}\,\{\hat{f}v_y\} = \frac{1}{L_x L_y}\frac{1}{B}\sum_\alpha f_\alpha(X)\frac{\partial \varepsilon_\alpha}{\partial X} = \frac{n}{B}\zeta' = \frac{en}{B}E_x \tag{2.5.47}$$

to lowest order in $E_x = \zeta'/e = [\zeta(L_x) - \zeta(0)]/eL_x$ [2.9g, 88].

Since the velocity $v_{y\alpha\alpha}$ is non-vanishing only near the edges a and b (2.5.45) the Hall current can be viewed as a surface (or, respectively, edge) phenomenon. As our calculation indicates, however, the net Hall current is determined by the momentum of the velocity $(v_{y\alpha\alpha}X)$ and is, thus, a volume (surface) quantity. The situation is much the same as for free electron diamagnetism.

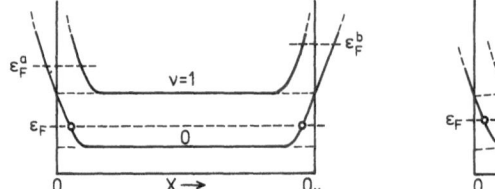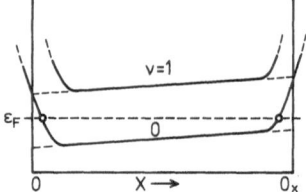

Fig. 2.50. Energy spectrum of a finite system in x direction for zero (*left*) and finite (*right*) electric field in the same direction, with different values of the Fermi energy

Let us assume that in the bulk the Fermi energy lies in the gap between two Landau levels. At the surface the Fermi energy cuts a number of energy levels and is thus pinned, Fig. 2.50. The excitation of electrons across the system [furnishing (2.5.44)] is now obvious [2.168–170].

The electron-impurity potential has no significant influence on the edge currents. Indeed, the dynamics of electrons near the edges of the system is determined by the very rapidly increasing surface potential (which motivates the vanishing boundary conditions) and the magnetic field. Consequently the surface (Hall) current will not be extinguished by localisation. If in the bulk the Fermi energy varies within a localisation regime, the Hall conductivity remains unchanged. Actually, if localisation takes place, the bulk of the system can be removed completely without affecting the Hall current.

These considerations may elucidate *Laughlin's* gauge argument [2.168]. Since a satisfactory proof is still lacking, the gauge argument should be considered as a heuristic principle or perhaps as a conjecture. Critical researchers may comprehend it as a reformulation of experimental evidence. Recent [2.172, 9h] and future numerical investigations may help to come to a final conclusion.

Laughlin's gauge argument should not be confused with the current formula (2.5.43) which remains on solid ground [2.173]. For an infinite ergodic system the corresponding formula for arbitrary temperatures

$$j_y = -\left(\frac{\delta F}{\delta A_y}\right)_{T,N}, \tag{2.5.48}$$

where F denotes the free energy [2.174], can be shown to be equivalent to the usual (average current) Kubo formula (2.3.26) [2.175]. For a sufficiently large but finite system ($L_x < \infty$) the current given by (2.5.48) vanishes because edge and bulk contributions compensate each other exactly [2.175]. Thus, to maintain the equivalence with the Kubo formula the edge states (carrying the edge current) have to be excluded from F.

g) The Influence of a Periodic Potential

In the previous sections we have considered an electron gas with some effective electron mass, free or interacting with randomly distributed fixed impurities. We turn now, very briefly, to the opposite limit of a real system: an ideal electron gas in a periodic (lattice) potential. The Coulomb interaction between the electrons will be neglected.

Assuming, for $B=0$, a well-separated conduction band with dispersion $\varepsilon(k)$ [$k=(k_x,k_y)$], we can apply the Peierls replacement approximation, Sect. 2.2. Since $\varepsilon(k)$ is a periodic function in k space, e.g.

$$\varepsilon(k)=\varepsilon_1 \cos (a_1 k_x)+\varepsilon_2 \cos (a_2 k_y), \tag{2.5.49}$$

the Peierls approximation leads to a differential equation of infinitely high order. The eigenvalue problem has been studied extensively [2.176–182] (for an introduction to the subject cf. *Hund* [2.23]). The spectrum, obtained numerically [2.179], is very complex. There are denumerable infinite energy gaps, the positions of which are determined by

$$N_1 =j+i\phi_1, \tag{2.5.50}$$

where $N_1 = N\Omega$ is the number of energy states per unit lattice cell below the gap, $\phi_1 = B\Omega/\phi_0$ is the number of flux quanta within the unit cell area Ω, and i and j are integers [2.180]. Inserting (2.5.50) into the Středa formula (2.5.34), we obtain the quantised values (2.5.23) [2.182] first derived by *Thouless* et al. [2.183] (cf. also [2.184, 186]). This result agrees with Laughlin's gauge argument, which is remarkable because the gaps determined by (2.5.50) do not correspond to integer filling.

The problem of motion in a periodic potential appears also in the case of a free electron gas with Coulomb interaction when the ground state is ordered (Wigner crystal or charge density wave) [2.185].

h) Fractional QHE

When studying high mobility ($\mu>4\times 10^5$ cm^2/Vs) GaAs-(GaAl)As hetero-structures at low temperatures in the quantum limit, *Tsui* et al. [2.187, 188] discovered fractional quantisation of the Hall resistivity. Plateaux in ϱ_{xy} and minima in ϱ_{xx} were observed at filling factors $\eta=1/3$ and $2/3$. In subsequent work [2.189] additional structures at $\eta=2/5$, $3/5$, $4/5$ and $2/7$ were found in ϱ_{xx} (Fig. 2.51). Also structures showed up at $\eta=4/3$ and $5/3$, none however at $\eta=1/2$ and $1/4$. These findings suggest that fractional quantisation occurs in multiple series. Although, for the time being, plateaux in ϱ_{xy} have been demonstrated only for $\eta=1/3$ and $2/3$, it is expected that the missing ones would appear if the temperature were lowered into the mK range.

Fractional quantisation has a remarkable consequence if the gauge argument is assumed to hold: since for unit elementary charge this argument yields integer quantisation, e.g.

$$\sigma_{xy}=-\frac{1}{3}\frac{e^2}{h} \tag{2.5.51}$$

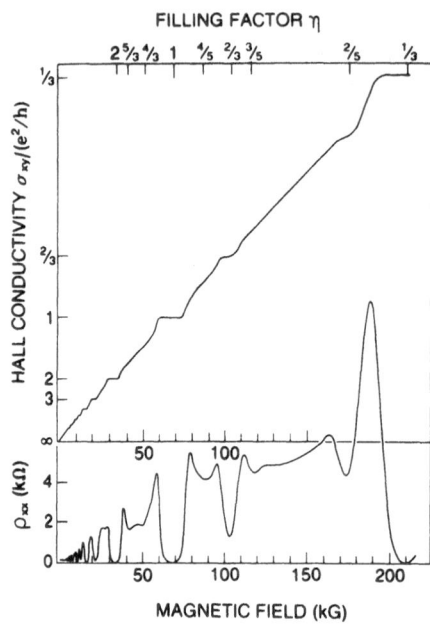

FILLING FACTOR η

Fig. 2.51. Fractional QHE in a GaAs-(GaAl)As heterojunction [2.189]

requires the production of a flux quantum $\phi = h/e$ to be accompanied by the transfer of a charge $-ie^*$. Thus, by (2.5.44),

$$\sigma_{xy} = -i\frac{e^*e}{h} \tag{2.5.52}$$

with $e^* = -e/3$. Generally, for $\eta = p/q$ elementary excitations with charge $e^* = -e/q$ must exist.

Laughlin [2.190] suggested a new kind of (many-electron) ground state for some partial filling which indeed is separated by a *gap* from (single-particle) excited states which carry *fractional charge*. He used the symmetrical gauge $A = (-By/2, Bx/2, 0)$ for which the free electron states of the $\nu = 0$ Landau level are (up to a normalisation factor)

$$|n> = z^n \exp(-|z|^2/4), \quad n \geq 0, \tag{2.5.53}$$

where $z = (x - iy)/l$. The state $|n\rangle$ is exponentially localised on a circle with radius $\sqrt{2n} \cdot l$. There is one (obviously antisymmetric) Laughlin state for each odd integer m,

$$\Psi_m = \prod_{i<j} (z_i - z_j)^n \exp\left(-\sum_k |z_k|^2/4\right), \tag{2.5.54}$$

$i, j, k \leq N$. The state Ψ_m is built up by single-particle states with $n \leq m(N-1)$ and corresponds, therefore, to filling $\eta = N/m(N-1) \simeq 1/m$ for large N. The effective radius of the electron distribution in the state is $\sqrt{2m(N-1)} \cdot l$. The m-fold vanishing of Ψ_m for $z_i \to z_j$ keeps the average Coulomb interaction energy at a low value.

One can construct a hole-like excitation at the origin of the system by pushing each constituent single-electron state one step outwards,

$$\Psi_{\text{hole}}^{(m)} = \prod_i z_i \prod_{j<k} (z_j - z_k)^m \exp \left(-\sum_l |z_l|^2/4\right). \tag{2.5.55}$$

Since, by construction, m orbitals belong to one electron, enlarging the system by one orbital means $1/m$ electron missing, so the charge of the hole excitation (2.5.55) is $e^* = e/m$. This excitation is separated from the ground state by a gap of the order of e^2/m^2l. Consequently, the gauge argument yields $\sigma_{xy} = -i(e^2/h)/m$ for filling $\eta = 1/m$. The plateaux can be understood by assuming the low-lying electron and hole excitations to be localised by the random impurity potential.

Laughlin was able to estimate the ground state energy and found that his quantum fluid states are more advantageous than the charge density wave states assumed previously, if $m \lesssim 10$. Thus, plateaux at $\eta = 1/5$ and $1/7$ could be seen, in principle. A painstaking attempt to detect the quantum liquid – Wigner crystal phase transition experimentally turned up a weak anomaly at $\eta = 1/5$ but absolutely nothing at $1/7$ and below [2.191].

While the 2/3 plateau is arguably related to the holes forming a 1/3 Laughlin state, there is no such simple explanation for plateaux at 2/5 and 2/7 fillings: further quantum fluid states are called for. There are several proposals as to how to construct them. *Anderson* [2.192] noted that the hole state $\Psi_{\text{hole}}^{(m)}$ is not just a local perturbation of Ψ_m but differs from it globally so that the Pauli principle does not prohibit filling the state Ψ_m with N electrons and $\Psi_{\text{hole}}^{(m)}$ with further N electrons. This is the construction of the $\eta = 2/m$ ground state. Similarly, we can build the $\eta = 3/m$ ground state by filling Ψ_m, $\Psi_{\text{hole}}^{(m)}$ and the state with two holes at the origin with N electrons each, etc. Though Anderson does not estimate the ground state energies, it stands to reason that this construction is a good suggestion since in the thermodynamic limit Ψ_m and $\Psi_{\text{hole}}^{(m)}$ have the same energy density. In this sense the $1/m$-Laughlin ground state possesses an m-fold degeneracy.

An N-electron state with holes at z_A and z_B is produced by multiplying Ψ_m by $\prod_i (z_i - z_A)(z_i - z_B)$. Since this is invariant against interchanging z_A and z_B we might say that the holes are bosons. *Haldane* [2.193] put forward a hierarchy of quantum fluid states based on the idea that one can change η by injecting holes (or their mirror images: fractionally charged particles) into the Laughlin states, and then requiring that the holes themselves form a correlated Laughlin-type state, but with *even* index q. This gives the filling

$$\eta = \frac{1}{m \pm \dfrac{1}{q}} \tag{2.5.56}$$

(+ and − for holes and electrons, respectively). For instance, $m=3$, $q=2$ give the 2/5 state with particles, and the 2/7 state with the holes. One can then require that the hole Laughlin liquid is itself a composite state; the filling factors of the ensuing hierarchy are given by a continued fraction, e.g.

$$\frac{3}{5} = 1\bigg/\bigg[1+1\bigg/\bigg[2-\frac{1}{2}\bigg]\bigg].$$

Since $\sigma_{xy} = -(p/q)(e^2/h)$ for $\eta = p/q$ is nothing but the free electron result, an appealingly simple explanation for the subplateaux is that in spite of the variation of the gate voltage, the actual filling is pinned to $\eta = p/q$ *and the η dependence of the gap by treating the electron-electron interaction as a perturbation. In this work, and in a revised and extended calculation by *Tao* [2.194], it was found that the gap is indeed exceptionally large at certain simple fractions, including those corresponding to the observed subplateaux. A puzzling feature is that though the largest gap corresponds to 1/3 and 2/3, there are fairly large gaps at fractions with even denominators as well, so in principle a 1/2-plateau, etc., could exist. It is not clear why plateaux with even denominators are not realised in nature.

i) Cyclotron Resonance

An approximate expression for the dynamic conductivity of a 2d system can be obtained by using (2.3.39, 40) in GBA [2.111]. The result, relevant for the transition $v=0 \rightarrow v=1$ is

$$M''(\omega) = \omega_c \frac{\Gamma_0^2}{\Gamma_0 + \Gamma_1} \frac{2x^2}{(1+x)^2} I(\omega) \tag{2.5.57}$$

$$I(\omega) = \frac{1}{2\pi l^2 n} \int \frac{d\omega'}{\pi} (\Gamma_0 + \Gamma_1) D_0(\omega' - \omega) D_1(\omega') \frac{f(\omega' - \omega) - f(\omega)}{\omega}. \tag{2.5.58}$$

The relaxation spectrum $M''(\omega)$ exhibits a resonance. The peak value occurs at $\omega = \omega_c$ and for $\hbar\omega_c > k_B T$ is given by

$$\Gamma_c \equiv M''(\omega_c) \simeq \frac{32}{3\pi} \frac{\Gamma_0^2}{\Gamma_0 + \Gamma_1} \frac{x^2}{(1+x)^2}; \tag{2.5.59}$$

the half-width is approximately $\Gamma_0 + \Gamma_1$. This prediction of the theory has been confirmed by *Allen* et al. [2.195]. They measured the dynamic conductivity of silicon inversion layers in a magnetic field of $B = 6.15\,$T at $T = 1.3\,$K. Extracting $M''(\omega)$ from the experimental data by Kramers-Kronig analysis and comparing it with (2.5.57) (Fig. 2.52) the strength parameter $U = \sqrt{1+x}\,\Gamma_0 = (n_i g^2/4\pi r^2)^{1/2}$

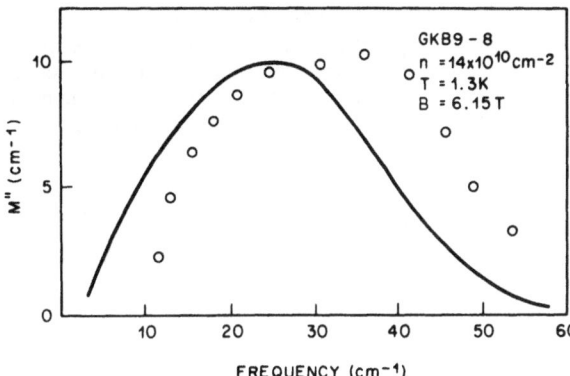

Fig. 2.52. Observed (O) and calculated (——) relaxation spectrum of a (100) Si inversion layer [2.195]

and the range r of the impurity potential were determined to be $U = 15 - 30$ K and $r = 50$ Å.

For $x \ll 1$ the shape of the cyclotron resonance line is Lorentzian with a linewidth Γ_c. For $x \gtrsim 1$, however, this is not the case since $M''(\omega)$ drops to zero within a frequency range of the order of Γ_c. In spite of this fact Γ_c still gives an estimate for the width of the cyclotron line which reaches a maximum for $x \simeq 1$ and behaves like $\sqrt{B}$ for $x \gg 1$ (also obtained by *Ando* [2.196] and by *Fujita* and *Prasad* [2.197]), in agreement with experiments on n-channel (100) Si MOSFETs and GaAs-(GaAl)As heterojunctions [2.198–200]. This indicates the dominant role of short-range impurity scattering in these systems.

In GaAs-(GaAl)As modulation doped heterojunctions with low electron concentration and very high mobilities *Englert* et al. [2.220a] observed oscillations of the cyclotron resonance linewidth between 0.07 and 0.32 T as a function of the resonance magnetic field. They demonstrated the correlation between filling factor of the Landau levels and the linewidth, which showed maxima at integer filling factors and minima when the levels were half filled. It was proposed that the oscillations were caused by screening effects. Subsequently a detailed calculation of the cyclotron resonance linewidth was performed by *Lassnig* and *Gornik* [2.220b]. The data of *Englert* et al. [2.220a] are well explained by ionized impurity scattering in conjunction with screening. Recently, much work has been devoted towards the understanding of the cyclotron resonance lineshape in GaAs-(GaAl)As heterojunctions with the goal, to get information about the quality of the GaAs-(GaAl)As interface. An up to date report has been given by *Gornik* et al. [2.220c].

In the limit $x \ll 1$ $\Gamma_c \propto 1/B^2$ [2.111] (in [2.197] $\Gamma_c \propto 1/B$ was obtained). Experiments on InSe accumulation layers by *Kress-Rogers* et al. [2.201] show a rapidly decreasing line width with increasing magnetic field in an intermediate field range and a tendency to saturation for higher magnetic field. As shown in Fig. 2.53, good agreement between theory and experiment can be achieved by taking into account both long range impurity scattering [2.111, 202, 116c] and interaction with acoustic phonons, *Prasad* et al. [2.116d], – the importance of

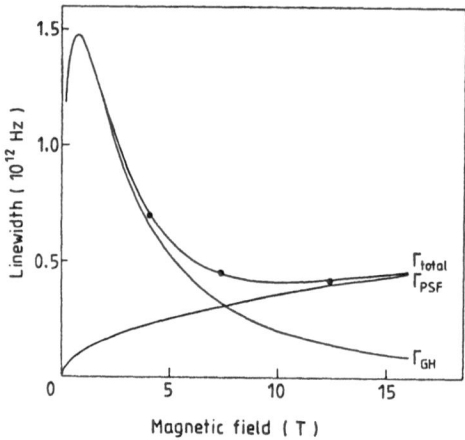

Fig. 2.53. Experimental data (●) and theoretical model calculations of the cyclotron linewidth as a function of the magnetic field. GH: Gaussian impurity scattering; PSF: Acoustic phonon scattering [2.201]

which even at low temperatures was emphasized by *Horovitz* and *Madhukar* [2.203]. The total line width is defined by $\Gamma_{\text{total}} = (\Gamma_{\text{imp}}^2 + \Gamma_{\text{ph}}^2)^{1/2}$. For different samples $r = 78$–250 Å and $U = 30$–620 K has been extracted [2.201].

Ando [2.196] has pointed out that in the case of point impurity scattering a SdH-type oscillation is superimposed on the cyclotron resonance line, caused by the rapid variation of the Fermi energy in two dimensions. This prediction was experimentally confirmed by *Abstreiter* et al. [2.204], Fig. 2.54. The specimen was a (100) silicon inversion layer in which the electron concentration could be changed by a factor of 2. A laser spectrometer with a wavelength of 311 μm was employed. The cyclotron resonance spectrum shown is a typical example of the absorption signal obtained in a Faraday configuration transmission experiment on a 2d system. Note the relatively large width and the asymmetry of the resonance line. Whereas the position of the maxima of the superimposed oscillations agree very well with the predictions of *Ando* [2.196, 205], their amplitude is considerably overestimated by the theory. The discrepancy is partially caused by the GBA and can be reduced by refined calculation (cf. Ref. [2.12], p. 555). Another reason might be inhomogeneity effects brought about by the large cross section of the specimen used in the experiment. The linewidth Γ_c was determined for different electron concentrations and compared with the $B = 0$ inverse relaxation time τ [2.204], Fig. 2.55. According to (2.5.14, 49) the ratio $\sqrt{\omega_c \tau}/(\hbar \omega_c / \Gamma_c)$ should be constant for point impurity scattering. That this is at least approximately the case may be seen in the upper part of Fig. 2.55. At low concentrations there are deviations, however.

Similarly to the 3d case, the most important application of cyclotron resonance is to determine the cyclotron mass as a function of electron concentration, temperature, uniaxial pressure and other variables. Whereas in n-type silicon inversion layers the cyclotron mass does not depend much on the

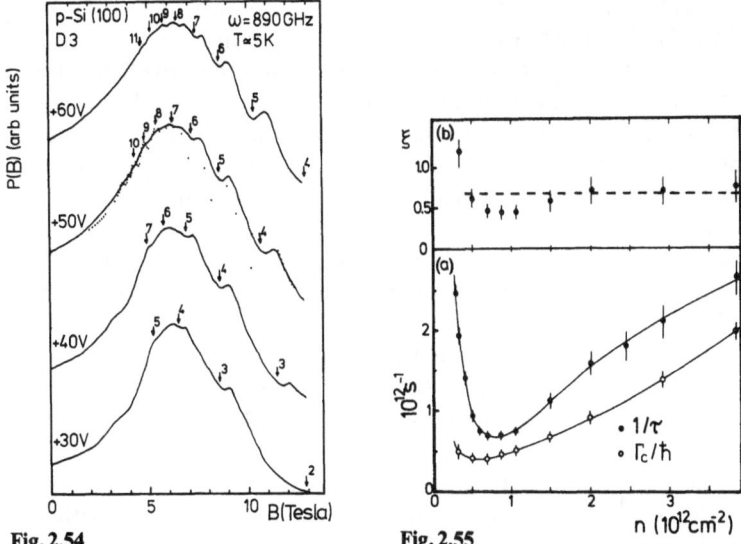

Fig. 2.54. Cyclotron resonance in an n-channel (100) Si inversion layer. The magnetic fields where the Fermi energy lies at the centre of a Landau level are denoted by arrows. ($\cdots$) represents the theoretical prediction by Ando. The electron concentration corresponding to the gate voltage of 10 V is $\sim 10^{12}\,\text{cm}^{-2}$ [2.204]

Fig. 2.55. Comparison of the zero-field collision rate $1/\tau$ with cyclotron linewidth Γ_c. The dashed line in the upper part (at 0.65) is the value of $\xi = \sqrt{\omega_c \tau/(\hbar\omega_c/\Gamma_c)}$ predicted for point scatterers in GBA [2.204]

surface electric field, this is not so for p-type layers. In this case the effective mass can increase substantially with increasing hole concentration in certain ranges.

It has been pointed out already that the subband structure of p-type inversion layers is relatively complicated since there are three valence bands, two of which are degenerate in zero electric field. High electric surface fields change the hole dispersion $\varepsilon(k_x, k_y)$ and the cyclotron mass considerably [2.101, 206]. A cyclotron resonance experiment with a 337 µm laser and magnetic fields up to 20 T [2.69d] to determine the cyclotron mass of holes confirmed the masses derived from SdH data in accordance with the self-consistent subband calculations [2.75a].

In the case of p-type silicon the band structure is not substantially modified by magnetic fields as high as 20 T. The situation is different, however, for p-type GaAs. This became obvious, when Landau levels were calculated, self-consistently for p-type GaAs-(GaAl)As heterojunctions by *Bangert* and *Landwehr* [2.208a] and by *Broido* and *Sham* [2.208b]. It was mentioned already, that experimental cyclotron resonance data only partially agree with theory [2.105a]. From the subband calculations cyclotron masses of $0.16\,m_e$ and $0.61\,m_e$ were deduced semiclassically from the constant energy contours, whereas the experimental data were interpreted in terms of two masses of $0.38\,m_e$ and $0.60\,m_e$ [2.104]. The calculation of Landau levels with incorporation of a high transverse magnetic field in the self-consistent solution of the Schrödinger and the Poisson equation reveal, that the band structure is considerably modified by the magnetic field. The cyclotron resonance spectra have to be interpreted in terms of Landau transitions, and the definition of a single effective cyclotron mass from experiments in high magnetic fields becomes meaningless.

In n-channel silicon MOSFETs it was observed that for electron concentrations below $10^{12}/cm^2$ the resonance field can drop sharply. The effect varies from sample to sample and has qualitatively been explained by electron localisation [2.209].

Whereas in bulk semiconductors the first cyclotron resonance experiments employed microwave radiation (cf. [2.210]), submillimeter waves were preferred in the study of 2d systems in order to meet the $\omega_c \tau \gg 1$ condition. A laser-source in experiments on silicon layers was first used by *Abstreiter* et al. [2.211]. An alternative method is to employ a Fourier transform spectrometer and record the transmission spectrum at constant magnetic field. The first data with this method were obtained by *Allen* et al. [2.212].

j) Thermomagnetic Effects

Few experiments and great conceptual confusion characterise the research on thermomagnetic effects. The first is fully understandable, the second, however, hardly. High precision thermomagnetic experiments are generally known to be difficult. The general aspects of the theory, on the other hand, have been clarified for 3d systems long ago (Sect. 2.4), while the translation to two dimensions is obvious [2.213–215].

The main subjects of confusion are (1) the correct description of the (diamagnetic) edge currents and their contribution to the observed quantities, and (2) the correct description of the thermodynamic state of the system. Instead of an exhaustive treatment of the problem we restrict ourselves to a few comments.

Following *Widom* [2.216] we define what he calls the Hall current by

$$j' = -\operatorname{curl} M, \tag{2.5.60}$$

with opposite sign convention, however [2.217, 219]. Assuming the system to be in a quasi-equilibrium state, the magnetisation M per unit area depends on the space coordinates only via the chemical potential ζ and the temperature T,

$$j' = -\left(\frac{\partial M}{\partial \zeta}\right)_{T,B} \times \nabla\zeta - \left(\frac{\partial M}{\partial T}\right)_{\zeta,B} \times \nabla T. \tag{2.5.61}$$

In a quasi-equilibrium transformation the thermodynamic potential $\Omega(\zeta, T, B)$ per unit area changes by

$$d\Omega = -s\,dT - n\,d\zeta - M\,dB, \tag{2.5.62}$$

$$s = -\left(\frac{\partial\Omega}{\partial T}\right)_{\zeta,B}, \quad n = -\left(\frac{\partial\Omega}{\partial\zeta}\right)_{T,B}, \quad M = -\left(\frac{\partial\Omega}{\partial B}\right)_{T,\zeta} \tag{2.5.63}$$

and the Maxwell relations

$$\left(\frac{\partial M}{\partial\zeta}\right)_{T,B} = \left(\frac{\partial n}{\partial B}\right)_{\zeta,T}; \quad \left(\frac{\partial M}{\partial T}\right)_{\zeta,B} = \left(\frac{\partial s}{\partial B}\right)_{\zeta,T} \tag{2.5.64}$$

are satisfied. The entropy density is $s = ns_1$. With (2.5.61, 64) the transport coefficients defined by (2.4.7) are

$$\sigma'_{xy} = -\sigma'_{yx} = -e\left(\frac{\partial n}{\partial B}\right)_{\zeta,T}, \quad \beta'_{xy} = -\beta'_{yx} = -\left(\frac{\partial s}{\partial B}\right)_{\zeta,T} \tag{2.5.65}$$

and the transverse magneto-Seebeck coefficient is

$$\alpha'_{xy} = \frac{\beta'_{xy}}{\sigma'_{xy}} = \frac{1}{e}\frac{(\partial s/\partial B)_{\zeta,T}}{(\partial n/\partial B)_{\zeta,T}} = \frac{1}{e}\left(\frac{\partial s}{\partial n}\right)_{\zeta,T}, \tag{2.5.66}$$

in contrast to the usual free electron results $\sigma_{xy} = -en/B$, $\beta_{xy} = -s/B$, $\alpha_{xy} = s/en$, (2.4.12, 13, 17).

The electron distribution function leading to (2.5.60) is barometric and describes an inhomogeneous density. The quasi-equilibrium *isothermal* current (2.5.61) is generally different from the current in an *isolated* system usually assumed to be realised in solid state experiments, and is calculated by solving an appropriate Boltzmann-type kinetic equation or by linear response theory (Sect. 2.4). As reported in the previous section, for an electron-impurity system the Kubo formula can be rearranged to give (2.5.31–34). Thus (at least for such systems at $T=0$) the Widom-Hall conductivity σ'_{xy} agrees with the usual one defined by the Kubo formula *only* if the Fermi energy lies in a gap. The case with β'_{xy} is similar [2.218].

It may be instructive to calculate $(\partial n/\partial B)_{T,\zeta}$ for free electrons. Using Landau states and Landau counting (!) $n = (eB/h)\sum_\nu f(\varepsilon_\nu)$

$$\left(\frac{\partial n}{\partial B}\right)_{T,\zeta} = \frac{n}{B} + \frac{eB}{h}\sum_\nu\left(-\frac{\partial f(\varepsilon_\nu)}{\partial\zeta}\right)\frac{\partial\varepsilon_\nu}{\partial B}. \tag{2.5.67}$$

Here the second term is $(\partial M_L/\partial\zeta)_{T,B}$, $(M_L = \chi_L B$ is the Landau diamagnetic moment per unit area) and vanishes (at $T=0$) if $\varepsilon_F \neq \varepsilon_\nu$. The free electron Hall conductivity is obtained by subtracting from (2.5.60) the contribution from the Landau diamagnetism [2.219].

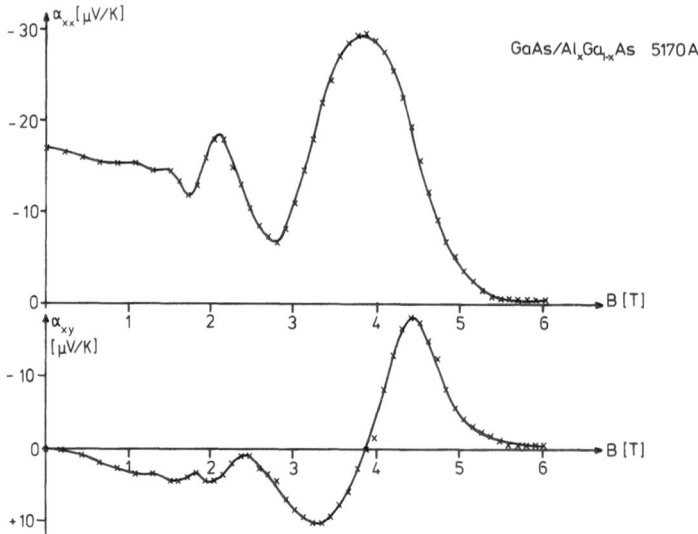

Fig. 2.56. Thermomagnetic coefficients α_{xx} and α_{xy} for a GaAs-(GaAl)As heterostructure as a function of the magnetic field at $T=4.2\,\mathrm{K}$ [2.223]

The coefficients (2.5.65) have been interpreted in different ways [2.216–221]. In any case, at $T=0$ they yield the correct values even in the presence of the impurities, as long as the Fermi energy lies in a gap. The free electron expressions quoted above hold for any values of Fermi energy and temperature but, in general, only in the limit of vanishing impurity concentration.

Figure 2.56 shows data of an experiment on GaAs-(GaAl)As heterostructures at 4.2 K with electron concentration $2.4 \times 10^{11}\,\mathrm{cm}^{-2}$ and mobility 24,000 cm²/Vs, performed by *Obloh* et al. [2.10b]. The vanishing of α_{xy} and α_{xx} at high magnetic fields indicates a localisation effect similar to the QHE. At present no theory is available which explains the observation. The peak values of α_{xx} predicted for free electrons

$$(\alpha_{xx})_{\mathrm{peak}} = -\frac{k_{\mathrm{B}}}{e}\frac{\ln 2}{\nu + \frac{1}{2}} \tag{2.5.68}$$

[2.213–215] have been confirmed fairly well [2.10b]. At low temperatures the level broadening has to be taken into account [2.10c, 105b].

The free electron formula $\alpha_{xy} = s/en$ seems to hold for $\omega_{\mathrm{c}}\tau \lesssim 1$ [2.222]. The reason for this may be the cancellations of the damping factors $(\omega_{\mathrm{c}}\tau)^2/[1 + (\omega_{\mathrm{c}}\tau)^2]$ appearing in the kinetic expressions of σ_{xy} and β_{xy} when taking the ratio defining α_{xy}. The Wiedemann-Franz law is valid for $k_{\mathrm{B}}T(\tau/\hbar) \gg 1$ [2.223a] but violated for $k_{\mathrm{B}}T(\tau/\hbar) \simeq 1$ [2.223b].

2.A Appendix: Dynamic Magnetoconductivity (LOCA)

Here we outline the derivation of the Mori formula for the dynamic magneto-conductivity and its evaluation in the lowest-order correlation approximation (LOCA). For simplicity $Vol=1$, $\hbar=1$.

2.A.1 Conductivity and Current Susceptibility

For an isotropic system the Kubo formula (2.3.26) for the electric conductivity can be written as

$$\sigma(z)=\frac{i}{z}\,[\chi(z)-\chi(0)], \tag{2.A.1}$$

where

$$\chi(z)=i\langle\!\langle j;j\rangle\!\rangle_z \tag{2.A.2}$$

$$\langle\!\langle A;B\rangle\!\rangle_z=\int_{-\infty}^{\infty}(\pm i)\theta(\pm t)\langle[A^+(t),B]\rangle\,e^{izt}dt,\ \mathrm{Im}\,\{z\}\gtrless 0 \tag{2.A.3}$$

is the current susceptibility (Green's function). Here $\langle\ldots\rangle$ denotes the thermal equilibrium and impurity average. In the transverse case $(j\perp B)$

$$j=\frac{1}{\sqrt{2}}\,(j_x+ij_y) \tag{2.A.4}$$

and the relations (2.3.37) hold, whereas in the longitudinal case $j=j_z$ and $\sigma(z)=\sigma_{zz}(z)$.

The susceptibility $\chi_{AB}(z)\equiv\langle\!\langle A;B\rangle\!\rangle_z$ is related to the correlation function

$$\phi_{AB}(t)=i(A(t)|B) \tag{2.A.5}$$

$$(A|B)=\int_0^{\beta}[\langle A^+B(i\lambda)\rangle-\langle A^+\rangle\langle B\rangle]d\lambda \tag{2.A.6}$$

$$A(t)=e^{iHt}Ae^{-iHt}, \tag{2.A.7}$$

the Laplace transform of which is

$$\phi_{AB}(z)=\int_0^{\infty}\phi_{AB}(t)e^{izt}dt. \tag{2.A.8}$$

Both χ and ϕ have Kramers-Kronig type spectral representation

$$F(z) = \frac{1}{\pi} \int \frac{F''(\omega)}{\omega - z} d\omega \qquad (2.A.9)$$

with

$$\chi''_{AB}(\omega) = \int \tfrac{1}{2} \langle [A^+(t), B] \rangle e^{i\omega t} dt \qquad (2.A.10)$$

and

$$\phi''_{AB}(\omega) = \int \tfrac{1}{2} (A(t)|B) e^{i\omega t} dt, \qquad (2.A.11)$$

respectively. These quantities are related by

$$\omega \phi''_{AB} = \chi''_{AB}(\omega) \qquad (2.A.12)$$

$$\phi''_{AB}(\omega) = \pi f_{AB} \delta(\omega) + \frac{\chi''_{AB}(\omega)}{\omega} \qquad (2.A.13)$$

as long as $\lim_{\omega \to 0} \chi/\omega$ exists. Consequently

$$\phi_{AB}(z) = -\frac{f_{AB}}{z} + \frac{1}{z} [\chi_{AB}(z) - \chi_{AB}(0)]. \qquad (2.A.14)$$

Since in the limit $z \to \infty$ $\chi_{AB} \to 0$ and $z\phi_{AB} = -(A|B) = -\chi^T_{AB}$, the isothermal susceptibility

$$\chi^T_{AB} = f_{AB} + \chi_{AB}(0). \qquad (2.A.15)$$

Ergodicity requires $f_{AB} = 0$.

2.A.2 Projectors and Mori Formula

Now we introduce the Liouville operator $\mathscr{L}$ defined by

$$\mathscr{L}A = -i\dot{A} = [H, A] \qquad (2.A.16)$$

$$A(t) = \exp(i\mathscr{L}t)A, \quad (A|\mathscr{L}B) = (\mathscr{L}A|B)$$

$$\phi_{AB}(z) = \left(A \left| \frac{1}{\mathscr{L} - z} \right| B \right). \qquad (2.A.17)$$

Furthermore we introduce the matrix

$$g_{ik} = (A_i | A_k) \tag{2.A.18}$$

and the projection operator

$$P = \sum_{ik} |A_i) g_{ik} (A_k| \equiv 1 - Q \tag{2.A.19}$$

$$P^2 = P, \quad PQ = QP = 0. \tag{2.A.20}$$

Using the identity

$$\frac{1}{x+y} = \frac{1}{x} - \frac{1}{x} y \frac{1}{x+y} \tag{2.A.21}$$

one easily obtains the Mori equation (in matrix notation)

$$(z - \omega g^{-1} - \hat{M} g^{-1}) \phi = g, \tag{2.A.22}$$

where

$$\omega : \omega_{ik} = (A_i | \mathscr{L} | A_k) \tag{2.A.23}$$

$$\hat{M} : M_{ik} = \left(Q \mathscr{L} A_i \left| \frac{1}{Q \mathscr{L} Q - z} \right| Q \mathscr{L} A_k \right) \tag{2.A.24}$$

$$\phi : \phi_{ik} = \left(A_i \left| \frac{1}{\mathscr{L} - z} \right| A_k \right). \tag{2.A.25}$$

If A_k is the transverse current

$$j = \frac{i}{ml} \sum_{\alpha} \sqrt{v+1} \, a_{\alpha+1}^+ a_\alpha \tag{2.A.26}$$

$\alpha = (v, k_y, k_z)$, $\alpha + 1 = (v+1, k_y, k_z)$ (a^+ and a are the usual Fermion creation and annihilation operators)

$$g = (j|j) = \frac{e^2 n}{m}, \quad \omega = (j|\mathscr{L}|j) = \frac{e^2 n}{m} \omega_c \tag{2.A.27}$$

$$\hat{M} = \left(Q \mathscr{L} j \left| \frac{1}{Q \mathscr{L} Q - z} \right| Q \mathscr{L} j \right)$$

$$= \frac{e^2}{m^2} \left(F \left| \frac{1}{Q \mathscr{L} Q - z} \right| F \right), \tag{2.A.28}$$

where $F = 1(F_x + iF_y)/\sqrt{2}$ is the transverse force due to the impurities ($F = -\text{grad } U$). (We have used $Q\mathscr{L}j = ieF/m$ and $(j|F) = 0$.). Approximating the denominator in (2.A.28) by $\mathscr{L} - z$, $\hat{M} = (e^2/m^2)\phi_{FF}$ and assuming ergodicity, $\sigma = i\phi$,

$$\sigma = i \frac{e^2 n/m}{z - \omega_c + M} \qquad (2.A.29)$$

$$M = \frac{1}{mn} \frac{\chi_{FF}(z) - \chi_{FF}(0)}{z}. \qquad (2.A.30)$$

These formulae were obtained in [2.26] by using the Götze-Wölfle method [2.44].

2.A.3 Lowest Order Correlation Approximation

In the single-site single-scattering approximation

$$\chi_{FF}(z) = n_I \sum_q |u(q)|^2 \tfrac{1}{2} q_\perp^2 X(q, \omega), \qquad (2.A.31)$$

where $u(q)$ is the Fourier transform of the impurity potential, $q_\perp^2 = q_x^2 + q_y^2$ and

$$X(q, z) = \langle\langle \varrho(q); \varrho^+(q) \rangle\rangle \qquad (2.A.32)$$

is the density susceptibility, and

$$\varrho(q) = \sum_{\alpha\beta} \varrho_{\alpha\beta}(q) a_\alpha^+ a_\beta \qquad (2.A.33)$$

with

$$\varrho_{\alpha\beta}(q) = (\exp -iqr)_{\alpha\beta} \qquad (2.A.34)$$

is the density fluctuation. We need the spectral density

$$X''(q, \omega) = \sum_{\alpha\alpha'} \sum_{\beta\beta'} \varrho_{\alpha\beta}(q) \varrho_{\alpha'\beta'}^*(q) f_{\alpha\beta,\alpha'\beta'}(\omega) \qquad (2.A.35)$$

where

$$f_{\alpha\beta,\alpha'\beta'}(\omega) = \int \tfrac{1}{2} \langle [a_\alpha^+(t) a_\beta(t), a_{\beta'}^+ a_{\alpha'}] \rangle e^{i\omega t} dt. \qquad (2.A.36)$$

In the lowest-order correlation approximation the correlation function occurring in (2.A.36) is factorised,

$$\langle [a_\alpha^+(t)a_\beta(t),\, a_{\beta'}^+ a_{\alpha'}] \rangle$$

$$= \langle a_\alpha^+(t)a_{\alpha'} \rangle \langle a_\beta(t)a_{\beta'}^+ \rangle - \langle a_{\alpha'}a_\alpha^+(t) \rangle \langle a_{\beta'}^+ a_\beta(t) \rangle. \qquad (2.A.37)$$

Assuming translational and rotational invariance of the impurity averaged quantities

$$\langle a_\alpha^+(t)a_\beta \rangle = \int D_\alpha(\omega)f(\omega)e^{i\omega t}d\omega\, \delta_{\alpha\beta} \qquad (2.A.38)$$

$$\langle a_\alpha a_\beta^+(t) \rangle = \int D_\alpha(\omega)[1-f(\omega)]e^{i\omega t}d\omega\, \delta_{\alpha\beta}, \qquad (2.A.39)$$

where $f(\omega)$ is the Fermi function, $D_\alpha(\omega)$ is the single-electron spectral function defined in (2.2.46) and

$$X''(q,\omega) = \pi \sum_{\alpha\beta} |\varrho_{\alpha\beta}(q)|^2 \int d\omega_1 \int d\omega_2 \delta(\omega + \omega_1 - \omega_2) \qquad (2.A.40)$$

$$\cdot D_\alpha(\omega_1)D_\beta(\omega_2)[f(\omega_1)-f(\omega_2)]. \qquad (2.A.41)$$

With this expression we obtain the relaxation spectrum $M''(\omega)$ as given by (2.3.40). The derivation of the LOCA longitudinal conductivity formula is similar.

2.A.4 Generalized Born Approximation

To derive the GBA equation for the electron Green's function by using the Mori formalism we introduce the more appropriate scalar product

$$(A|B) = \langle A^+ B + BA^+ \rangle \qquad (2.A.42)$$

[2.44]. The average Green's function

$$G_{\alpha\beta}(z) = \int (\mp i)\theta(\pm t)(a_\alpha(t)|a_\beta)e^{izt}dt,\ \mathrm{Im}\{z\} \gtrless 0$$

$$= -\left(a_\alpha \left| \frac{1}{\mathscr{L}-z} \right| a_\beta \right) = G_\alpha \delta_{\alpha\beta} \qquad (2.A.43)$$

obeys the Mori equation

$$(z - \omega g^{-1} - \Sigma g^{-1})G(z) = g, \qquad (2.A.44)$$

where

$$g : g_{\alpha\beta} = (a_\alpha|a_\beta) = \delta_{\alpha\beta} \qquad (2.A.45)$$

$$\omega:\omega_{\alpha\beta}=(a_\alpha|\mathscr{L}|a_\beta)=\varepsilon_\alpha\delta_{\alpha\beta} \tag{2.A.46}$$

$$\Sigma:\Sigma_{\alpha\beta}=-\left(Q\mathscr{L}a_\alpha\left|\frac{1}{Q\mathscr{L}Q-z}\right|Q\mathscr{L}a_\beta\right)$$

$$=\Sigma_\alpha\delta_{\alpha\beta}. \tag{2.A.47}$$

Thus,

$$G_\alpha(z)=\frac{1}{z-\varepsilon_\alpha-\Sigma_\alpha(z)}. \tag{2.A.48}$$

The Hamilton operator (2.2.1),

$$H=\sum_\alpha \varepsilon_\alpha a_\alpha^+ a_\alpha+\sum_{\alpha\beta} U_{\alpha\beta}a_\alpha^+ a_\beta \tag{2.A.49}$$

with $\langle U\rangle_I=0$ has been used. In LOCA the expression for Σ_α factorises

$$\Sigma_\alpha=-\sum_\beta\left(U_{\alpha\beta}a_\beta\left|\frac{1}{Q\mathscr{L}Q-z}\right|U_{\alpha\gamma}a_{\alpha\gamma}\right)$$

$$=\sum_\beta\langle|U_{\alpha\beta}|^2\rangle_I G_\beta. \tag{2.A.50}$$

Equations (2.A.48, 49) lead to the GBA equation for self-energy

$$\Sigma_\alpha=\sum_\beta\frac{\langle|U_{\alpha\beta}|^2\rangle_I}{z-\varepsilon_\beta-\Sigma_\beta} \tag{2.A.51}$$

used in the text.

2.A.5 Comments

A great advantage of the Mori formalism is its flexibility. For instance, we can choose the partial currents j_v, defined by $j=\sum_v j_v$ and (2.A.26) as the A_i operators (instead of the current itself). This choice allows a traditional kinetic approximation [2.9c] and leads in LOCA to the straightforward generalisation (GBA) of (2.3.24). Similarly, in this formulation the Hall coefficient in LOCA is not identical to the free electron expression (cf. the discussion at the end of Sect. 3.3) but is a straightforward generalisation of the result obtained by solving a Boltzmann-Bloch type kinetic equation.

Acknowledgements. We are indebted to Dr. Patrik Fazekas for his help in formulating Chap. 2.5.4h, Prof. Wlodek Zawadzki for his useful comments and Dr. Suzanne Krebs for her skillful preparation of the manuscript.

Selected Bibliography

A *monograph* on quantum magneto-transport still does not exist.

Some *review articles* are:

E.N.Adams, R.W.Keyes: In *Progress in Semiconductors*, Vol. 6, ed. by Gibson, Burgess, Kröger (Heywood, London 1962) p. 85

L.M.Roth, P.N.Argyres: In *Semiconductors and Semimetals* 1, 159 (Academic, New York 1966)

G.Landwehr: In *Physics of Solids in Intense Magnetic Fields*, ed. by Haidemenakis (Plenum, New York 1969) p. 415

C.R.Pidgeon: In *Handbook on Semiconductors*, Vol. 2, ed. by Balkanski (North-Holland, Amsterdam 1980) p. 223

T.Ando, A.B.Fowler, F.Stern, Rev. Mod. Phys. **54**, 437 (1982)

For various *theoretical aspects* cf.

R.Kubo, S.M.Mijake, N.Nashitsume: In *Solid State Physics* 17, 269 (Academic, New York 1965)

J.R.Barker: In *Hot Electrons in Semiconductors*, ed. by D. G. Seiler and A. E. Stephens (Pergamon, Oxford 1978) p. 197 (Solid-State Electr. **21**, 1–323 (1978))

J.R.Barker: In *Handbook on Semiconductors*, Vol. 1, ed. by W. Paul (North-Holland, Amsterdam 1982) p. 617

J.Hajdu: In *Narrow Gap Semiconductors* – Physics and Applications, ed. by W. Zawadzki, Lecture Notes Phys. **133**, 219 (Springer, Berlin, Heidelberg, New York 1980)

J.Hajdu: In *Theoretical Aspects and New Developments in Magneto-Optics*, ed by J. T. Devreese (Plenum, New York 1980) p. 195

Some *conference proceedings* and *summer school notes* are:

E.D.Haidemenakis (ed.): *Physics of Solids in Intense Magnetic Fields* (Plenum, New York 1969)

G.Landwehr (ed.): *The Generation of High Magnetic Fields and their Application in Solid State Physics* (unpublished lecture notes, Würzburg 1972)

G.Landwehr (ed.): *The Application of High Magnetic Fields in Semiconductor Physics* (unpublished lecture notes, Würzburg 1974)

G.Landwehr (ed.): *The Application of High Magnetic Fields in Semiconductor Physics* (unpublished lecture notes, Würzburg 1976)

J.F.Ryan (ed): *The Application of High Magnetic Fields in Semiconductor Physics* (unpublished lecture notes, Oxford 1978)

S.Chikazumi, M.Miura (eds.): *Physics in High Magnetic Fields*, Springer Ser. Solid-State Sci., Vol. 24 (Springer, Berlin, Heidelberg, New York 1981)

G.Landwehr (ed.): *Application of High Magnetic Fields in Semiconductor Physics*, Lecture Notes Phys. Vol. 177 (Springer, Berlin, Heidelberg, New York 1983)

A.J.Freeman (ed.): *Solids and Plasmas in High Magnetic Fields* (North-Holland, Amsterdam 1979) [J. Magn. Magn. Mat. **11**, 1 (1979)]

J.J.Quinn, P.J.Stiles (eds.): *Electronic Properties of Quasi-Two-Dimensional Systems* (North-Holland, Amsterdam 1976) [Surf. Sci. **58**, 1 (1976)]

G.Dorda, P.J.Stiles (eds.): *Electronic Properties of Two-Dimensional Systems* (North-Holland, Amsterdam 1978) [Surf. Sci. **73**, 1 (1978)]

S.K.Kawaji (ed.): *Electronic Properties of Two-Dimensional Systems* (North-Holland, Amsterdam 1980) [Surf. Sci. **98**, 1 (1980)]

F.Stern (ed.): *Electronic Properties of Two-Dimensional Systems* (North-Holland, Amsterdam 1982) [Surf. Sci. **113**, 1 (1982)]

R.J.Nicholas (ed.): *Electronic Properties of Two-Dimensional Systems* (North-Holland, Amsterdam 1984) [Surf. Sci. **142**, 1 (1984)]

G.Bauer, F.Kuchar, H.Heinrich (eds.): *Two-Dimensional Systems, Heterostructures and Superlattices*, Springer Ser. Solid-State Sci., Vol. 53 (Springer, Berlin, Heidelberg, New York, Tokyo 1984)

References

2.1   L.Shubnikov, W.J. de Haas: Leiden Commun. **207**a, c, d, **210**a (1930)

2.2   L.M.Bliek, G.Landwehr, M. von Ortenberg: Proc. IX Intern. Conf. on Phys. Semiconductors (Nauka, Moscow 1969) p. 710

2.3   L.D.Landau: Z. Physik **64**, 629 (1930)

2.4   R.B.Dingle: Proc. Roy. Soc. (London): **A211**, 517 (1952)

2.5   G.Landwehr: In *Physics of Solids in Intense Magnetic Fields*, ed. by E.D.Haidemenakis (Plenum, New York 1969) p. 415

2.6   A.B.Fowler, F.F.Fang, L.Howard: Proc. of VIII Intern. Conf. Phys. Semiconductors, Kyoto 1966, p. 331

2.7   K. von Klitzing, G.Dorda, M.Pepper: Phys. Rev. Lett. **45**, 494 (1980); K. von Klitzing: unpublished

2.8   E.Braun, E.Staben, K. von Klitzing: PTB-Mitteilungen **90**, 350 (1980)

2.9   G.Landwehr (ed.): In *Application of High Magnetic Fields in Semiconductor Physics*, Lecture Notes Phys. Vol. 177 (Springer, Berlin, Heidelberg, New York 1983)

2.9a   Th. Englert: ibid p. 87

2.9b   E.Gornik: ibid p. 248

2,9c   J.Hajdu, U.Paulus: ibid p. 415

2.9d   K.Heift, J.Hajdu: ibid p. 139

2.9e   S.Kawaji, J.Kawaguchi: ibid p. 53

2.9f   H.Fukujama: ibid p. 47

2.9g   J.Hajdu: ibid p. 23

2.9h   A.Aoki: ibid p. 11

2.10   R.J.Nicholas (ed.): *Electronic Properties of Two-Dimensional Systems* (North-Holland, Amsterdam 1984) [Surf. Sci. **142**, 1 (1984)]

2.10a   H.L.Störmer: ibid p. 130

2.10b   H.Obloh, K. von Klitzing, K.Ploog: ibid p. 236

2.10c   W.Zawadzki, R.Lassnig: ibid p. 225

2.11   L.M.Roth, P.N.Argyres: *Semiconductors and Semimetals* **1**, 159 (Academic, New York 1966)

2.12   T.Ando, A.B.Fowler, F.Stern: Rev. Mod. Phys. **54**, 437 (1982)

2.13   P.G.Harper, J.W.Hodby, R.A.Stradling: Rep. Prog. Phys. **36**, 1 (1973)

2.14   R.L.Peterson: In *Semiconductors and Semimetals* **10**, 221 (Academic, New York 1975)

2.15a   G.Bauer: In *Springer Tracts Mod. Phys.* **74**, 1 (Springer, Berlin, Heidelberg 1974)

2.15b   D.G.Seiler, A.E.Stephens (eds.): Hot Electrons in Semiconductors (Pergamon, Oxford 1978) [Solid State Electr. **21**, 1–323 (1978)]; D.K.Ferry, J.Barker, A.Jacoboni (eds.):*Physics of Nonlinear Transport in Semiconductors* (Plenum, New York 1980)

2.16   See, e.g. *Solids and Plasmas in High Magnetic Fields*, ed. by R.L.Aggarwal, A.J.Freeman, B.B.Schwartz (North-Holland, Amsterdam 1979)

2.17   R.Peierls: Z. Physik **80**, 763 (1933)

2.18   R.Kubo, H.Hasegawa, N.Hashitsume: J. Phys. Soc. Jpn. **14**, 56 (1959)

2.19   L. Onsager. Philos. Mag. (7) **43**, 1006 (1952)

2.20   E.O.Kane: J. Phys. Chem. Solids **1**, 249 (1957); *Semiconductors and Semimetals* **1**, 75 (Academic, New York 1966)

2.21   W.Zawadzki, B.Lax: Phys. Rev. Lett. **16**, 1001 (1974)

2.22   J.Schnakenberg: Z. Physik **175**, 445 (1963); G.Eilenberger: Z. Phys. **175**, 445 (1963)

2.23   F.Hund: *Theorie des Aufbaus der Materie* (Teubner, Stuttgart 1961)

2.24   Y.Ono, J.Hajdu: Z. Physik **B33**, 61 (1979)

2.25   R.R.Gerhardts, J.Hajdu: Z. Physik **245**, 126 (1971)

2.26   W.Götze, J.Hajdu: J. Phys. **C11**, 3993 (1978)

2.27   T.Ando, Y.Uemura: J. Phys. Soc. Japan **36**, 959 (1974)

2.28   K.Heift: Dissertation, Köln University (1981);
K.Heift, Y.Ono, J.Hajdu: In Lecture Notes Phys., Vol. 152, ed. by E.Gornik, H.Heinrich, L.Palmetshofer (Springer, Berlin, Heidelberg, New York 1982) p. 373

2.29   U.Paulus, J.Hajdu: Solid State Commun. **20**, 687 (1967)
2.30   D.Blochinzew, L.Nordheim: Z. Physik **84**, 168 (1933)
2.31   R.Peierls: *Quantum Theory of Solids* (Oxford U. Press, London 1955)
2.32   A.A.Abrikosov: Solid State Phys. Suppl. **12**, 1 (1972)
2.33   V.S.Titeica: Ann. Physik (Leipzig) **22**, 129 (1935)
2.34   E.N.Adams, T.D.Holstein: J. Phys. Chem. Solids **10**, 254 (1959)
2.35   D.Bergers, J.Hajdu: Solid State Commun. **20**, 683 (1976)
2.36   A.H.Kahn: Phys. Rev. **119**, 1189 (1960);
        V.G.Skobov: Sov. Phys. JETP **10**, 1039 (1960); **11**, 941 (1960);
        Yu.A.Bychkov: Sov. Phys. JETP **12**, 483 (1961)
2.37   R.Kubo, S.I.Miyake, N.Hashitsume: Solid State Phys. **17**, 269 (1965)
2.38   E.Bangert: Z. Physik **215**, 177, 192 (1968)
2.39   B.Davydov, I.Pomeranchuk: J. Phys. USSR **2**, 147 (1940)
2.40   L.M.Bliek: Dissertation, TH Braunschweig (1969) unpublished
2.41   P.N.Argyres: J. Phys. Chem. Solids **4**, 19 (1958)
2.42   R.Kubo: J. Phys. Soc. Jpn. **12**, 570 (1957)
2.43   H.Mori: Prog. Theor. Phys. **34**, 423 (1965)
2.44   W.Götze, P.Wölfle: J. Low Temp. Phys. **5**, 575 (1971);
        Phys. Rev. **B6**, 1226 (1972)
2.45   R.R.Gerhardts: Z. Physik **B22**, 327 (1975)
2.46   H.Keiter, J.Hajdu: Phys. Status Solidi **38**, 757 (1970)
2.47   J.Rauluszkievicz, M.Gorska, E.Kaczmarek: In *Physics of Narrow Gap Semiconductors* (PAN, Warsaw 1978)
2.47a   W.Zawadzki: ibid p. 281
2.47b   A.Raymond, J.L.Robert, C.Bernard, C.Bousquet, A.Aulombard: ibid p. 303
2.48   R.A.Stradling: Private communication
2.49   I.M.Tsidilkovski: Preprint (1979)
2.50   S.Narita, K.Suizu: Suppl. Prog. Theor. Phys. **57**, 187 (1975)
2.51   L.S.Dubinskaya: Sov. Phys. JETP **29**, 436 (1969)
2.52   R.R.Gerhardts, J.Hajdu: Solid State Commun. **9**, 1607 (1971)
2.53   N.N.Bertschenko, M.W.Paschkofski: Sov. Phys. Solid State **15**, 2497 (1974)
2.54   C.S.Whitsett: Phys. Rev. **138A**, 829 (1965)
2.55   A.Freudenberger: Diplomarbeit, U. Würzburg (1975)
2.56   J.Kossut, J.Hajdu: Solid State Commun. **27**, 1401 (1978)
2.57   O.Beckmann, E.Hanamura, L.J.Neuringer: Phys. Rev. Lett. **18**, 773 (1967);
        Kh.J.Amirkhanov, R.S.Bashirov: Sov. Phys. Semicond. **1**, 558 (1967);
        S.T.Pavlow, R.V.Parfenev, Yu.A.Firsov, S.S.Shalyt: Sov. Phys. JETP **21**, 1049 (1965);
        R.Dornhaus, G.Nimtz, W.Schlabitz, P.Zaplinski: Solid State Commun. **15**, 495 (1974);
        Yu.N.Gavrilyuk, S.G.Grasanzade, E.A.Salkov, G.A.Shepelskii: Sov. Phys. Semicond. **31**, 64 (1979);
        R.Dornhaus, G.Nimtz: Solid State Commun. **22**, 41 (1977)
2.58   E.H.Putley: In *Semiconductors and Semimetals* **1**, 286 (Academic, New York 1966)
2.59   G.Nimtz, B.Schlicht, E.Tyssen, R.Dornhaus, L.D.Haas: Solid State Commun. **32**, 669 (1979);
        G.Nimtz, B.Schlicht: In *Festkörperprobleme* **20**, 369 (Vieweg, Braunschweig 1980)
2.60   E.P.Wigner: Phys. Rev. **46**, 1002 (1934)
2.61   W.G.Kleppmann, R.J.Elliott: J. Phys. **C8**, 2729 (1975)
2.62   R.R.Gerhardts: Solid State Commun. **36**, 397 (1980)
2.63   G.Nimtz, B.Schlicht, H.Lehmann, E.Tyssen: Appl. Phys. Lett. **35**, 640 (1979)
2.64   G.I.Guseva, P.S.Zyryanov: Phys. Status Solidi **25**, 775 (1968)
2.65   R.R.Gerhardts: Solid State Commun. **10**, 107 (1972)
2.66   A.Bastin, C.Lewiner, O.Betbeder-Matibet, P.Nozieres: J. Phys. Chem. Solids **32**, 1811 (1971)
2.67a   B.Lax, J.G.Mavroides: Solid State Phys. **11**, 261 (1960) (Academic, New York 1960)
2.67b   C.R.Pidgeon: In *Handbook on Semiconductors*, Vol. 2, ed. by M.Balkanski, T.S.Moss (North-Holland, Amsterdam 1980) p. 223

2.68   N.Miura, G.Kido, S.Chikazumi: Solid State Commun. **18**, 885 (1975);
         G.Kido, N.Miura, M.Akihiro, H.Katayama, S.Chikazumi: In *Physics in High Magnetic Fields*, ed. by S.Chikazumi, N.Miura, Springer Ser. Solid-State Sci., Vol. 24 (Springer, Berlin, Heidelberg, New York 1981) p. 72
2.69   G.Landwehr (ed.): *The Application of High Magnetic Fields in Semiconductor Physics* (unpublished conference lecture notes, Würzburg 1976)
2.69a   F.Herlach: ibid p. 84
2.69b   N.Miura, G.Kido, K.Suzuki, S.Chikazumi: ibid p. 441
2.69c   M.Heuser, J.Hajdu: ibid p. 208
2.69d   J.Kotthaus: ibid p. 663
2.70   M. von Ortenberg, K.J.Button, D.Fischer, G.Landwehr: Phys. Rev. **B6**, 2100 (1972)
2.71   B.D.McCombe, R.Kaplan, R.J.Wagner, E.Gornik, W.Mueller: Phys. Rev. **B13**, 2536 (1976)
2.72   M. von Ortenberg: Habilitationsschrift, U. Würzburg (1974)
2.73   M. von Ortenberg: In *Infrared and Millimeter Waves*, Vol. 3, ed. by K.J.Button (Academic, New York 1980) p. 275
2.74   M. von Ortenberg, K.Schwarzbeck, G.Landwehr: *Physique sous Champs Magnétiques Intenses* (CNRS, Paris 1975) p. 305
2.75   J.Q.Ramage, R.A.Stradling, R.J.Tidey, J.R.Burke: In *Proc. XII. Int. Conf. Phys. Semicond.* (Teubner, Stuttgart 1974) p. 531
2.76   E.Tyssen, R.Dornhaus, G.Haider, P.Kokoschinegg, K.-H.Mueller, G.Nimtz, H.Happ: Solid State Commun. **17**, 1459 (1975)
2.77   J.R.Apel, T.O.Poehler, C.R.Westgate, R.I.Joseph: Phys. Rev. **B4**, 436 (1971)
2.78   M.Heuser, J.Hajdu: Solid State Commun. **20**, 313 (1976)
2.79   E.E.H.Shin, P.N.Argyres, B.Lax: Phys. Rev. **B7**, 3572 (1973)
2.80   A.J.Kawabata: J. Phys. Soc. Jpn. **23**, 999 (1967)
2.81   A.Lodder, S.Fujita: J. Phys. Soc. Jpn. **25**, 774 (1968); Phys. Lett. **A46**, 381 (1974)
2.82   V.K.Arora, M.A.Al-Massari, M.Prasad: Physica **106B**, 311 (1981)
2.83   J. van Royen, J.T.Devreese: Solid State Commun. **40**, 947 (1981)
2.84   S.J.Miyake: Prog. Theor. Phys. Suppl. **69**, 311 (1981)
2.85   Yu.N.Obraztsov: Sov. Phys. Solid State **6**, 331 (1964); **7**, 455 (1965); **8**, 506 (1966)
2.86   P.S.Zyryanov, V.P.Silin: Sov. Phys. JETP **19**, 366 (1964);
         Sov. Phys. Solid State **8**, 503 (1966)
2.87   A.I.Anselm, Yu.N.Obraztsov, R.G.Tarkhanyan: Sov. Phys. Solid State **7**, 2293 (1966)
2.88   M.Heuser, J.Hajdu: Z. Physik **270**, 289 (1974)
2.89   R.Kubo, M.Yokota, S.Nakajima. J. Phys. Soc. Jpn. **12**, 1203 (1957);
         S.Nakajima: Prog. Theor. Phys. **20**, 948 (1958)
2.90   L.Smrčka, P.Středa: J. Phys. **C10**, 2153 (1977)
2.91   J.Hajdu, S.Fischer: Z.Physik **181**, 479 (1964);
         V.G.Baryakhtar, S.V.Peletminskii: Sov. Phys. JETP **21**, 126 (1965);
         Sov. Phys. Solid State **7**, 356 (1965);
         S.V.Peletminskii: Sov. Phys. Solid State **7**, 2157 (1966)
2.92   B.Schroeder, G.Landwehr: Solid State Commun. **22**, 589 (1977)
2.93   D.G.Seiler, R.R.Galazka, W.B.Becker: Phys. Rev. **B3**, 4274 (1971); also in *The Physics of Semimetals and Narrow Gap Semiconductors*, ed. by D.L.Carter, R.T.Bate (Pergamon, Oxford 1971)
2.94   G.Dorda: Electron. Fis. Apl. **17**, 203 (1974);
         H.L.Störmer, A.Pinczuk, A.C.Gossard, W.Wiegmann: Appl. Phys. Lett. **38**, 691 (1981)
2.95   R.Dingle, H.L.Störmer, A.C.Gossard, E.Wiegmann: Surf. Sci. **98**, 90 (1980)
2.96   F.Stern, W.E.Howard: Phys. Rev. **163**, 816 (1967)
2.97   F.Stern: Proc. X Int. Conf. Phys. Semicond. (Cambridge, Mass. 1970) p. 451
2.98   T.Ando: unpublished. Cf. also Ref. [2.101]
2.99   J.M.Luttinger, W.Kohn: Phys. Rev. **97**, 869 (1958)
2.100   E.Bangert, K. von Klitzing, G.Landwehr: in *Proc. XII. Int. Conf. Phys. Semicond* (Teubner, Stuttgart 1974) p. 714

2.101 G.Landwehr: In *Festkörperprobleme* **15**, 49 (Vieweg, Braunschweig 1975)
2.102 F.J.Ohkawa, Y.Uemura: Prog. Theor. Phys. Suppl. **57**, 164 (1975)
2.103 J.J.Quinn, P.J.Stiles (eds.): *Electronic Properties of Quasi-Two-Dimensional Systems* (North-Holland, Amsterdam 1976) [Surf. Sci. **58**, 1 (1976)]
2.103a E.Bangert, G.Landwehr: ibid p. 138
2.103b R.Gerhardts: ibid p. 227
2.103c T.Ando: ibid p. 128; Phys. Rev. **B13**, 3468 (1976)
2.104 H.L.Störmer, Z.Schlesinger, A.C.Gossard, W.Wiegmann: Phys. Rev. Lett. **51**, 126 (1983)
2.105 G.Bauer, F.Kuchar, H.Heinrich (eds.): *Two-Dimensional Systems, Heterostructures and Superlattices*, Springer Ser. Solid-State Sci., Vol. 53 (Springer, Berlin, Heidelberg, New York, Tokyo 1984)
2.105a G.Landwehr, E.Bangert: ibid p. 40
2.105b W.Zawadzki: ibid p. 79
2.106 T.Ando, Y.Uemura: J. Phys. Soc. Jpn. **36**, 959 (1974)
2.107 T.Ando: J. Phys. Soc. Jpn. **37**, 622 (1974)
2.108 R.Gerhardts: Z. Phys. **B21**, 275, 285 (1975)
2.109 F.Wegner: Z. Phys. **B51**, 279 (1983)
2.110 S. Das Sarma: Solid State Commun: **36**, 357 (1980)
2.111 W.Götze, J.Hajdu: Solid State Commun. **29**, 89 (1979)
2.112 T.Ando, Y.Matsumoto, Y.Uemura, M.Kobayashi, K.F.Komatsubara: J. Phys. Soc. Jpn. **32**, 859 (1972)
2.113 T.Ando: J. Phys. Soc. Jpn. **37**, 1233 (1974)
2.114 Th.Englert: Dissertation, U. Würzburg (1977) unpublished
2.115 F.J.Ohkawa, Y.Uemura: J. Phys. Soc. Jpn. **43**, 907, 917 (1977)
2.116 G.Dorda, P.J.Stiles (eds.): *Electronic Porperties of Quasi-Two-Dimensional Systems* (North-Holland, Amsterdam 1978) [Surf. Sci. **73**, 1 (1978)]
2.116a L.J.Sham, M.Nakayama: ibid p. 272; Phys. Rev. **B20**, 734 (1979)
2.116b Th.Englert, K. von Klitzing: ibid p. 70
2.116c M.Prasad, S.Fujita: ibid p. 494
2.116d M.Prasad, T.K.Srinivas, S.Fujita: ibid p. 505
2.117 R.Kümmel: Z. Physik **B22**, 223 (1975)
2.118 G.Dorda, H.Gesch, I.Eisele: Solid State Commun. **20**, 429, 677 (1976);
       I.Eisele, H.Gesch, G.Dorda: Solid State Commun. **22**, 185 (1977);
       Th.Englert, G.Landwehr, K. von Klitzing, G.Dorda, H.Gesch: Phys. Rev. **B18**, 794 (1978)
2.119 P.Stallhofer, J.P.Kotthaus, J.F.Koch: Solid State Commun. **20**, 519 (1976)
2.120 T.Neugebauer, K. von Klitzing, G.Landwehr, G.Dorda: Solid State Commun. **17**, 295 (1975)
2.121 A.A.Lakhani, P.J.Stiles: Phys. Lett. **A51**, 117 (1975)
2.122 M.J.Kelly, L.M.Falicov: Phys. Rev. **B15**, 1974 (1977);
       J.Phys. **C10**, 4735 (1977)
2.123 U.Paulus: Z. Phys. **B30**, 165 (1978);
       Y.Ono, U.Paulus: Z. Phys. **B34**, 11 (1979)
2.124 D.C.Tsui, G.Kaminsky: Phys. Rev. Lett. **42**, 595 (1979)
2.125 F.F.Fang, P.J.Stiles: Phys. Rev. **174**, 823 (1968)
2.126 T.Ando, Y.Uemura: J. Phys. Soc. Jpn. **37**, 1044 (1974)
2.127 K.F.Komatsubara, K.Narita, Y.Katayama, N.Kotera: J. Phys. Chem. Sol. **35**, 123 (1974)
2.128 J.Wakabayashi, S.Kawaji: J: Phys. Soc. Jpn. **44**, 1839 (1978)
2.129 T.Englert: Unpublished
2.130 P.W.Anderson: Phys. Rev. **109**, 1492 (1958)
2.131 N.F.Mott: *Metal-Insulator Transitions* (Taylor and Francis, London 1974)
2.132 J.M.Ziman: *Models of Disorder* (*Cambridge Press, Cambridge* 1979)
2.133 Y.Nagaoka, H.Fukujama (eds.): *Anderson Localization*, Springer Ser. Solid-State Sci., Vol. 39 (Springer, Berlin, Heidelberg, New York 1982)
2.133a T.Ando: ibid p. 176
2.133b Y.Ono: ibid p. 207

2.134 G.Landwehr (ed.): *Impurity Bands in Semiconductors* (conference lecture notes, Würzburg 1979) in Phil. Mag. **42B** 725 (1980)
2.135 W.Götze: Philos. Mag. **43**, 219 (1981);
       W.Götze: In *Modern Problems in Solid State Physics*, Vol. 1, ed. by Yu.E.Lozovik, A.A.Maradudin (North-Holland, Amsterdam 1984)
2.136 L.Bányai: In *Physique des Semiconducteurs*, ed. by M.Hulin (Dunod, Paris 1964) p. 417
2.137 M.Tsukada: J. Phys. Soc. Jpn. **41**, 1466 (1976)
2.138 H.Aoki, H.Kamimura: Solid State Commun. **21**, 45 (1977)
2.139 H. Aoki: J. Phys. **C10**, 2583 (1977); **C11**, 3823 (1978)
2.140 H.Levine, S.B.Libby, A.M.M.Pruisken: Phys. Rev. Lett. **51**, 1915 (1983)
2.141 H.Aoki: J. Phys. **C15**, L 1227 (1982)
2.142 D.Yoshioka, Y.Ono, H.Fukuyama: J. Phys. Soc. Jpn. **50**, 3419 (1981); **51**, 340 (1982); Y.Ono, D.Yoshioka, H.Fukuyama: J. Phys. Soc. Jpn. **50**, 2143 (1981)
2.143 E.Abrahams, P.W.Anderson, D.C.Licciardello, T.V.Ramakrishnan: Phys. Rev. Lett. **42**, 673 (1979);
2.144 M.Tsukada: J. Phys. Soc. Jpn. **40**, 1515 (1976); **42**, 391 (1977)
2.145 R.Gerhardts: Phys. Rev. **B24**, 1339; 4068 (1981)
2.146 R.Gerhardts, Y.Kuramoto: Z. Phys. **B44**, 301 (1981)
2.147 Y.Kuramoto, R.Gerhardts: J. Phys. Soc. Jpn. **51**, 3810 (1982)
2.148 H.Fukuyama, P.M.Platzman: Phys. Rev. **B25**, 2943 (1982)
2.149 J.Kosch: Diplomarbeit, U. Köln (1984)
2.150 T.Ando, Y.Matsumoto, Y.Uemura: J. Phys. Soc. Jpn. **39**, 279 (1975)
2.151 S.Kawaji, T.Igarashi, J.Wakabayashi: Prog. Theor. Phys. Suppl. **57**, 176 (1975)
2.152 S.K.Kawaji (ed.): *Electronic Properties of Two-Dimensional Systems* (North-Holland, Amsterdam 1980) [Surf. Sci. **98**, 1 (1980)])
2.152a J.Wakabayashi, S.Kawaji: ibid p. 299
2.153 D.C.Tsui, A.C.Gossard: Appl. Phys. Lett. **38**, 550 (1981)
2.154 T.Englert: Private communication
2.155 K. von Klitzing, G.Ebert: Physica **B117-118**, 682 (1983)
2.156 H.L.Störmer, Z.Schlesinger, A.Chang, D.C.Tsui, A.C.Gossard, W.Wiegmann: Phys. Rev. Lett. **51**, 126 (1983)
2.157 T.Aoki, H.Ando: Solid State Commun. **38**, 1079 (1981)
2.158 R.E.Prange: Phys. Rev. **B23**, 4802 (1981);
       R.E.Prange, R.Joynt: Phys. Rev. **B25**, 2943 (1982)
2.159 W.Brenig: Z. Physik **B50**, 305 (1983)
2.160 J.Chalker: J. Phys. **C16**, 4297 (1983)
2.161 D.J.Thouless: J. Phys. **C14**, 3475 (1981)
2.162 P.Středa: J. Phys. **C15**, L717 (1982)
2.163 G.A.Baraff, D.C.Tsui: Phys. Rev. **B24**, 2274 (1981)
2.164 S.V.Iordansky: Solid State Commun. **43**, 1 (1982)
2.165 R.F.Kazarinov, S.Luryi: Phys. Rev. **B25**, 626 (1982)
2.166 S.A.Trugman: Preprint (1983); Bull. Am. Phys. Soc. **28**, 365 (1983)
2.167 A.MacKinnon: Private communication
2.168 R.B.Laughlin: Phys. Rev. **B23**, 5632 (1981)
2.169 Y.Imry: J. Phys. **C15**, L1227 (1982)
2.170 B.I.Halperin: Phys. Rev. **B25**, 2185 (1982)
2.171 E.Teller: Z.Physik **67**, 311 (1931)
2.172 H.Aoki: J. Phys. **C16**, 1893 (1983)
2.173 N.Byers, C.N.Yang: Phys. Rev. Lett. **7**, 46 (1961)
2.174 L.D.Landau; E.M.Lifshitz: *Electrodynamics of Continuous Media* (Pergamon, Oxford 1960)
2.175 J.Hajdu, U.Gummich: Solid State Comm. **52**, 985 (1984)
2.176 P.G.Harper: Proc. Phys. Soc. (London) **A68**, 874, 879 (1955)
2.177 A.D.Brailsford: Proc. Phys. Soc. (London) **A70**, 275 (1961)
2.178 G.E.Zil'berman: Sov. Phys. JETP **5**, 208 (1957); **6**, 299 (1958)
2.179 D.R.Hofstadter: Phys. Rev. **B14**, 2239 (1976)

2.180   F.H.Claro, G.W.Wannier: Phys. Rev. **B19**, 6068 (1969); G.H.Wannier: Phys. Status Solidi **B88**, 757 (1978)
2.181   A.H.MacDonald: Phys. Rev. **B28**, 6713 (1983); **B29**, 3057 (1984)
2.182   P. Středa: J. Phys. **C15**, L 1299 (1982)
2.183   D.J.Thouless, M.Kohmoto, M.P.Nightingale, M. den Nijs: Phys. Rev. Lett. **49**, 405 (1982)
2.184   R.Rammal, G.Toulouse, M.T.Jaekel, B.I.Halperin: Phys. Rev. **B27**, (1983)
2.185   D.Yoshioka: Phys. Rev. **B27**, 3637 (1983)
2.186   A.MacKinnon, L.Schweitzer, B.Kramer: J. Phys. C (to appear)
2.187   D.C.Tsui, H.L.Störmer, A.C.Gossard: Phys. Rev. Lett. **48**, 1559 (1982)
2.188   H.L.Störmer; D.C.Tsui, A.C.Gossard, J.C.M.Hwang: Physica **117-118B**, 688 (1983)
2.189   H.L.Störmer, A.Chang, D.C.Tsui, J.C.M.Hwang, A.C.Gossard, W.Wiegmann: Phys. Rev. Lett. **50**, 1953 (1983)
2.190   R.B.Laughlin: Phys. Rev. Lett. **50**, 1395 (1983)
2.191   E.E.Mendey, M.Heiblum, L.L.Chang, L.Esaki: Phys. Rev. **B28**, 4886 (1983)
2.192   P.W.Anderson: Phys. Rev. **B28**, 2264 (1983)
2.193   F.D.M.Haldane: Phys. Rev. Lett. **51**, 605 (1983)
2.194   R.Tao: Preprint 1983
2.195   S.J.Allen, Jr., B.A.Wilson, D.C.Tsui: Phys. Rev. **B26**, 5590 (1982)
2.196   T.Ando: J. Phys. Soc. Jpn. **38**, 989 (1975)
2.197   S.Fujita, M.Prasad: J. Phys. Chem. Sol. **38**, 1351 (1977)'
2.198   R.J.Wagner, T.A.Kennedy, B.D.McCombe: Phys. Rev. **B22**, 945 (1980)
2.199   P.Voisin, Y.Guldner, Y.P.Vieren, M.Voos, P.Delesculuse, N.T.Linh: Appl. Phys. Lett. **39**, 982 (1981)
2.200   J.C.Portal, R.J.Nicholas, M.A.Brummell, A.Y.Cho, K.Y.Cheng, T.P.Persall: Solid State Commun. **43**, 907 (1982)
2.201   E.Kress-Rogers, R.J.Nicholas, Y.Chevy: J. Phys. **C16**, 2439 (1983)
2.202   M.Prasad, S.Fujita: Phys. Lett. **A63**, 147 (1977)
2.203   B.Horovitz, A.Madhukar: Solid State Commun. **32**, 695 (1979)
2.204   G.Abstreiter, J.P.Kotthaus, J.F.Koch, G.Dorda: Phys. Rev. **B14**, 2480 (1976)
2.205   T.Ando: Phys. Rev. Lett. **36**, 1383 (1976)
2.206   G.Landwehr, E.Bangert, K. von Klitzing, T.Englert, G.Dorda: Solid State Commun. **19**, 1031 (1976)
2.207   G.Landwehr, E.Bangert, K. von Klitzing: In *Physique sous Champs Magnétiques Intenses* (CNRS, Paris 1975) p. 177;
        F.J.Ohkawa, Y.Uemura: Prog. Theor. Phys. Suppl. **57**, 164 (1975)
2.208   E.Bangert, G.Landwehr: Proc. Int. Conf. on Superlattices, Microstructures and Micro-devices, Champaign-Urbana (1984) in print
        D.A.Broido, L.J.Sham: Proc. XVIIth Int. Semicond. Conf. San Francisco (1984) to be published by Springer, New York
2.209   J.Mikeska, H.Schmidt: Z. Phys. **B20**, 43 (1975)
2.210   R.L.Aggarwal, A.J.Freeman, B.B.Schwartz (eds.): *Solids and Plasmas in High Magnetic Fields* (North-Holland, Amsterdam 1979) (J. Magn. Magn. Mat. **11**)
2.211   G.Abstreiter, J.F.Koch: Phys. Rev. Lett. **32**, 104 (1974)
2.212   S.J.Allen, Jr., D.C.Tsui, J.V.Dalton: Phys. Rev. Lett. **32**, 107 (1974)
2.213   S.P.Zelenin, A.S.Kondrat'ev, A.Kuchma: Sov. Phys. Semicond. **16**, 355 (1982)
2.214   P.Středa: J. Phys. **C16**, L369 (1983)
2.215   S.M.Girvin, M.Jonson: J. Phys. **C15**, L1147 (1982); M.Jonson, S.M.Girvin: Phys. Rev. **B29**, 1939 (1984)
2.216   A.Widom: Phys. Lett. **A90**, 474 (1982)
2.217   H.Oji: Phys. Lett. **A98**, 127 (1983)
2.218   P.Středa, L.Smrčka: J. Phys. **C16**, L895 (1983);
        P.Středa, H.Oji: Phys. Lett. **102A**, 201 (1984)
2.219   J.Hajdu, U.Gummich: Phys. Lett. **A99**, 396 (1983)
2.220   Th.Englert, J.C.Maan, Ch.Uihlein, D.C.Tsui, A.C.Gossard: Physica **117B**, **118B**, 631 (1983);

R.Lassnig, E.Gornik: Solid State Commun. **47**, 959 (1983);
E.Gornik, W.Seidenbusch, R.Lassnig, H.L.Störmer, A.C.Gossard, W.Wiegmann: In *Two-Dimensional Systems, Heterostructures, and Superlattices*, ed. by G.Bauer, F.Kuchar, H.Heinrich, Springer Ser. Solid-State Sci. Vol. 53 (Springer, Berlin, Heidelberg 1984) p. 60

2.221   A.Widom: Phys. Rev. **B28**, 4858 (1983)
2.222   H.Obloh: Private communication
2.223a P.Středa: Phys. Stat. Sol. **B125**, 849 (1984)
2.223b H.Oji: J. Phys. C to appear; Phys. Rev. **B29**, 3148 (1984)
2.224   A.D.Wieck, E.Batke, D.Heitmann, J.P.Kotthaus, E.Bangert: Phys. Rev. Lett. **53**, 493 (1984)
2.225   Th.Englert, K. von Klitzing, R.J.Nicholas, G.Landwehr, G.Dorda, M.Pepper: Phys. Stat. Sol. **B99**, 237 (1980)
2.226   F.Stern (ed.): Electronic Properties of Two-Dimensional Systems (North-Holland, Amsterdam 1982) [Surf. Sci. **113**, 1 (1982)]
2.226a T.Ando: ibid p. 182
2.226b T.Englert, D.C.Tsui, A.Gossard, Ch.Uihlein: ibid p. 295
2.226c R.B.Laughlin: ibid p. 22
2.227   According to M.Kubisa and W.Zawadzki, the 0⁻ peak in Fig. 2.14 has to be associated with the "bump" at about 3.5 T. This is in agreement with Ref. 2.25, p. 52

References Concerning the QHE Added at Proof

Several papers in Ref. 2.10 and 2.105.

Review

S.Kawaji: Proc. Int. Symp. Foundation of Quantum Mechanics, Tokyo (1983) p. 327

Localisation

Y.Ono: J. Phys. Soc. Jpn. **53**, 2342 (1984)
T.Ando: Technical Report of ISSP Ser. A No. 1443 (1984)
L.Schweitzer, B.Kramer (eds.): Supplement, Proc. Int. Conf. on Localisation, Interaction and Transport Phenomena (PTB, Braunschweig 1984)

Field Theory

M.H.Friedman, J.B.Sokoloff, A.Widom, Y.N.Srivastava: Phys. Rev. Lett. **52**, 1587 (1984)
H.Levine, S.B.Libby, A.M.M.Pruisken: Nucl. Phys. **B240**, 30, 49, 71 (1984)

Periodic Potential

D.J.Thouless: Phys. Rev. **B27**, 6083 (1983)
Q.Niu, D.J.Thouless: J. Phys. **A17**, 2453 (1984)
J.E.Avron, R.Seiler, B.Simon: Phys. Rev. Lett. **51**, 51 (1983)

Scattering Theory

R.Joynt, R.E.Prange: Phys. Rev. **B29**, 3303 (1984)

Gauge Argument

R.Tao, Yong-Shi Wu: Phys. Rev. **B30**, 1097 (1984)
J.Avron, R.Seiler: Preprint 1984

Open System

O.Heinonen, P.L.Taylor: Phys. Rev. **B28**, 6119 (1983)
T.Toyoda, V.Gudmundsson, Y.Takahashi: Phys. Lett. **102A**, 130 (1984)

3. High-Field Magnetism

Mitsuhiro Motokawa, Louis W. Roeland, and Cornelis J. Schinkel

With 13 Figures

The different elementary interactions in a crystal containing magnetic atoms are discussed and it is shown that some of the interactions correspond to very strong magnetic fields. As this chapter deals mostly with magnetization measurements in strong pulsed magnetic fields, this method of measurement is described in practical detail. Experiments are concerned with crystal fields, exchange interactions and related phase transitions, spin waves, itinerant electron systems, relaxation effects, de Haas-van Alphen effect, and magnetic impurities. Other experimental techniques for the study of cooperative magnetic systems are magnetic resonances (mainly in the far infrared at these very high fields), magnetoresistance measurements, the Faraday effect and optical spectroscopy.

3.1 Background

The study of magnetism involves primarily the interactions between elementary magnetic moments in solids. Originally more concerned with the development and practical applications of magnetic materials, it has now evolved into a vast field of sophisticated research, both theoretical and experimental. A system of interacting magnetic dipoles is an excellent model for problems in statistical physics: well-defined interactions can be combined in various ways to obtain all kinds of ordered states, phase transitions and critical phenomena which are well suited for theoretical treatment and experimental observation.

In most materials, the quantum mechanical exchange interaction dominates over the magnetic dipole interaction. The relative strength of these interactions can change as a function of temperature and crystal structure. The local distribution of electric fields in a crystal, usually referred to as the "crystal field", can have a strong influence on the anisotropy. An extreme case is given by an Ising system where the magnetic moments can be parallel or antiparallel to only one specific direction in the crystal. A system with isotropic interactions is called a Heisenberg magnet, and if the direction of the magnetic moments is confined to a particular plane, this is called an $x-y$ system. The spatial arrangement of the magnetic atoms in the crystal gives rise to another different kind of reduced dimensionality: they can be arranged in linear chains where the distance between the chains is so large that interactions are confined to the one-dimensional chains, or the interactions can be confined to planes in the crystal structure. In recent years, many types of crystals have been synthesised in which the one or other of these features is strongly enhanced; in some cases this is a sensitive function of temperature and magnetic field.

An interesting theoretical problem is given by the application of locally random magnetic fields to a system of interacting magnetic moments. This cannot, of course, be realised in practice but it can be studied for the analogous case of a diluted antiferromagnet with random configuration of the magnetic moments.

In the last two decades, much progress has been made in the development of sophisticated experimental techniques, e.g. neutron diffraction, equipment for obtaining very low temperatures, superconducting magnets and lasers. More recently, very high pulsed magnetic fields have become available allowing studies in magnetism under extreme conditions. This may bring out fundamental effects that are normally masked by others at lower field levels.

The elementary magnetic moments in a solid are mainly due to the spin and molecular orbits of electrons. The magnetic moments of nuclei contribute only second-order effects, e.g. hyperfine splitting. Some particular effects, such as the de Haas-van Alphen effect and Landau diamagnetism, are due to quasi-free electrons moving under the influence of the Lorentz force. Most magnetic phenomena are due to electrons in partially filled inner shells such as the d and f shells and to the electrons that take part in the bonding, e.g. valence electrons and conduction electrons. The electrons in the filled inner shells respond to an applied magnetic field very weakly due to the strong internal interactions which correspond to fields of the order 10^5 T. The electrons in an incompletely filled outer shell couple according to well-known rules to produce a quantum state carrying a magnetic moment μ. They are subjected to thermal agitation with energy $k_B T$ at temperature T and to the interaction with the external magnetic field B with energy μB, where k_B is the Boltzmann constant. As the field increases it will align the moments against the thermal agitation. The magnetisation as a function of field and temperature follows the Brillouin law

$$M = NgJ\mu_B B_J(gJ\mu_B \cdot B/k_B T), \tag{3.1}$$

where N is the number of the magnetic moment, g the g factor, J the total quantum number, and μ_B the Bohr magneton. By comparing energies it follows that $B = k_B T/\mu$ is the critical order of magnitude of the field; for a magnetic moment μ of 1 Bohr magneton: 9.27×10^{-24} Joule/Tesla this is $1.5 \cdot T$ T. Indeed, such a system is very close to saturation at $B/T = 3$ T/K and can be studied in moderate fields at low temperatures.

There are other interactions which necessitate the use of higher magnetic fields, such as the internal electrical field produced by charges surrounding the atom carrying the magnetic moment, i.e. the crystalline electric field. Let us consider a magnetic atom (or rather ion) with total quantum number J. In general the crystalline electric field will split the degenerate state into a number of crystalline electric field (CEF) states, spanning an energy interval proportional to the field strength.

In the rare earths the partially filled $4f$ shell is responsible for the magnetic moment. The $4f$ orbits have a relatively small extension and consequently they experience a weak crystalline electric field. Even so the overall splitting may amount to several hundred kelvin and thus several hundred tesla may be necessary to quench the CEF effect and to induce saturation. Most of these systems are anisotropic; this estimate applies to a "hard" direction. The $3d$ transition elements experience larger crystalline electric fields. The internal LS coupling also requires excitation fields that are an order of magnitude larger. If the temperature is sufficiently low the exchange coupling can give rise to an ordered state. Besides ferromagnetic and antiferromagnetic systems there are complicated spiral and modulated structures, determined by the nature of the exchange coupling. The anisotropy results from the crystalline electric field, magnetoelastic coupling, etc. To compete with this ordering, an external field has to be of the order of magnitude of the exchange field, which can be estimated from the ordering temperature. This is of the order of several hundred kelvin, corresponding to a field of the order of 100 T.

In metals there is magnetism due to itinerant electrons. Superimposed on the Landau diamagnetism, which we shall not consider here, there are oscillatory contributions, the de Haas-van Alphen effect, due to the passage of Landau levels through the Fermi energy. An applied magnetic field makes a distinction between spin-up and spin-down electrons, occupying two subbands up to the Fermi level. These bands are shifted with respect to one another and electrons will spill from one spin state to the other. The resulting net moment is, in a first approximation, proportional to the field and independent of temperature. This, so-called, Pauli spin paramagnetic susceptibility is given by $X = n(E_F) \mu_B^2$, $n(E_F)$ being the total (spin-up + spin-down) density of states at the Fermi level. Strictly speaking this expression is valid only if $n(E)$ is a constant in the region of interest around the Fermi level. A better approximation is given by the Stoner-Edwards-Wohlfarth theory. If, for instance, the band was such that $n(E)$ is linear in E, the Pauli contribution and the deviations from it are of comparable magnitude when $\mu_B B \simeq n(E_F)/[dn(E)/dE]_{E_F}$. Thus higher fields will give some information about the density of states as a function of energy. In the presence of exchange interactions the bands may be spontaneously split but the same quantitative argument applies for the incremental magnetisation in an applied field.

For a general introduction to magnetism, several textbooks are available ([3.1, 2]). Low-dimensional magnets have specifically been discussed by *de Jongh* and *Miedema* [3.3], who indicate the importance of these systems in statistical mechanics. There are three major series of conferences on magnetism: The International Conference on Magnetism (. . . Amsterdam 1976, München 1979, Kyoto 1982 [3.4]), The Conference on Magnetism and Magnetic Materials, in short MMM (. . . Atlanta 1981, Montreal 1982 [3.5], Pittsburgh 1983) and for the more practical applications and engineering problems the related INTERMAG Conference (. . . Grenoble 1981, Montreal 1982 [3.6], Philadelphia 1983 and Hamburg 1984). Regarding the highest fields, there have been sessions on magnetism at several conferences on high magnetic fields [3.7–12], and the first conference dedicated to this was held in Osaka in 1982 [3.13]. This latter field is extended in the following section which reviews current research in magnetism at the highest available magnetic fields.

3.2 Techniques for Measuring Magnetisation in Pulsed High Fields

Most magnetisation measurements are performed either by measuring the force on the sample in an inhomogeneous field, or by measuring the voltage induced in a pick-up coil containing the sample [3.14]. While these methods are fairly straightforward in stationary fields (where the sample is moved for the induction method), some difficulties are encountered in very strong fields which are available only in the form of impulses of short duration.

The force method was adapted to pulsed fields by *Kapitza* in 1931 [3.15]: the force on the sample was transferred to a diaphragm and the small displacement of this was hydraulically and electrically amplified in an ingenious way. More recently, *Belov* et al. [3.16] used piezoelectric transducers for this purpose, but on the whole it is felt that this method is not practical for fast transient fields.

In the pulsed field induction method the time derivative of the magnetic flux is measured by the voltage induced in a pick-up coil containing the sample. A large part of this voltage is due to the variation of the applied field and only a minute part comes from the magnetisation of the sample. The contribution due to the applied field can be removed by a separate compensation coil wound in the opposite direction. This simple method may give satisfactory results for slowly varying fields, but for pulsed fields of short duration it is inadequate, because the eddy currents induced in various metal parts of the experimental assembly result in changes of the spatial field distribution which bring the compensation out of equilibrium.

A more sophisticated compensation method was introduced by *Gersdorf* et al. [3.17] who showed that a coil assembly consisting of confocal concentric ellipsoidal coils can be arranged to generate no external field while the interior field is homogeneous. Thus there is no mutual inductance with an external magnet and coupling with a magnetic moment placed at the interior is independent of the sample's position. This would be an ideal system but it has the drawback that the sample volume is not easily accessible. This is avoided in the approximation shown in Fig. 3.1a. The inner coil and the three outer ones approximate the ellipsoidal coils. The field at the centre is expanded in spherical harmonics and the parameters determining the system are chosen such that the zeroth-, second- and fourth-order terms in the expansion are zero; to this order pick up of an external field is eliminated. The output signal of such a coil can be electronically integrated to yield the magnetisation. In practice, there is a leftover component of the external field due to the limited machining accuracy, in particular for the large zero-order contribution. An external signal derived from a separate pick-up coil is used to compensate this deficiency. The remaining signal is due to the imperfect adjustment of this compensation and to the

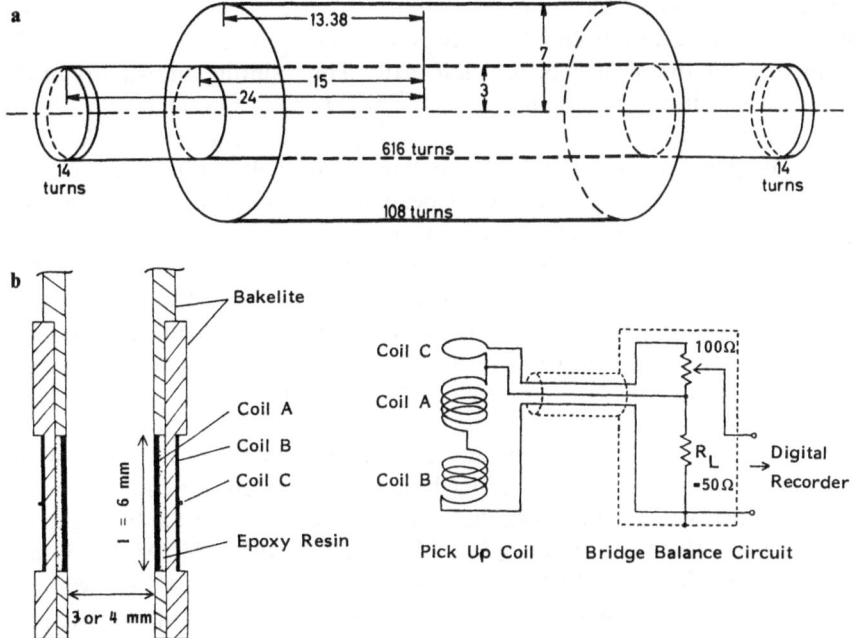

Fig. 3.1. (a) Example of a pick-up coil as used at the Amsterdam 40 T installation. The dimensions are in mm. **(b)** Pick-up coil and method of flux compensation used at Osaka University [3.18]

magnetic response of the coil system itself. The construction used at the Amsterdam centre [3.17] is made of glass cloth impregnated with polyester resin. At the Leiden centre, epibond is used which is reported to have a somewhat more favourable magnetic response [3.19]. Internal strains in the construction, particularly after thermal cycling, will alter the delicate balance between the coils resulting in an irreproducible signal proportional to the applied field. Subtraction of signals obtained at long time intervals (due to long cooling times of the coil) may therefore be wrong in this respect. This error is known to be proportional to the field and can usually be corrected or reduced by averaging over several runs. At the Amsterdam installation, curvatures in a magnetisation versus field plot can be determined much more accurately than is warranted by the accuracy of the slope [3.20, 21], because of the use of specially programmed pulse shapes. The system can be calibrated with a known magnetic moment like $ZrGd_2S_5$ (Leiden) or the zero-field saturation magnetisation of pure nickel (Amsterdam).

A number of disturbing effects are related to the pulse duration. Relaxation effects, magnetocaloric effects, and eddy currents set up in a metallic sample will render the magnetisation results in short pulses inaccurate. To some extent one can cope with this problem by averaging the data with the field going up and down. (With short pulses the compensation of a coil is also more difficult because of phase shifts between the coils.) It is preferable to work with long pulses with constant height. The Amsterdam installation offers a pulse duration of 80 ms, long enough for eddy currents to decay even in very good conductors, and field stability during that time is better than $1:10^3$. The resulting uncertainty of the magnetisation data is less than $\pm 2 \times 10^{-5}\,\mathrm{Am^2}$ $(2 \times 10^{-2}\,\mathrm{G\,cm^3})$.

Higher fields can be obtained by the discharge of a fast capacitor bank (Chap. 6.2.1). This generates a field pulse in the shape of one half-period of a sine wave. The short pulse duration causes time-dependent inhomogeneities in the field due to eddy currents. As the sample space is not large enough to accommodate a pick-up coil with a complicated configuration, a simple pick-up coil with a coaxial compensation coil and a correction coil for fine adjustment is used although some higher order uncompensated components remain. *Ozhogin* et al. reported [3.23] that a sensitivity of $4 \times 10^{-6}\,\mathrm{emu/cm^3}$ for the susceptibility is obtained by using a phase correction coil [3.22] in a field up to 50 T with 15 ms duration. They consider it important to fix the solenoid magnet rigidly to the floor of the laboratory and suspend the measuring system on a pillar of the building to avoid noise due to coil vibration. For precise orientation of the crystal against the field direction, the sample axis can be adjusted by tilting the cryostat from the vertical by $\pm 3°$ with an error within 5′. At the Osaka University High Magnetic Field Laboratory where fields up to 70 T are available [3.24], a pick-up coil system, as shown in Fig. 3.1b, is used [3.18]. Copper wire doped with 2% tin of 0.1 mm diameter is used for the pick-up coil. The large residual resistance eliminates eddy current effects in the probe wire, which would otherwise cause a considerable uncompensated component. This also maintains the condition of good compensation, as the temperature is changed; with the pure copper wire the compensation condition changes drastically at low temperatures due to change of electric resistivity and consequent change of the eddy current effect. The voltage induced in the pick-up coil is recorded by digital recorders, once with a sample and then without it. The two signals are subtracted from each other by a microcomputer, eliminating the small residual uncompensated components contained in both signals. The sensitivity of the susceptibility measurement finally obtained in this manner is $5 \times 10^{-6}\,\mathrm{emu/cm^3}$.

The magnitude of the magnetisation is calibrated by the saturation magnetisation of the well-known antiferromagnet $CuCl_2 \cdot 2H_2O$. When a conductive sample is measured, it must be in the form of a thin plate or fine powder which is cast with epoxy resin, to avoid skin effect and Joule heating.

The vibrating coil (or sample) method, which was developed by *Foner* [3.25] for use in a steady field, has been adopted for a pulsed field up to 50 T with 10 ms duration time at the ISSP (University of Tokyo) [3.26]; satisfactory results are obtained especially for samples which show remanence (cf. Sect. 6.5.2).

3.3 Crystal Fields

In a solid the magnetic ion is subjected to the electric field of the surrounding charges. The strength of this interaction is mainly determined by the proximity of the external charges. The transition ion $3d$ electrons have a large spatial extension and therefore are usually subjected to strong electric fields. On the other hand, the rare earth $4f$ orbits are confined to a smaller volume and the crystal fields can in these cases be treated on the same level as an external magnetic field, that is as a perturbation on a spin orbit split J level.

The crystal field exhibits the symmetry of the lattice and it will usually split the free ion J multiplets in crystalline electric field states, CEF states, which are labelled by the appropriate irreducible representations of the crystal point group. The precise way in which a level is split is determined by the strength of the crystalline electric field.

Theoretically the crystal field Hamiltonian is most elegantly given in terms of effective crystal field operators that act within the ground state J multiplet

$$H_{\mathrm{CEF}} = \sum_{l,n} A_l^n < J|\alpha_l^n|J > O_l^n(J) \tag{3.2}$$

The $O_l^n(J)$ are linear combinations of tensor operators of order l that are invariant under the point group of the crystal; n labels these combinations if more than one exists. The $\langle J|\alpha_l^n|J \rangle$ are the so-called Steven coefficients. The A_l^n are the crystal field parameters (CFP) which give the strength of the corresponding field component. Selection rules limit the summation to terms with l smaller than or equal to $2J$, $2L$ or $2l'$, which is the smallest, where J, L, l' are the free ion total, orbital and configurational quantum number.

For example, Tb^{+++} $[(4f)^8, S=3, L=3, J=6]$ has a crystal field Hamiltonian up to $l=6$. (However, Gd^{+++} has no first-order crystal field effect because $L=0$, and Eu^{+++} none because $J=0$).

Thus the CEF Hamiltonian of Tb^{+++} in a hexagonal field will be

$$H_{\mathrm{CEF}} = B_2^0 \alpha O_2 + B_4^0 \beta O_4 + B_6^1 \gamma O_6^1 + B_6^2 \gamma O_6^2. \tag{3.3}$$

In this case $O_2 = J_z^2 - J(J+1)/3$ and similar expressions hold for the other tensor operators.

The crystal field strength parameters B_l^n can be easily computed in a model in which the surrounding charges are well localised on the ions outside the $4f$ cloud. However, this simple point charge model hardly ever gives results that agree with experiment. Many contributions to the CEF have to be considered; at present the state of the art is limited to treating the crystal field strength parameters as adjustable parameters.

Experimentally, this topic can be studied by a variety of methods; neutron spectroscopy is commonly used. Inelastic transitions between CEF levels are observed to obtain the energy differences. The initial susceptibility of mono-crystals as a function of temperature can be analysed in terms of the CEF Hamiltonian. The first information about the CEF level scheme was provided by Schottky anomalies in the specific heat. However, very often the analysis of all these data cannot be unambiguous since several sets of parameters give satisfactory fits to the data. High field magnetisation measurements provide a useful tool for a cross-check. The strength of the method is that high field magnetisation data include higher levels, whereas the other analyses are based on the low-lying states. This applies not only to their energies, as for specific heat data, but also to their wave functions.

The first reports on high field magnetisation data were of a qualitative nature. At the 1961 high magnetic fields conference *Henry* [3.27] presented many pieces of high field data on rare earth metals with the comment that crystal field effects were probably responsible for the lack of saturation.

When monocrystals of the rare earth metals became available these effects could be studied in much more detail. Especially the light rare earths, where the crystal field and exchange interactions are of the same order of magnitude, were promising candidates for a more quantitative treatment. Unfortunately, as will be explained for praseodymium, this is prevented by the complex crystallographic structure of the RE metals. A qualitative analysis for compounds was carried out by *Busch* et al. [3.28] for the rare earth pnictides in fields up to 20 T in 1966.

As a first example we give the high field magnetisation for Eu metal [3.29]. No crystal field effects are expected for this metal because the Eu ions are in a divalent state and consequently have a ground state of $^8S_{7/2}$. This is borne out by the data; in several other directions the data are indistinguishable from those of Fig. 3.2 for the [110] direction. In the same figure the data for Nd metal [3.29] are given and a strong anisotropy between [110] and [001] is observed.

The most spectacular manifestation of crystal field effects is observed in Pr metal (Fig. 3.2) [3.29–31]. The crystallographic structure of Pr is dhcp. There are

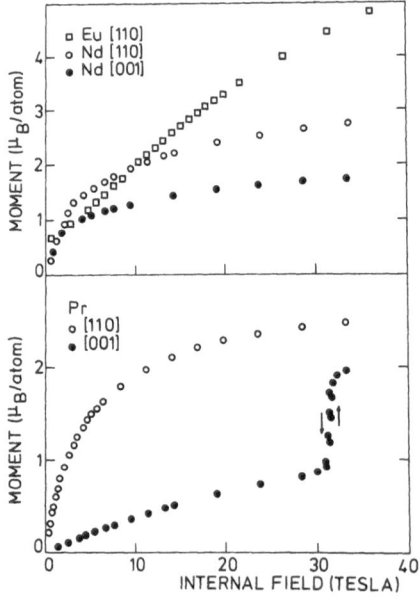

Fig. 3.2. High field magnetisation data for Eu, Nd and Pr metal monocrystals at 4.2 K

two sorts of sites; in one the local environment is cubic, in the other hexagonal. On both sites the $^3H_4$ ground state of Pr^{+++} is split by the crystal field and from neutron scattering data it is concluded that in low magnetic fields along the c axis the hexagonal sites do not carry a magnetic moment. It is therefore assumed that the crystal field ground state for the hexagonal sites is purely $|4, M=0\rangle$. The expectation value of the magnetic moment of this state is zero when the magnetic field is applied along the c axis, so consequently the state does not change. However, excited states like $|4, \pm 1\rangle$ are Zeeman split by a magnetic field and if the lowest of these crosses the $|4,0\rangle$ ground state it will be populated. At low temperature this causes a discontinuous change in the magnetic moment for this site and hence in the bulk magnetisation.

In Fig. 3.2 the magnetisation is shown both with the magnetic field in the basal plane perpendicular to the c axis, and along the c axis. Whilst the basal plane magnetisation rises smoothly, a discontinuity in the c-axis magnetisation is observed at 32 T. It may be pointed out that in both cases the magnetisation is considerably below the saturation value of $3.2\,\mu_B$/atom, indicating that the highest field in this plot is still not strong enough to overcome the total crystal field splitting.

Although the qualitative explanation in terms of a crossing of CEF states is satisfactory, it is not possible to extract quantitative information from this on a simple basis. For instance, the assumption that on both sites the CEF are the same is incompatible with the data [3.30]. A system that is more manageable is provided by dilute alloys of rare earth atoms in hosts like Y, Sc, Lu and Mg. These hosts are in a hcp phase and therefore have just one crystallographic site. Moreover, the exchange interactions are small because the alloys are so strongly diluted ($\sim 3\%$).

A number of such alloys has been studied by *Touborg* et al. in low fields [3.32]. The low field susceptibility on monocrystals has been measured as a function of temperature and data have been fitted to a crystal field Hamiltonian, treating the exchange on the basis of a molecular field. Although the exchange is small, it is often big enough to induce some kind of ordering at low temperatures. Care must be taken to measure the real crystal field susceptibility. This being done, they fitted their data with a crystal field Hamiltonian. The results permit a calculation of the high field magnetisation. The magnetic field is taken into account on the same level as the crystal field, that is $H_{CEF}+H_{Zeeman}$ is diagonalised in the *LSJ* ground level.

The measurements were performed at the 40 T facility of the University of Amsterdam. Small spheres of 3 mm diameter single crystals were used in gradually decreasing fields. The contribution of the pure host was subtracted. The results for Y-based alloys are given in Figs. 3.3–5. The agreement between the predicted and measured magnetisation is satisfactory, supporting the reliability of the crystal field parameters. Typical crossings of crystal field levels are often observed.

To get a more fundamental knowledge of crystal field effects, it is of interest to look for a systematic behaviour of the parameters, as is done for a number of

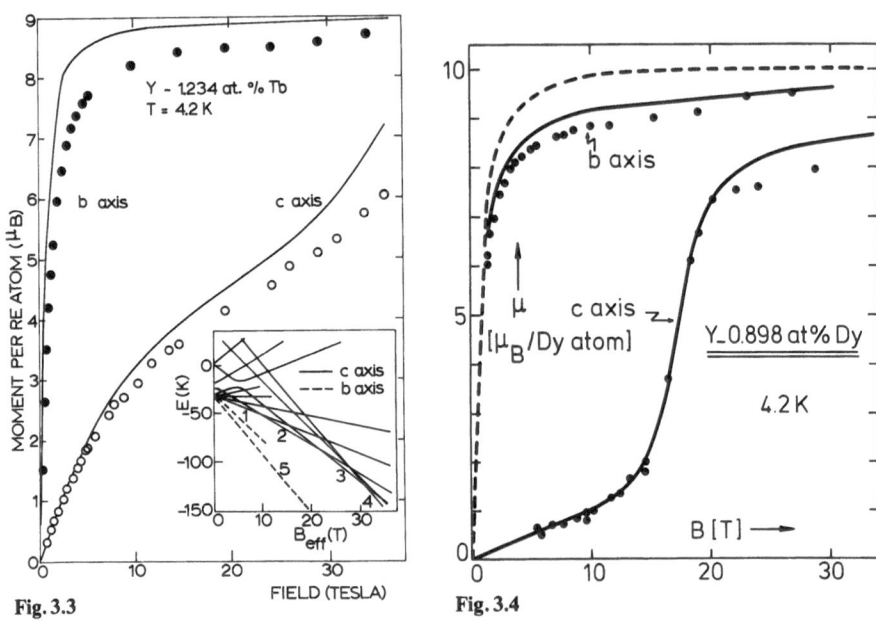

Fig. 3.3

Fig. 3.4

Fig. 3.3. High-field magnetisation data for Y-1.2 at. % Tb. (——) are calculated with crystal field parameters from [3.32]. The insert shows the dependence of the CEF levels as a function of the magnetic field. (○○○) are the experimental data for the *c* axis, (●●●) for the *b* axis

Fig. 3.4. High field magnetisation data for Y-0.9 at. % Dy. (——) are calculated with crystal field parameters from [3.32]

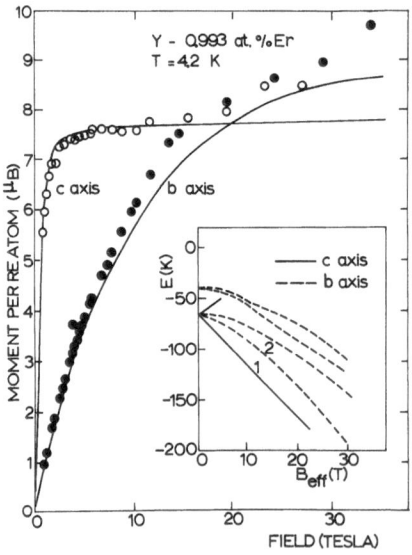

Fig. 3.5. As for Fig. 3.3, for Y-1.0 at. % Er

dilute alloys of rare earths in Y, Lu, Sc and Mg [3.32, 33]. The parameters appear to vary less than expected going from one host to another and also within the rare earth series for each of the hosts. Hence one cannot be sure of the details of the theoretical description and, for instance, a field dependence of the parameters B in the CEF Hamiltonian cannot be excluded a priori. This is phenomenologically the first correction to be applied to the CEF Hamiltonian. Consequently it is interesting to investigate whether field-dependent CEF parameters can lead to a more appropriate description of crystal field effects.

3.4 Spin Waves

With interacting magnetic moments in a solid, lowering the temperature will eventually lead to magnetic order. Below the ordering temperature the thermal motion is a perturbation of the perfect order. It is possible to analyse these perturbations in terms of elementary excitations, in much the same way as the thermal disturbance on the position of the atoms is analysed in terms of phonons. For magnetic moments these are called spin waves, quantised collective motions of the spins.

If we associate a saturation magnetisation with the ferromagnetic fully aligned systems at 0 K, then the excitation of spin waves at elevated temperatures will decrease the magnetisation. The reduction will be determined by the energy of the spin waves, which are characterised by a wave vector. The dispersion relation $E(k)$ is determined by a large number of microscopic parameters such as the anisotropy and the elastic and magnetoelastic properties. At this level of sophistication precise data obtained in different ways are necessary to check such theories.

The dispersion relation itself is measured as usual by spectroscopy, in this case by neutron spectroscopy [3.34]. These data can then be fitted to a Hamiltonian containing all the above-mentioned ingredients, and the reliability of the fit can be checked by high field magnetisation measurements. At first sight the necessity of strong fields is not obvious; one could simply measure the magnetisation as a function of temperature. However, with the large number of parameters an additional degree of experimental freedom is desirable and the high field susceptibility provides valuable additional information. Moreover, low field values are unreliable because of crystal imperfections and contaminations. Strong fields overcome the influence of such defects. Apart from that, if the field is to act as a parameter on the same footing as the temperature, it must be of the same order of magnitude, regarding energy.

Already at the 1961 high field conference *Gaskell* and *Motz* [3.35] reported magnetisation data on Gd metal as a function of magnetic field and temperature. However, at that time the technology, both of the equipment and of the sample preparation, precluded any sort of accuracy. Not much work in this field was reported until 1975, when *Roeland* et al. [3.36, 37] undertook the high field measurements on Gd and Tb. An important point was the sample purity. It

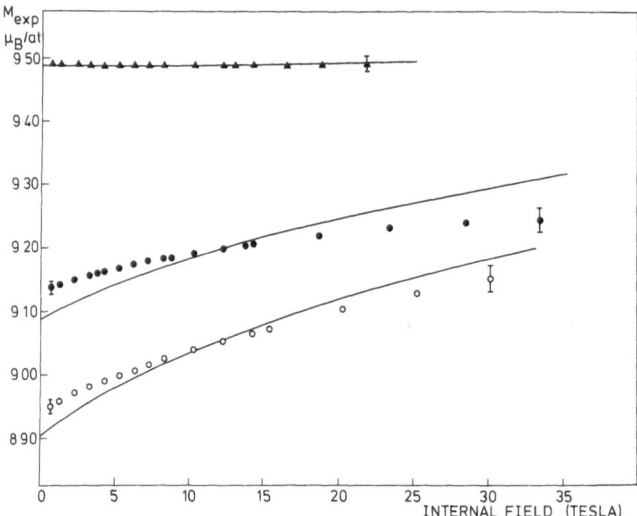

Fig. 3.6. Magnetisation of Tb metal in the easy (b) direction as a function of internal field at 4.2 K (▲), 64.5 K (●), and 77 K (○). (——) are calculated as explained in [3.36]

turned out that commercially available rare earth metals were substantially contaminated with oxides. Solid state electrolysis removed this impurity content by more than one order of magnitude. The saturation magnetisation of Gd at 4.2 K was found to be 7.630 μ_B, in contrast to the previous value of 7.55 μ_B. The relative difference may not appear to be large, but one has to bear in mind that only 7.0 μ_B can be ascribed to the 4f electrons. The remainder must be due to the conduction electrons, presumably the 5d electrons. With respect to this contribution the difference is substantial, and of major importance for band structure and Fermi surface calculations.

For Tb the magnetisation was measured not only at very low temperatures, but also at temperatures high enough to expect a reduction of the magnetisation due to spin waves [3.38]. Figure 3.6 shows that a magnetic field quenches the spin waves and tends to restore full saturation. The full lines are calculated on the basis of a spin wave theory by *Lindgård* [3.36]. The parameters in the Hamiltonian were found by fitting the energy gap in the spin wave spectrum as a function of field strength, field direction and temperature. Although the differences between the calculated high field magnetisation and the experimental data are small, they are significant and indicate that the theory is not satisfactory and must be supplemented with, e. g., anisotropic exchange. At this point we can mention the work that *Pauthenet* reported recently [3.39], which was done in a dc field. Very accurate high field magnetisation data of monocrystalline nickel, iron and cobalt show an excellent fit to the phenomenological relation

$$\sigma(T, B_i) = \sigma(T, 0) + A(T) B_i^{1/2} + \chi B_i, \tag{3.4}$$

where B_i is the internal field and $A(T)$ is proportional to T. The first two terms on the rhs have been obtained by *Holstein* and *Primakoff* on the basis of spin wave theory [3.40]. The term χ, which is called "superimposed susceptibility" [3.39], is actually the Pauli paramagnetism of the $3d$ and $4s$ electrons and orbital contribution of the $3d$ electrons.

Another important aspect of spin waves is antiferromagnetism. In an antiferromagnet we can divide the spin system into two sublattices. The magnetic moment of each sublattice is not at saturation even at $0\,K$ because the fully aligned state, called the "Néel state", is not a good quantum state. Zero point excitation of spin waves necessarily exists and it reduces the total magnetic moment of each sublattice. Since this is a quantum effect, the zero point spin reduction increases with decreasing spin number and dimensionality. The one-dimensional $S = 1/2$ spin system has the largest effect. When an external field is applied to an antiferromagnet, magnetic moments first are aligned perpendicular to the applied field, if the anisotropy is neglected, and then tilt towards it as the field increases, which results in an increase of the induced magnetisation. The induced magnetisation finally saturates at $B = 2\,B_E$, where B_E is the exchange field associated with the exchange interaction between spins on different sublattices. At this field all the spins align parallel to the field and spin reduction is no longer expected. The initial susceptibility χ_i is less than the molecular field susceptibility $\chi_m = M_0/2\,B_E$, where M_0 is the saturation moment at $2\,B_E$; then the magnetisation curve is non-linear with respect to the applied field. From this experiment one can determine the magnitudes of the exchange interaction and spin reduction precisely. The first theoretical argument was given in 1964 by Griffiths but experiments on low-dimensional Cu compounds were undertaken only recently as stronger fields became available [3.41, 42].

3.5 Itinerant Electron Systems

In the preceding sections we have dealt with materials whose magnetic properties are determined mainly by localised magnetic moments, that is magnetic moments due to electrons with wave functions that can be approximated best starting from the ionic wave functions. Another class of materials is given by the itinerant electron magnets. In these materials the magnetic moments are ascribed to electrons with Bloch wave functions and for which the electron energy band picture is appropriate.

The response to a magnetic field of such an itinerant electron system is in first approximation determined by the Pauli susceptibility, which is temperature independent and proportional to the density of states at the Fermi energy. It is assumed that the density of states around the Fermi level does not vary over a region of the order of $k_B T$ and/or $\mu_B B$. In most metals the (s, p) conduction electron band is so wide and consequently so flat that these conditions are fulfilled for realistic values of temperature T and magnetic field B. However, if the bands are formed by d or f electrons they can be so narrow and have so much

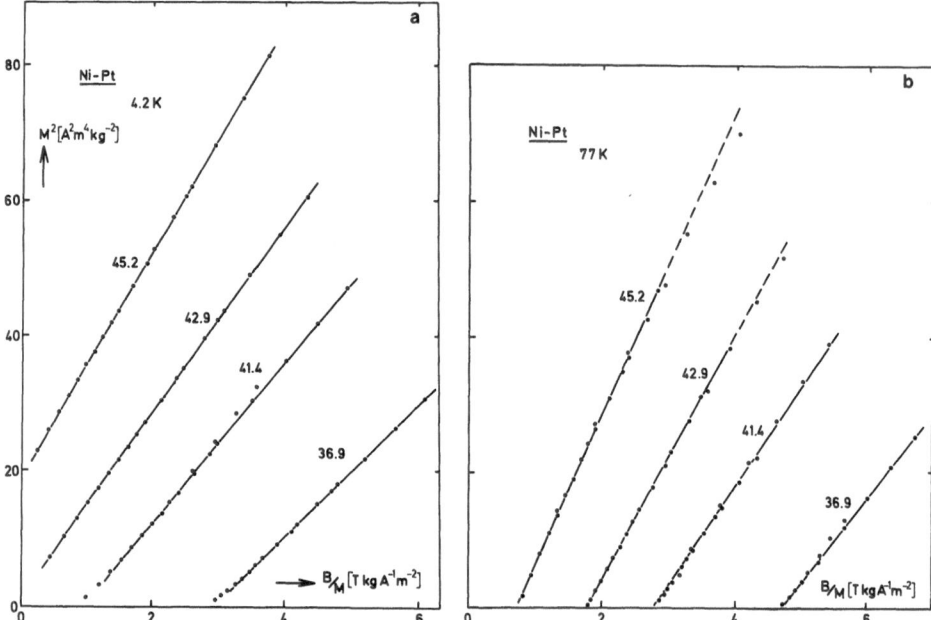

Fig. 3.7a, b. Magnetisation data for Ni-Pt alloys [3.44]. The curves are labelled with the nickel content in at.%, (**a**) at 4.2 K; (**b**) at 77 K

fine structure that the corresponding susceptibility is neither independent of temperature nor of magnetic field, particularly when the applied field is enhanced by ferromagnetic interactions between the electrons.

Edwards and *Wohlfarth* [3.43] have given a detailed treatment of itinerant electron systems with ferromagnetic interactions. The molecular field approximation is used to take care of the exchange interactions, following Stoner, and a series expansion of the density of states in the vicinity of the Fermi level up to second order to account for structure of the density-of-states curve. The Stoner-Edwards-Wohlfarth model is applicable only to weak itinerant magnets in which a spontaneous splitting of the bands due to large exchange interactions does not lead to one completely filled spin-up band or completely emptied spin-down band. The resulting expression for the magnetic isotherms is

$$B = (a + cT^2)M + bM^3, \tag{3.5}$$

where a is determined by the Fermi density of states and the exchange interaction parameter. The coefficients b and c contain the density of states n and its first two derivatives n' and n'' at the Fermi energy. An expression of this form follows from a straightforward series expansion of the energy in terms of the order parameter M up to M^4. The merits of the Stoner-Edwards-Wohlfarth model are that the coefficients can be calculated exactly. On the other hand, these coefficients can be determined from experiment, for instance by plotting the magnetisation curves as M^2 versus B/M, the so-called Arrott plots. In the paramagnetic regime the inverse susceptibility is given by $a + cT^2$ ($\sigma^3 \to 0$). Thus if one knows the Fermi density of states from low-temperature specific heat data, one should be able to solve for n' and n''.

There are a number of systems that obey (3.1) fairly well, for example palladium metal [3.20, 44], the intermetallic compound Ni_3Ga [3.45] and the Ni-Pt alloys [3.44] (Fig. 3.7). However, a complete and quantitative agreement with the SEW model could never be obtained. The origin of this difficulty most likely lies in the temperature dependence of the exchange interaction parameter [3.44, 45]. This would imply that low-temperature magnetic isotherms, although not decisive for the band parameters, are of great help for band structure calculations, because the coefficient b in (3.1) contains only band parameters and not the exchange interaction. Anyhow, by means of high field magnetisation data at least a first distinction can be made between systems which are possibly homogeneous itinerant magnets and those which are certainly not.

Another point of interest in these systems arises from the first-order transition, i.e. the so-called metamagnetic transition, in a high magnetic field [3.46]. This occurs as a consequence of the influence of a magnetic field on the energy band structure and must be distinguished from usual metamagnetic transitions due to the strong anisotropy. According to the model mentioned above, *Wohlfarth* found that ferromagnetism arises at a certain critical field [3.46]

$$B_c^* = \frac{\sqrt{2}}{3\sqrt{3}} \left(\frac{1}{\chi^3 |b|} \right)^{1/2} \tag{3.6}$$

under the condition $b < 0$, which means

$$\frac{n''}{n} = -3 \frac{(n')^2}{n^2} > 0, \tag{3.7}$$

where χ is the paramagnetic susceptibility.

To observe a first-order transition in a practical field, these results require a high paramagnetic susceptibility and a high positive curvature of n. Correspondingly, a specimen must show a maximum in χ with respect to temperature. From these points of view, a few materials are discussed by *Wohlfarth* [3.46]. The susceptibility of Pd has been measured in a field up to 38 T and 77 K, but at this temperature b is positive. At 4.2 K b is negative but B_c^* is estimated to be ~ 300 T theoretically. Recently Gersdorf et al. reported a non-linear increase of the magnetisation up to 35 T. At 6 T $TiBe_2$ is expected to show the transition, but the magnetisation transition is not sharp although the specific heat in a field above 6 T apparently shows ferromagnetic behaviour. The reason is ascribed to the position of the Fermi energy in a sharp peak of the density of states and very close to a dip within this peak [3.46].

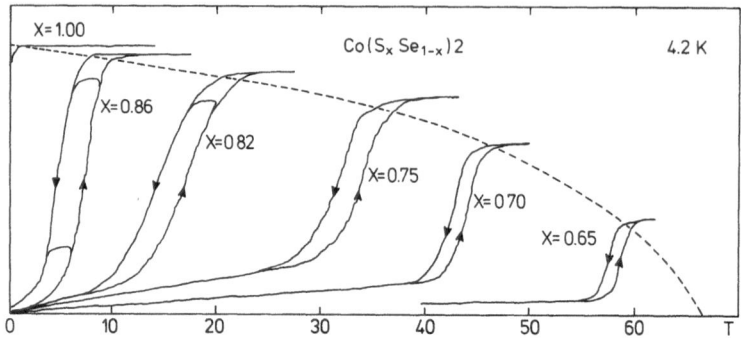

Fig. 3.8. Magnetisation curves of Co(S$_x$Se$_{1-x}$)$_2$ at 4.2 K [3.48]

The B_c^* of YCo$_2$ is estimated to be 100 T. The magnetisation of this compound has been measured up to 38 T by *Schinkel* [3.47]. An upward deviation from linear behaviour is found but no clear transition as yet.

Evidence of this type of transition is obtained in the system Co(S$_x$Se$_{1-x}$)$_z$ [3.48, 49]. This shows ferromagnetism for $0.88 < x < 1$ and is paramagnetic for $0 < x < 0.88$. A spectacular metamagnetic transition is found in the paramagnetic alloys in the range of $0.65 < x < 0.88$, Fig. 3.8. *Takahashi* and *Tano* explained this phenomenon by a calculation of free energy based on the density of states derived from the band structure calculation [3.50], and thus they also stress that the positive curvature in the density of states near the Fermi level is the origin of the metamagnetism. Some other recent topics on itinerant electron magnetism in high magnetic field are discussed by *Franse* [3.51].

3.6 Exchange

In the previous sections we have considered the competition of the applied magnetic field with the electrostatic crystal field in the case of localised moments, with the lattice potential in the case of itinerant electrons, and with the thermal agitation (spin waves). Another important competitor can be the exchange field in the case of interacting magnetic moments. For an illustration, let us consider a simple system: two magnetic moments with $S = 1/2$ (e. g., two electrons), coupled via an isotropic Heisenberg interaction $+ 2 JS_1 \cdot S_2$. For J positive an energy level scheme results, with a ground state singlet $S = S_1 + S_2 = 0$ and an excited triplet at a distance J with $S = S_1 + S_2 = 1$. The application of a magnetic field leads to a Zeeman splitting of the excited level, and at a field $B = J/g\mu_B$ to a crossing of the lower Zeeman level with the original ground level. The resulting magnetisation curve at zero temperature is a step function and thus the exchange parameter J can be determined from the experiments.

A somewhat more complex but still very simple example is a trinitroxide [3.44] in which three exchange coupled electrons reside on each molecule, very weakly coupled to the electrons on the other molecules. In this case the magnetisation at zero temperature should show a jump from $1\,\mu_B$ to $3\,\mu_B$ per molecule. At finite temperatures, in the liquid helium region the curve is smoothed by thermal effects but still allows for a determination of the exchange interaction.

In a system of mutually interacting magnetic moments, for instance a crystal containing transition metal atoms, the situation is more complex, because then the problem becomes a many-particle problem. However, in many cases the appropriate Hamiltonians can be solved, either for a particular crystal structure, giving rise to a one- or two-dimensional magnetic system, or when the molecular field approximation can be adapted. A number of examples can be found in the results of the Leiden group, obtained by means of a 50 T pulse coil [3.19] as well as of the Osaka group and others [3.52–54]. The merits of intense magnetic fields in studying magnetic exchange interactions become evident when the interactions cannot be given in a so-called bilinear form, i.e. $J(S_i \cdot S_j)$, but contain higher order terms such as a biquadratic form, i.e. $j(S_i \cdot S_j)^2$. A number of microscopic mechanisms can be considered [3.55, 56] which would result in higher order exchange, but we restrict ourselves to a phenomenological description in terms of the molecular field approximation. The isotropic Heisenberg exchange $J(S_1 \cdot S_2)$ can be accounted for by introducing an effective field, the molecular field (exchange field) B_E, which is proportional to the bulk magnetisation M. The magnetic behaviour of each magnetic moment in the crystal is then determined by $B_{total} = B_{applied} + B_E = B_{applied} + \lambda M$. Higher order exchange interactions are represented by terms of higher order in M in the molecular field, which for symmetry reasons contains only odd powers of M. Hence:

$$B_{tot} = B_{appl} + \lambda_1 M + \lambda_2 M^3 + \lambda_3 M^5 + \ldots \ . \tag{3.8}$$

The presence and importance of higher order terms were demonstrated by *Cooper* et al. on (Tb, Y) Sb [3.57] and (Ce, La) Sb [3.58]. Other more recent examples are $Mn(CH_3COO)_2 \cdot 4H_2O$ [3.53] and $CeTl_3$ [3.59]. In this latter case the magnetic field evidently modifies the wave function of the ground state of the crystalline electric field level scheme, which in its turn changes the exchange parameter J. Although the experimental data cannot be described on a microscopic scale at the moment, intense magnetic fields will be a valuable tool to clarify this field via magnetisation measurements on monocrystals.

From the point of view that four-spin exchange interaction must be introduced, in addition to the biquadratic term, C_6Eu is an interesting material [3.60]. This compound has intercalated layers of Eu in hexagonal graphite and Eu ions are on a two-dimensional triangular lattice. The spin structure is a triangular array at zero field. The magnetisation is measured in a field up to 40 T and a curious result is obtained when the field is parallel to the layer (Fig. 3.9). As the field is increased, the triangular spin state changes to a ferromagnetic state at H_{c0}. Above H_{c1} it assumes ferrimagnetic spin-canted order and finally it

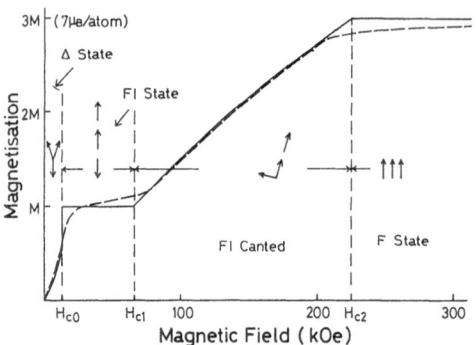

Fig. 3.9. Theoretical (——) and experimental (– – –) magnetisation curves of C_6Eu and the corresponding model of spin structure [3.60]

saturates as a ferromagnetic state at H_{c2}. This behaviour cannot be explained by a model of the usual bilinear exchange interaction between Eu spins even if long-range interactions are taken into account. *Date* introduced a biquadratic exchange interaction, mentioned above, and also a "four-spin exchange interaction" expressed as

$$U_k = k\,[(S_iS_j)\,(S_kS_l) + (S_iS_l)\,(S_jS_k) - (S_iS_k)\,(S_jS_l)] \tag{3.9}$$

in this spin system. The idea of "four-spin exchange" was first introduced to explain the spin structure of solid ^{3}He and applied to NiS_2 to account for the complicated magnetic structure. Usually this interaction has very complicated forms but for C_6Eu the model is simplified because of the two-dimensional spin state. The magnetisation curve obtained is successfully explained, Fig. 3.9.

Another example of the complicated interaction is the Dzyaloshinsky interaction which is expressed as $D(m_xl_y + m_yl_x)$, where m and l are the sum and difference of two sublattice moments, respectively. *Ozhogin* et al. measured the magnetisation of rutile-type CoF_2 up to 50 T and found this interaction to be important here [3.23].

Another topic of recent interest is magnetisation measurements of "spin-Peierls" materials. A few materials with an antiferromagnetically coupled linear chain spin system show the so-called spin-Peierls transition [3.61, 62]. This transition occurs at a certain temperature where a displacement of atoms takes place so that each atom moves alternately closer to the next one and farther from the opposite one, keeping the total length constant. This leads to alternating exchange interactions so consequently gives rise to a dimerisation of spins which results in a non-magnetic state in this spin system. However, these dimerised spins are different from isolated spin pairs, mentioned briefly in the beginning of this section, because even after dimerisation the dimers couple with their neighbours. When an external field is applied to this system, the induced magnetisation is not step-like as expected from isolated paired spins, but rather as shown in Fig. 3.10 [3.61–63]. This behaviour is found in TTF-CuBDT, TTF-Au BDT and MEM-$(TCNQ)_2$. The critical field where a steep increase of the

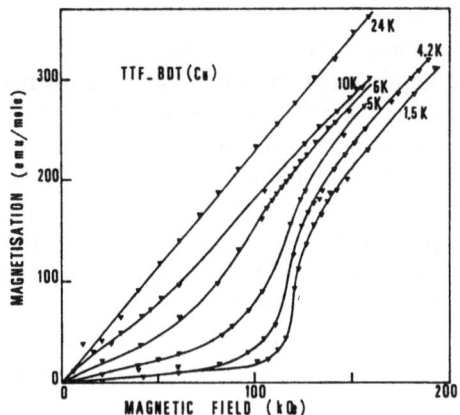

Fig. 3.10. Magnetisation curves of TTF-Cu BDT [3.61–63]

magnetisation appears is 12.5 T at 1.5 K, 2.0 T at 1.10 K and 19.2 T at 4.2 K, respectively. This spin system is non-magnetic below H_c, but above H_c it enters a new phase. To explain this new phase, three models are proposed [3.62]. One is that the spin system becomes incommensurate above H_c and the second is an infinite series of m-merisation with commensurate phase, where $m = 2, 3, 4, \ldots$ are dimerisation, trimerisation, quadramerisation and so on, since incommensurate distortion of the lattice is usually unstable. The last model is a domain formation model: some parts in the dimer chain turn into a normal antiferromagnetic array, resulting in domains forming at H_c and the domains grow with increasing field. However, conflicts among the explanations still remain and the character of the field-induced transition is not clear.

3.7 Relaxation Effects

A special problem in magnetism is the question how the magnetisation relaxes to its new equilibrium value when the applied magnetic field is changed. One can think of several mechanisms, each one giving its specific dependence of the relaxation time on the temperature and magnetic field [3.64].

A number of methods exists to measure the relaxation time. Conceptually it is most straightforward to measure directly the decay of the magnetisation after the change in applied field [3.65]. A more sophisticated method is to measure the ac complex susceptibility in an applied field [3.66]. This quantity is related to the relaxation time (if there is only one) through the *Casimir-Du Pré* relations [3.67].

Another direct way of measuring relaxation times is the observation of the change in level population through the change in optical absorption of Zeeman lines [3.68] or through the change in a resonance intensity like in EPR [3.69].

The temperature dependence of the relaxation times has been studied extensively in the past, but the field dependence has not been, because of greater experimental complexity. It may be of interest to do these experiments in higher fields and we want to discuss briefly some of the problems involved.

In principle it is required that for the direct observation of the relaxation time the rise or decay time of the field be smaller than or comparable with the relaxation time. This may be a severe requirement for high field experiments because the relaxation time tends to decrease strongly with fields. *Rimai* et al. [3.69] have shown that it is possible to analyse data obtained with a resonance technique, even if this requirement is not met. The resonance is observed in the rising and falling field and the relaxation of the population is described as a result of the mechanism acting just prior to passing the resonance. As a result, the strength of the resonance depends on the slope of the field. On the other hand, the ac susceptibility method requires that the high fields be sustained long enough for the system to attain its equilibrium state in the field. Another difference between the ac and the direct methods is that the latter tends to discriminate against fast relaxation processes in favour of slow ones, whilst in the former this is vice versa.

In the past all these techniques have been used in conjunction with high field facilities (although not always in high fields). The pulsed field EPR was used in fields up to 10 T [3.69]. The resonance is observed in the field going up and down and for long times the relaxation is found as the integrated effect of the field region inbetween. The analysis for very short times has already been mentioned.

Measurements of the relaxation times directly from the magnetisation have been performed by *Breur* and *Fain* [3.65] at the 40 T Amsterdam installation. In the systems they studied the relaxation times became prohibitively short above about 2 T. Nevertheless the high field installation was used because of its flexibility in generating a great variety of pulse shapes. There is no reason why these measurements could not be extended to higher fields as long as the relaxation times are of the order of the rise time, or decay time of the field.

The ac complex susceptibility method was used by *van Duyneveldt* et al. [3.70] both in a steady field and in a quasi-static field. Relaxation times of the order of 10^{-7} s were measured in fields up to 8 T. Where the data overlap there is good agreement between the two sets of measurements.

Recently *Misu* [3.68] reported measurements of relaxation times of the order of 1 ms in the field range 0–12.8 T. He observed the change in level population straightforwardly by the optical absorption in Zeeman split spectral lines.

It can be concluded that the measurement of relaxation effects in fields in excess of 10 T is feasible with a variety of techniques.

3.8 De Haas – Van Alphen Effects

When a magnetic field is applied to a metal, the conduction electrons will follow spiral paths around the direction of the field. A semiclassical treatment of this problem shows that in reciprocal space this motion is represented by a k vector which moves along a line of constant energy in a plane perpendicular to the field. The Bohr-Sommerfeld quantisation rules imply that the areas of these orbits in reciprocal space are quantised. The density of states is then periodic in $1/B$ and this results in an oscillating contribution to the magnetisation of the form $M_0 \cos[(2\pi F/B) + \phi]$, where F is determined by the Fermi surface topology and the field direction [3.71]. By measuring F as a function of the field direction the dHvA effect provides a powerful tool for investigating the Fermi surface.

The effect was first observed in 1930 by *de Haas* and *van Alphen* (dHvA) in the susceptibility (defined as the ratio of magnetisation and field) of a Bi monocrystal [3.72]. The magnetisation was measured with a Faraday balance which implies the use of an inhomogeneous field. Usually F/B is a very large number, of the order of 1000. Therefore if B is low and varies ever so little over the volume of the sample, the contributions of different parts of the sample will interfere destructively. The small ellipsoidal parts of the Fermi surface of Bi provided the condition that allowed the effect to be observed so early despite the inhomogeneity of the field.

Even when larger parts of the Fermi surface are studied in homogeneous fields (e.g. by the torque method), a problem is given by the small distance ΔB between successive zeroes of the oscillation. At 3 T, ΔB is of the order 10^{-4} T; as it is proportional to B^2 this can be considerably improved by using higher fields.

Shoenberg [3.73] was the first to measure dHvA effects in pulsed fields up to 10 T, generated by a capacitor discharge with a half-period of 30 ms. The magnetisation was measured with a balanced pair of pick-up coils. Both the field and the derivative of the magnetisation were displayed on an oscilloscope and the frequency was determined by counting the number of oscillations between two known field values.

Although this method has yielded valuable information, and still does, it was subsequently replaced by a modulation method [3.74], where a static field is modulated by a small oscillating field. Because of the great sensitivity of this technique, combined with improved magnetic field facilities, this low field method is now the most widely used [3.75, 76].

At the 1978 conference on high magnetic fields, *Shoenberg* [3.77] discussed the use of high magnetic fields for the dHvA effect. An important point is the fact that the amplitude M_0 contains a factor $\exp(-2\pi^2 m^* k_B T_D / ehB)$, where m^* is the cyclotron mass, and T_D the, so-called, Dingle temperature which is a measure of the inverse relaxation time of the electrons (or holes). In the case of large m^* (d electrons for instance) or large T_D (fast relaxation, as in alloys) a high value for B is needed to give M_0 an observable magnitude. On the other hand, in very strong fields which are inevitably transient, a modulation technique cannot be applied and therefore the detection of the dHvA oscillations and the accurate determination of the corresponding frequencies and Dingle temperatures becomes a problem. Consequently the advantage of a high field is counteracted by the lack of sensitivity of the detection system.

However, a set-up with a long pulse duration, like the one at the Amsterdam centre, combines strong fields with considerable accuracy, both with respect to frequencies and to amplitudes. It is worth noting that even in powders of intermetallic compounds dHvA oscillations were clearly observed [3.78, 79].

At Nijmegen some interesting results have been obtained on magnetic materials like the dense Kondo system RB_6 (R: La, Ce, Pr and Nd) [3.80] and the itinerant electron magnetic systems $ZrZn_2$ [3.81] and $TiBe_2$ [3.82] by means of pulsed magnetic fields up to 35 T.

There is another reason for using high fields in the dHvA effect. To observe the effect in high T_c, high H_c superconductors it is first of all necessary to quench the superconducting state. For the A15 compounds Nb_3Sn and V_3Si fields of about 20 T are necessary. In 1977 the dHvA effect in these materials was observed for the first time [3.83] and a rough visual analysis was made.

In the near future experimental improvements can be foreseen and lots of experimental data on these hard superconductors and also on oscillations with a higher T_D and m^* will thus be collected, yielding information that cannot be obtained otherwise. This may be considered as one of the more important developments in solid state physics using high magnetic fields.

3.9 Miscellaneous

Strong field magnetisation measurements can give valuable information regarding more problems than are mentioned in the preceding sections.

A typical example is a para- or diamagnetic material contaminated with magnetic impurities having a large initial susceptibility. Application of a strong magnetic field saturates the impurity contribution and consequently the intrinsic susceptibility of the material can be determined accurately. A similar application is the determination of saturation magnetisation, e.g. on powdered ferromagnets with a strong anisotropy.

For magnetically ordered substances (ferromagnets excepted) the applied magnetic field might be strong enough to induce a complete alignment of the magnetic moments. The magnetisation behaviour, transition fields and in general the magnetic phase diagram, contain much information on the magnitudes of the magnetic moments, the magnetic interactions and the electronic structures. However, it is not always possible to evaluate the information because the most powerful tool for determining magnetic structures – neutron diffraction – is up till now used only in fields below 10 T. Illustrations of this kind of research are the experiments on FeRh [3.84], the uranium pnictides [3.85, 86] and the rare earth metals [3.87]. For iron group alloys, the high field susceptibilities of Ni-based alloys [3.88] and Fe-Cr and Co-Mn alloys [3.89] are measured.

Another application of strong fields, which actually falls under the heading "exchange" but is not explicitly mentioned there, is to systems of magnetic moments in a non-periodic structure. Here we refer to magnetically dilute systems, e.g. $(Mn, Mg)Y_2S_4$ [3.90], spin glasses, such as Pd(Mn) [3.91], and amorphous ferromagnetic alloys of which Fe-B is a well-known example [3.92]. In the former two classes of materials the magnetisation data can yield values for the exchange parameters, if the distribution of the magnetic moments is assumed to be random. In the amorphous magnets band structure, saturation moments and anisotropy can be extracted from the experimental data with a varying degree of accuracy.

Finally we mention a novel experiment with the random spin system $Fe_{1-x}Zn_xF_2$ [3.93]. The problem of random spin originates from *Imry* and *Ma*'s random field effect theory [3.94]; it was shown that the effective random field is induced by a uniform field applied to a diluted antiferromagnet [3.95]. The most interesting point is that the random field effect reduces the spatial dimensionality of the crystal by 1 or 2 in the Ising spin system, i.e. no long-range order is expected in a magnetic field even in the three-dimensional substance if 2 is right. Although there is still a conflict between theory and experiment, three-dimensional $Fe_{1-x}Zn_xF_2$ shows a long-range order in a weak field. When a strong field is applied, however, the magnetisation process reveals an order-disorder transition at a field B_c which is shown by shaded circles in Fig. 3.11 and

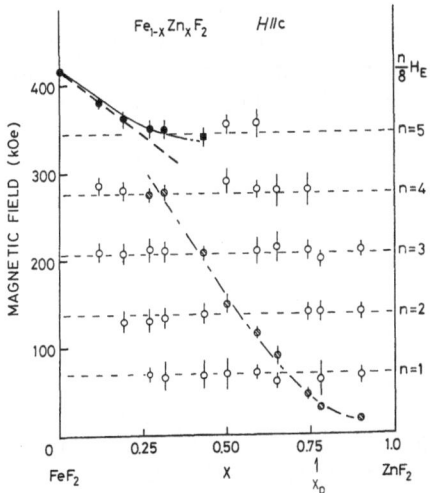

Fig. 3.11. Transition fields of $Fe_{1-x}Zn_xF_2$ as a function of concentration (see text) [3.93]

above which long-range order is broken. Open circles in the figure represent the fields where multiple-step-wise changes of the magnetisation are observed due to "spin-flip" at $B_S = (n/Z)B_E$, where B_E is the exchange field in pure FeF_2, $Z = 8$ is the number of the nearest-neighbour sites, and $n = 1, 2, 3, 4, 5$ depending on the number of the nearest-neighbour spins. Because of the strong anisotropy, even in the disordered state spin must be parallel or antiparallel to the field, so the spin-flip is still observed above B_c. For low concentrations of Zn, spin flop, solid circles in Fig. 3.11, occurs as well as in pure FeF_2.

3.10 Measurement of Magnetoresistance in Pulsed High Magnetic Fields

Magnetoresistance experiments are common in the research of metals, alloys and semiconductors; this technique is equally useful to investigate magnetic properties of magnetic conductors. One point of interest in this field is "spin fluctuations" of itinerant magnetic systems. In a spin system of a weakly ferromagnetic metal, the electric resistivity is strongly enhanced by the spin fluctuation at low temperature. By the application of a strong external field the spin fluctuation is suppressed and therefore negative magnetoresistance is expected. A good example is given by the intermetallic compound MnSi [3.96]. The electric resistivity of MnSi is expressed as $\varrho = \varrho_0 + \varrho_2 T^2$, where T is the temperature, ϱ_0 is the constant residual resistance. *Ueda* calculated the field dependence of ϱ_2 on the basis of self-consistent renormalisation theory of spin fluctuations [3.97, 98] and showed that ϱ_2 is strongly enhanced by the spin fluctuation effect and is significantly decreased by the field. Figure 3.12 shows the negative magnetoresistance of MnSi, which is in satisfactory agreement with the theoretical prediction that ϱ_2 becomes proportional to $B^{-1/3}$ in the strong field limit.

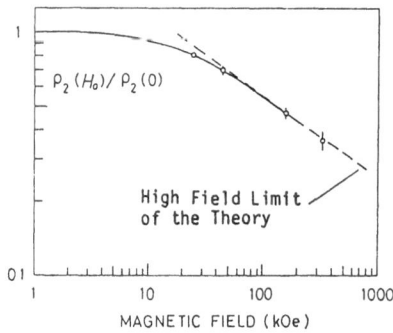

Fig. 3.12. Negative magnetoresistance of MnSi as a function of applied field [3.96]

Some other topics regarding magnetoresistance of magnetic materials are discussed in the proceedings of the International Symposium on High Field Magnetism [3.99, 100].

A direct magnetoresistance measurement was used to determine the very high upper critical field H_{c2} (over 60 T at 0 K) of Chevrel phase superconductors $Pb_xMo_6S_8$ [3.101] and the related compound $Pb_{1.2-x}Eu_xMo_6S_8$ [3.102].

In the case of magnetic materials the magnetocaloric effect must be considered [3.96], which inevitably arises intrinsically due to the change of magnetic entropy. Therefore it is important to obtain a good thermal contact between the spin system and the coolant. The demagnetisation effect must be reduced for ferromagnetic materials [3.100], e.g. by using a needle-shaped sample oriented parallel to the magnetic field.

3.11 Magnetic Resonance in Pulsed High Magnetic Fields

Magnetic resonance is an excellent method to determine the energy levels associated with the crystalline electric and magnetic fields [3.103]. At present, only two groups are engaged in this work, using pulsed high magnetic field, and concerning electron spin resonance (ESR) including electron paramagnetic resonance (EPR), antiferromagnetic resonance (AFMR) and ferromagnetic resonance (FMR). Nuclear magnetic resonance has not yet been used in this context although it is commonly used to investigate magnetic materials and might give interesting results, if adequate experimental methods can be developed.

To observe ESR in a high magnetic field, e.g. at 30 T, electromagnetic radiation of 840 GHz is required for a paramagnetic substance with $g = 2$. This frequency is well in the submillimeter or far infrared (FIR) region. In contrast to usual FIR spectroscopy, grating spectrometers or interferometers cannot be used in a pulsed field. By taking advantage of the sweep of the magnetic field, a monochromatic oscillator can be used, which is much stronger than the thermal sources, and results in a set-up comparable to microwave ESR. Fast detectors are available at the expense of sensitivity. In the final analysis, the important point is the signal-to-noise ratio, which presently depends more on the light source than on the detector. Light sources, backward-travelling wave tubes (BTWT, carcinotron) and FIR lasers are available; in particular the optically pumped (with a CO_2 laser) FIR laser provides many lines in the submillimeter wave region at intensities of the order of 1 mW. Stability is still a problem because the intensity of the FIR output depends critically on several factors: the tuning of the FIR cavity, the beam transport from the CO_2 laser to the FIR cavity which is sensitive to the mode pattern of the CO_2 laser, the intensity and the precise frequency of the pumping radiation. The BTWT is a convenient light source but it is still limited in high frequency and it is economically less attractive because of the narrow frequency range of each of the expensive tubes [3.104]. Much is expected from the free electron laser which is now under development for FIR spectroscopy; this has not yet been considered for combination with pulsed fields.

In most of the pulsed field EPR experiments transmission through the sample is observed. For fast detectors, the photoconductive effect of InSb or impurity doped Ge is generally used. An example of an absorption signal is shown in Fig. 3.13. At the Osaka University High Magnetic Field Laboratory [3.105], pulsed lasers are synchronised to the field pulse, with a pulse duration that is only slightly shorter than the 400 μs of the field pulse. The frequencies 0.891 and 0.965 THz are available from a HCN laser, 1.74 THz from a D_2O laser and 2.52 THz from an H_2O laser. The sample temperature is controlled by a feedback system, either by evaporating helium or by a heater. The part of the light pipe that extends into the magnet is made of a thick teflon tube, avoiding damage due to electromagnetic implosion caused by interaction of the magnetic field with eddy currents in the light pipe. At the University of Leuven, cw lasers are used,

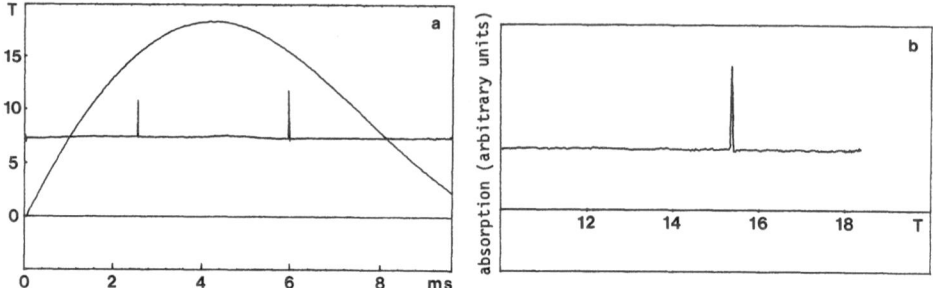

Fig. 3.13a, b. Example of experimental data of pulsed field ESR. (**a**) Intensities of magnetic field and laser beam as a function of time. Two spikes correspond to resonance absorptions in increasing and decreasing field; (**b**) resonance absorption as a function of magnetic field converted from (**a**) by minicomputer. The absorption is due to paramagnetic resonance of an organic compound MEM-(TCNQ)$_2$, which shows "spin-Peierls" transitions at low temperature, at $v = 428.63$ GHz and at room temperature

HCN as well as an optically pumped system [3.104]. The sample is cooled down to 4.2 K by a flow of cold helium gas. A stainless steel light pipe can be used in this apparatus because the pulse duration of 10 ms is long enough to allow the penetration of the field into the thin-walled tube. The ESR of metals has been measured at 0.891 THz with a stripline [3.106].

The ESR is a convenient technique for precise calibration of the magnetic field measuring system (inductive probe plus integrator). This has even been done in destructive fields, at 90 T with the ESR of ruby [3.26, 107].

The first FIR ESR experiment in pulsed fields over 30 T was a measurement of so-called exchange-splitting in CuCl·2H$_2$O [3.108]. To observe this phenomenon, the resonance field must be high enough to satisfy the condition $\Delta g \mu_B B / 2J \gg 1$. With pulsed fields, this can be satisfied for materials with fairly strong exchange interactions. Similar experiments were done with Cu(NH$_3$)$_4$·H$_2$O(CTS) and (NH$_2$)$_2$CuCl$_4$·2H$_2$O [3.109, 110] not only at high temperature but also under the condition $g \mu_B B / kT \gg 1$ and $\simeq 1$; the spectra are analysed by the molecular field approximation for the low-temperature data. For the region of intermediate temperatures, no theoretical treatment is presently available.

Kuroda et al. reported on the SH^3 term in Co-Tutton salt at high magnetic field [3.111]. The spin Hamiltonian contains the SH^3 term in principle but it is usually neglected because it is small. This term derives from a cross-over between ground orbital Zeeman levels and excited ones and is considered to be appreciable at high field where the Zeeman energy and orbital energy become comparable.

Motokawa applied ESR at 2.52 THz to Fe^{2+} in CdPS$_3$ and found a D value of -24 cm^{-1}, the highest known so far [3.105].

Many transitions including intermultiplet transitions were observed in LiErF$_4$ at 0.891 THz and up to 32 T by *Janssen* et al. [3.112]. In this case, one of the advantages of high field ESR is in reducing the dipolar broadening of the resonance line, due to the alignment of all spins in the same direction.

This method is useful to investigate antiferromagnets which have large spin wave energy gaps. A good example is AFMR of the two-dimensional antiferromagnet Rb_2NiF_4, where the temperature dependence of the width and shift of the absorption lines were measured [3.105].

One of the applications of high field ESR is the determination of the Néel temperature T_N [3.105]. Usually the Néel temperature of an antiferromagnet is determined by measuring specific heat, magnetic susceptibility, NMR, birefringence, ultrasonic attenuation, etc. However, none of these methods can be used in a pulsed field. As magnetic resonance can be observed in this case, the Néel temperature may be determined by observing an anomaly in the linewidth at T_N due to "critical slowing down". Using this method, the Néel temperatures of TMMC and Cu-formate have been determined at 32 and 29 T, respectively, and a large difference from T_N at zero field is found. This shift to higher temperatures at high field is qualitatively explained by the idea that thermal spin fluctuations in the paramagnetic state are suppressed by the magnetic field. In other words, this is a cross-over effect associated with a reduction in spin dimensionality, i.e. from Heisenberg type to XY and from XY to Ising type, with the anisotropy caused by the magnetic field. For TMMC, the increase of T_N at high field is quantitatively explained [3.113]; the reduction in the shift above 10 T is due to soliton excitation [3.114]. No theoretical explanation regarding two-dimensional Cu-formate has been given.

The high field ESR of metals, first investigated in brief by *de Groot* et al. [3.106], will be an interesting and important research topic in the future.

3.12 Optical Measurements in Pulsed High Magnetic Fields

Faraday rotation and the spectroscopy of absorption or emission of magnetic materials in the visible or infrared region are useful techniques to investigate magnetic properties.

As the angle of Faraday rotation is proportional to the magnetisation induced by an applied field, it provides a magnetisation measurement of magnetic materials. This indirect measurement without pick-up coil has the advantage that there is no pick-up noise and no need for compensation. Therefore this method is particularly useful for destructive high magnetic fields in which the compensation adjustment of a pick-up coil is almost impossible. Magnetisations of EuS, GdIG, TbIG, DyIG and YIG have been measured in this manner by means of destructive fields up to 160 T [3.26, 115].

Absorption or emission spectra are observed by a common grating spectrometer with a flash lamp synchronised to the peak of the pulsed field. The optical multichannel analyser is a piece of sophisticated equipment which is useful for detection and analysis [3.116]. The Paschen-Back effect of Cr^{3+} in ruby and the magnon sideband of antiferromagnetic Rb_2CoF_4 have been investigated.

References

3.1   D.C.Mattis: *The Theory of Magnetism* I, Springer Ser. Solid-State Sci., Vol. 17 (Springer, Berlin, Heidelberg 1981)

3.2   R.M.White: *Quantum Theory of Magnetism*, Springer Ser. Solid-State Sci., Vol. 32 (Springer, Berlin, Heidelberg 1983)

3.3   L.J. de Jongh, A.R.Miedema (eds.): *Experiments on Simple Magnetic Model Systems* (Taylor and Francis, London 1974)

3.4   P.F. de Chatel, J.J.M.Franse (eds.): Proc. Intern. Conf. on Magnetism, Amsterdam, September 6–10, 1976; Physica **86B–88B** (1977);
W.Zinn, G.M.Kalvius, E.Feldtkeller (eds.): Proc. Intern. Conf. on Magnetism, München, September 3–7, 1979; J. Mag. Magn. Mater. **13, 14** (1980);
K.Adachi (ed.): Proc. Intern. Conf. on Magnetism, Kyoto, September 6–10, 1982; J. Mag. Magn. Mater. **31–34** (1983)

3.5   R.Hasegawa, J.F.Janak (eds.): Proc. Third Joint Intermag-MMM Conference, J. Appl. Phys. **53**, 7631 (1982)

3.6   G.Bate, B.MacNeal (eds.): Proc. Third Joint Intermag-MMM Conference IEEE Trans. Magn. **MAG-18**, 1046 (1982)

3.7   H.Kolm, B.Lax, F.Bitter, R.Mills (eds.): *High Magnetic Fields*, Proc. Intern. Conf. on High Magnetic Fields, Cambridge, MA, 1961 (Wiley, New York 1962)

3.8   *Les Champs Magnétiques Intenses*, Colloq. Int. C.N.R.S. No. 166 (1966)

3.9   *High Magnetic Fields and their Applications*, Proc. Intern. Conf. on High Magnetic Fields and their Applications, Nottingham, UK, 1969 (The Institute of Physics, London 1969)

3.10   *Physique sous Champs Magnétiques Intenses*, Colloq. Int. C.N.R.S. No. 242 (1974)

3.11   A.J.Freeman, R.L.Aggarwal, B.B.Schwartz (eds.): *Solids and Plasmas in High Magnetic Fields*, Proc. Intern. Conf. on Solids and Plasmas in High Magnetic Fields, Cambridge, MA, 1978; J. Magn. Magn. Materials **11** (1979)

3.12   S.Chikazumi, N.Miura (eds.): *Physics in High Magnetic Fields* Springer Ser. Solid-State Sci., Vol. 24 (Springer, Berlin, Heidelberg 1981)

3.13   M.Date (ed.): *High Field Magnetism* (North-Holland, Amsterdam 1981)

3.14   H.Zijlstra: *Experimental Methods in Magnetism*, Part 2, Selected Topics in Solid State Physics, Vol. IX (North-Holland, Amsterdam 1967)

3.15   P.K.Kapitza: Proc. R. Soc. (London) **A131**, 224–321 (1931)

3.16   K.P.Belov, R.Z.Levitin, B.K.Ponomarev: Sov. Phys. JETP **24**, 1101–1104 (1967)

3.17   R.Gersdorf, F.A.Muller, L.W.Roeland: In [Ref. 3.8, pp. 185–188]

3.18   T.Sakakibara, H.Mollymoto, M.Motokawa, M.Date: In [Ref. 3.13, pp. 299–303]

3.19   J.J.Smit: Thesis, Univ. of Leiden, Netherlands (1979)

3.20   F.A.Muller, R.Gersdorf, L.W.Roeland: Phys. Lett. **31A**, 424–425 (1970)

3.21   L.W.Roeland, G.J.Cock, F.A.Muller, D.Shoenberg: Physica **79B**, 95–101 (1975)

3.22   Y.Allain, J. de Gunzbourg, J.P.Krebs, A.Miedan-Gros: Rev. Sci. Instrum. **39**, 1360–1365 (1968)

3.23   V.I.Ozhogin, K.G.Gurtovoj, A.S.Lagutin: In [Ref. 3.13, pp. 267–275]

3.24   A.Yamagishi, M.Date: In [Ref. 3.13, p. 289–298]

3.25   S.Foner: Rev. Sci. Instrum. **30**, 548 (1959)

3.26   G.Kido, N.Miura, K.Nakamura, H.Miyajima, K.Nakao, S.Chikazumi: In [Ref. 3.13, p. 309–318]

3.27   W.Henry: In [Ref. 3.7, pp. 552–560]

3.28   G.Busch, O.Vogt, P.Schwob: In [Ref. 3.8, pp. 401–407]

3.29   K.A.McEwen, G.J.Cock, L.W.Roeland, A.R.Mackintosh: Phys. Rev. Lett. **30**, 287–290 (1973)

3.30   G.J.Cock: Thesis, Univ. of Amsterdam, Netherlands (1976)

3.31   R.Gersdorf, F.R. de Boer, J.C.Wolfrat, F.A.Muller, L.W.Roeland: In [Ref. 3.13, p. 277–287]

3.32   P.Touborg: Phys. Rev. **B16**, 1201–1211 (1977)

3.33   J.Bijvoet, M.H. de Jong, H.Hölscher, P.F. de Chatel: Proc. Intern. Conf. on Cryst. Electr. Field and Struct. Eff. in f-Electr. Systems, Philadelphia 1979 (Plenum, New York 1980)
3.34   A.R.Mackintosh, H.Bjerrum Møller: In *Magnetic Properties of Rare Earth Metals*, ed. by R.J.Elliott (Plenum, New York 1972) Chap. 5
3.35   C.S.Gaskell, H.Motz: In [Ref. 3.7, pp. 561–567]
3.36   L.W.Roeland, G.J.Cock, P.-A.Lindgård: J. Phys. **C8**, 3427–3428 (1975)
3.37   L.W.Roeland, G.J.Cock, F.A.Muller, A.C.Moleman, K.A.McEwen, R.G.Jordan, D.W.Jones: J. Phys. **F5**, L233–237 (1975)
3.38   L.W.Roeland, F.A.Muller: J. Magn. Magn. Mater. **8**, 206–209 (1978)
3.39   R.Pauthenet: In [Ref. 3.12, pp. 326–335]; and in [Ref. 3.13, pp. 77–86]
3.40   T.Holstein, H.Primakoff: Phys. Rev. **58**, 1098 (1940)
3.41   H.Mollymoto, E.Fujiwara, M.Motokawa, M.Date: J. Phys. Soc. Jpn. **48**, 1771–1772 (1980)
3.42   T.Yosida: In [Ref. 3.13, pp. 305–308]
3.43   D.M.Edwards, E.P.Wohlfarth: Proc. R. Soc. (London) **A303**, 127–137 (1968)
3.44   C.J.Schinkel: In [Ref. 3.10, pp. 25–29]
3.45   C.J.Schinkel, F.R. de Boer, B. de Hon: J. Phys. **F3**, 1463–1469 (1973)
3.46   E.P.Wohlfarth: In [Ref. 3.13, p. 69–75]
3.47   C.J.Schinkel: J. Phys. **F8**, L87 (1978)
3.48   K.Adachi, M.Matsui, Y.Omata, H.Mollymoto, M.Motokawa, M.Date: J. Phys. Soc. Jpn. **47**, 675–676 (1979)
3.49   K.Adachi, M.Matsui: In [Ref. 3.13, pp. 51–54]
3.50   Y.Takahashi, M.Tano: J. Phys. Soc. Jpn. **51**, 1792–1798 (1982)
3.51   J.J.M.Franse: In [Ref. 3.4, pp. 819–828]
3.52   H.Mollymoto, M.Motokawa, M.Date: J. Phys. Soc. Jpn. **49**, 108–114 (1980)
3.53   M.Matsuura, Y.Okada, M.Morotomi, M.Mollymoto, M.Date: J. Phys. Soc. Jpn. **46**, 1031–1032 (1979)
3.54   Magnetisation data of many kinds of materials are found in [3.13]
3.55   R.J.Eliott, M.F.Thorpe: J. Appl. Phys. **39**, 802–807 (1968)
3.56   R.J.Birgenau, M.T.Hutchings, J.M.Baker, J.D.Riley: J. Appl. Phys. **40**, 1070–1079 (1969)
3.57   B.R.Cooper, I.S.Jacobs, C.D.Graham, O.Vogt: J. Physique Coll. **32**, **C1**-356–358 (1971)
3.58   B.R.Cooper, O.Vogt: J. Physique **32**, C1-1026–1027 (1971)
3.59   R.A.Elenbaas, C.J.Schinkel, S.Storm van Leeuwen, C.J.M. van Deudekom: J. Magn. Magn. Mater. **15–18**, 1218–1220 (1980)
3.60   M.Date, T.Sakakibara, K.Sugiyama, H.Suematsu: In [Ref. 3.13, p. 41–50]
3.61   J.W.Bray, L.V.Interrante, I.S.Jacobs, J.C.Bonner: *Extended Linear Chain Materials*, Vol. III ed. by J.S.Miller (Plenum, New York 1982) p. 353
3.62   D.Block, J.Voilon, L.J. de Jongh: In [Ref. 3.13, p. 19–36]
3.63   I.S.Jacobs, J.W.Bray, L.V.Interrante, D.Block, J.Voiron, J.C.Bonner: In *Physics In One Dimension*, ed. by J.Bernasconi, T.Schneider, Springer Ser. Solid-State Sci., Vol. 23 (Springer, Berlin, Heidelberg 1981) pp. 173–176
3.64   A.Abraham, B.Bleany: *Electron Paramagnetic Resonance of Transition Ions* (Clarendon, Oxford 1970)
3.65   W.Breur, S.C.Fain: Int. J. Magn. **2**, 145–151 (1972
3.66   A.J. de Vries, J.W.M.Livius: Appl. Sci. Res. **17**, 31–64 (1967)
3.67   H.B.G.Casimir, F.K.DuPré: Physica **5**, 505–511 (1938)
3.68   A.Misu: J. Magn. Magn. Mater. **11**, 161–163 (1979)
3.69   L.Rimai, R.W.Bierig, B.D.Silverman: Phys. Rev. **146**, 222–232 (1966)
3.70   A.J. van Duyneveldt, C.L.M.Pouw, W.Breur: Phys. Status Solidi **B55**, K63–65 (1973)
3.71   B.Lengeler: In *Springer Tracts Mod. Phys.* **82** (Springer, Berlin, Heidelberg 1978)
3.72   W.J. de Haas, P.M. van Alphen: Leiden Commun. 208d, 212a (1930)
3.73   D.Shoenberg: *Progress in Low Temperature Physics*, Vol. II, ed. by C.J.Gorter, (North-Holland, Amsterdam 1957) pp. 226–265
3.74   D.Shoenberg, P.J.Stilles: Proc. R. Soc. (London) **A281**, 62–91 (1964)
3.75   G.W.Crabtree, D.H.Dye, D.P.Karim: J. Magn. Magn. Mater. **11**, 236–246 (1979)
3.76   K. van Hulst, C.J.M.Aarts, A.R. de Vroomen, P.J.Wyder: J. Magn. Magn. Mater. **11**, 317–320 (1979)

3.77   D.Shoenberg: J. Magn. Magn. Mater. **11**, 216–220 (1979)
3.78   R.T.W.Meijer, L.W.Roeland, F.R. de Boer, J.C.P.Klaasse: Solid State Commun. **12**, 923–924 (1973)
3.79   J.C.P.Klaasse, R.T.W.Meijer, F.R. de Boer: Solid State Commun. **33**, 1001–1002 (1980)
3.80   A.P.J. van Deursen, Z.Fisk, A.R. de Vroomen: Solid State Commun. **44**, 609 (1982)
3.81   J.M. van Ruitenbeek, W.A.Verhoef, P.G.Mattocks, A.E.Dixon, A.P. van Deursen, A.R. de Vroomen: J. Phys. **F12**, 2919–2928 (1982)
3.82   A.P.J. van Deursen, A.R. de Vroomen: Solid State Commun. **36**, 305–307 (1980)
3.83   A.J.Arko, D.H.Lowndes, F.A.Muller, L.W.Roeland, J.Wolfrat, A.T. van Kessel, H.W.Myron, F.M.Mueller, G.W.Webb: Phys. Rev. Lett. **40**, 1590–1593 (1978)
3.84   C.J.Schinkel, R.Hartog, F.H.A.M.Hochstenbach: J. Phys. **F4**, 1412–1422 (1974)
3.85   C.J.Schinkel, R.Troć: J. Magn. Magn. Mater. **9**, 339–342 (1978)
3.86   O.Vogt, H.Bartholin: J. Magn. Magn. Mater. **15–18**, 1247–1248 (1980)
3.87   K.A.McEwen: In *Handbook on the Physics and Chemistry of Rare Earths*, Vol. 1, ed. by K.A.Gschneidner, L.R.Eyring (North-Holland, Amsterdam 1978)
3.88   T.Sakakibara, M.Date, K.Okuda: In [Ref. 3.4, pp. 83–64]
3.89   N.Kunitomi, S.Tsuge: In [Ref. 3.13, pp. 87–96]
3.90   H.H.Heikens, R.S.Kuindersma, C.F. van Bruggen, C.Haas: Phys. Status Solidi (A) **46**, 687–695 (1978)
3.91   J.J.Smit, G.J.Nieuwenhuis, L.J. de Jongh: Solid State Commun. **30**, 243–247 (1979)
3.92   I.Vincze: J. Magn. Magn. Mater. **15–18**, 1336–1338 (1980)
3.93   A.R.King, V.Jaccarino, T.Sakakibara, M.Motokawa, M.Date: Phys. Rev. Lett. **47**, 117 (1981); J. Appl. Phys. **53**, 1874–1878 (1982); J. Magn. Magn. Mater. **31–34**, 1119–1120 (1983)
3.94   Y.Imry, S.K.Ma: Phys. Rev. Lett. **35**, 1399–1409 (1975)
3.95   S.Fishman, A.Aharony: J. Phys. **C12**, L729–773 (1979)
3.96   T.Sakakibara, H.Mollymoto, M.Date: J. Phys. Soc. Jpn. **51**, 2439–2445 (1982); and in [Ref. 3.13, pp. 167–170]
3.97   K.Ueda, T.Moriya: J. Phys. Soc. Jpn. **39**, 605–615 (1975)
3.98   K.Ueda: Solid State Commun. **19**, 965–968 (1976)
3.99   F.R. de Boer, J.J.M.Franse, P.H.Frings, W.C.M.Mattens, P.F. de Chatel: In [Ref. 3.13, pp. 157–166]
3.100   Yang Fu-Min, Wu Yong-Sheng, Wang Yi-Zhong, Zhao Xi-Chao, Shen Bao-Gen, Liu Zhi-Yi, Pan Shiano-Thur: In [Ref. 3.13, pp. 121–126]
3.101   K.Okuda, M.Kitagawa, T.Sakakibara, M.Date: J. Phys. Soc. Jpn. **48**, 2157–2158 (1980)
3.102   K.Okuda, S.Noguchi, M.Honda, K.Sugiyama, M.Date: In [Ref. 3.13, pp. 143–146] and in [Ref. 3.4, pp. 517–518]
3.103   S.Foner: Phys. Rev. **107**, 683–685 (1957); Phys. Rev. **130**, 183–197 (1963)
3.104   F.Herlach, P. de Groot, P.Janssen, G. De Vos, J.Witters: In [Ref. 3.13, pp. 229–235]
3.105   M.Motokawa: In [Ref. 3.13, pp. 219–227]
3.106   P. de Groot, P.Janssen, F.Herlach, J.Witters, G. De Vos: In [Ref. 3.13, pp. 245–246]
3.107   G.Kido, N.Miura: Appl. Phys. Lett. **41**, 569–571 (1982)
3.108   M.Date, M.Motokawa, A.Seki, S.Kuroda, K.Matsui, H.Nakazato, M.Mollymoto: J. Phys. Soc. Jpn. **39**, 898–904 (1975)
3.109   M.Date, M.Motokawa, H.Hori, S.Kuroda, K.Matsui: J. Phys. Soc. Jpn. **39**, 257–258 (1975)
3.110   M.Motokawa, S.Kuroda, M.Date: J. Appl. Phys. **50**, 7762–7767 (1979)
3.111   S.Kuroda, M.Motokawa, M.Date: J. Phys. Soc. Jpn. **44**, 1797–1803 (1978)
3.112   P.Janssen, P. de Groot, G. De Vos, F.Herlach, J.Witters: In [Ref. 3.13, pp. 241–243]
3.113   K.Takeda, T.Koide, T.Tonegawa, I.Harada: J. Phys. Soc. Jpn. **48**, 1115–1122 (1980)
3.114   I.Harada, K.Sasaki, H.Shiba: Solid State Commun. **40**, 29–32 (1981)
3.115   M.Suekane, G.Kido, N.Miura, S.Chikazumi: In [Ref. 3.4, pp. 589–590]
3.116   H.Hori: In [Ref. 3.13, pp. 209–212]

4. Biomolecules and Polymers in High Steady Magnetic Fields

Georg Maret and Klaus Dransfeld

With 26 Figures

This is a review of recent progress in our knowledge about the behaviour of synthethic and biological macromolecules and larger biological particles in dilute solution and in the liquid crystalline state, when exposed to strong magnetic fields. Emphasis is on the observation of magnetic orientation as resulting mostly from an anisotropic molecular diamagnetic susceptibility, and on its interpretation in terms of molecular structure, flexibility and inter-molecular orientational correlation. Some biological applications of magnetic fields are briefly discussed, including the separation of macromolecules in inhomogeneous fields, the field-induced modification of cellular growth and the field sensitivity of animals.

4.1 Overview

In this chapter we first review briefly the dia- and paramagnetic properties of some organic molecules in Sect. 4.2. Subsequently, Sect. 4.3, we describe recent experiments on the magnetic orientation of various biological materials, polymers, liquid crystals and membranes [4.1]. For example, the magnetic orientation of *long* DNA *chains* (which amounts to about 1 % in a field of 20 T) and of other polymers gives direct information about the persistence length or radius of curvature of these macromolecules in solution [4.2]. Shorter segments of DNA or other polyelectrolytic rods (as, for example, virus particles) above a critical concentration form liquid crystalline phases with interparticle distances of up to a few hundred Å. These *polyelectrolytic liquid crystals* can be fully oriented in a strong magnetic field [4.3, 4] and are interesting anisotropic objects for the study both of their dynamics and their structure by scattering of light or neutrons. Furthermore, by evaporating the water from a polyelectrolytic liquid crystal – for example, of virus particles – exposed to a high magnetic field, dry filaments of fully aligned virus molecules have been produced which represent excellent oriented targets for the x-ray structure analysis of the virus itself [4.5].

Biological membranes show strong orientation in a magnetic field due to their intrinsic anisotropic structure. Whether the membrane orients itself parallel or perpendicular to the field depends on the proteins incorporated in the lipid membrane, and the degree of orientation varies whenever phase transitions occur in the membrane. In fact some lipid phase transitions can be observed most clearly by recording the magnetic orientations as a function of temperature [4.6]. Complex biological structures built up by folded membranes, such as retinal rods [4.7], can be fully oriented in relatively low fields.

High-field magnetic orientation turned out to be useful also for studying molecular structure and phase transitions in classical and polymeric liquid crystals as well as in microemulsions. When a polymer is formed from an oligomer by the addition of monomers *(polymerisation process)* in the presence of a strong magnetic field, the interesting question arises, whether the resulting polymers are oriented. In Sect. 4.4 evidence is presented that both the reaction rate [4.8] and the orientation [4.9] of the final polymer can be enhanced by strong magnetic fields.

We should also mention certain magnetic field dependent *chemical reactions* of great interest [4.10–24]; they are of very different origin and cannot be described here in detail. The reaction rates show a field dependence only at very low fields – between zero field and a few kilogauss at most – but are field independent at higher fields. In most of these reactions, recently reviewed by *Atkins* [4.25, 26] and *Turro* and *Kraeutler* [4.27], the nuclear hyperfine fields of the reacting molecules (10–100 G) are the rate-determining parameters at zero external field but not at much higher fields. Consequently the reaction rate in many cases changes with an applied field only between zero field and a field of several hundred gauss, with no further field dependence at higher fields. Recent developments in magnetochemistry are described in [4.28, 29].

If the magnetic field is strongly *inhomogeneous* diamagnetic macromolecules are expelled from (or attracted into) the regions of the maximum magnetic field, depending on the magnitude of their diamagnetic susceptibility (relative to the solvent). Thus in strong field gradients a separation of macromolecules, proteins or red blood cells from the rest of the solution has been achieved [4.30–32] as described in Sect. 4.5. Perhaps the changes of certain enzymatic activities in an inhomogeneous field observed by *Haberditzl* [4.33] are related to diamagnetic separation of reacting components from each other.

Since high magnetic fields affect the orientation, motion and polymerisation of biomolecules it is not too surprising that a direct *influence of magnetic fields on living cells* also exists. "Magnetobiology" has a long history [4.34–37], but many experimental results in the available fields were contradictory and therefore most of the conclusions appear rather speculative. It is established that certain animals are able to sense the earth's magnetic field and to use it for their orientation (survey in [4.38–42]). In Sect. 4.6 we briefly summarise the significant progress which has been made recently towards better understanding of the detection of such weak magnetic fields by bacteria [4.43, 39], insects [4.38, 44, 45], fish [4.41] and birds [4.40, 42, 46]. Convincing evidence exists that the roots of plants exposed to an *inhomogeneous* magnetic field show an oriented growth towards the low field region [4.47, 48]. We shall also discuss the first observation [4.49] that oriented growth (of monocellular pollen tubes) takes place in *homogeneous* strong fields. We believe that further study of oriented biological growth in high magnetic fields may lead to a better understanding of the growth process itself, including the mechanism of cell division. Therefore biological experiments in high fields represent in our view a very strong motivation for generating fields in excess of 30 T, particularly since most of the phenomena discussed here vary as H^2.

It goes beyond the scope of this chapter to review in detail the importance of high magnetic fields for high-resolution nuclear magnetic resonance spectroscopy of biological substances. This field of research has rapidly developed since 1960 when the first superconducting 5 T magnets were introduced. Now NMR spectrometers with fields up to 9.4 T have become commercially available and proved to be a powerful tool for structural analysis of many proteins and peptides [4.50–53]. Using high field NMR it has recently been possible to analyse quantitatively the large internal motion within protein molecules [4.54, 55]. In an attempt to improve the resolution and sensitivity of this technique further by working at even higher magnetic fields, a NMR spectrometer using a Nb_3Sn coil for a field of 14 T has recently been tested [4.56]. If one analyses large anisotropic molecules at such high fields they are no longer oriented at random in the solution but show alignment relative to the field. In fact, the complete magnetic orientation of suspended microcrystals of the paramagnetic protein metmyoglobin (Fig. 4.1) made it possible to monitor pseudo single-crystal NMR spectra; thus new structural features could be accurately determined [4.57].

The effects of alternating fields and microwaves on living cells are not discussed here; the reader is referred to the current literature [4.59].

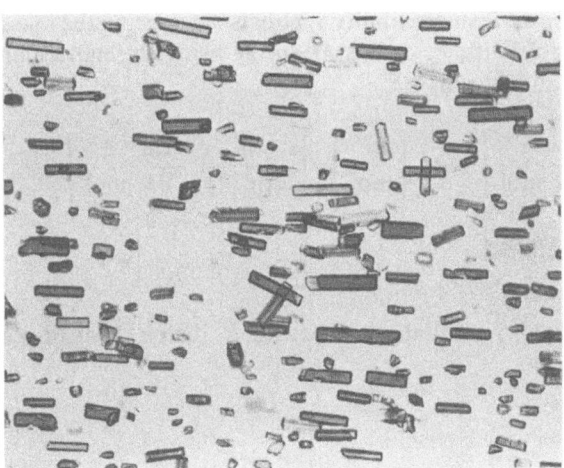

Fig. 4.1. Microcrystals (10–100) µm) of the protein met-aquo-myoglobin suspended in aqueous ammonium sulphate solution are almost completely oriented by a homogeneous magnetic field of only 0.3 T applied for 5 min. The protein carries a high-spin iron in the planar porphyrin group which is responsible for substantial anisotropy of the paramagnetic susceptibility due to some spin-orbit coupling even at room temperature. The observed magnetic orientation is very probably of *para*magnetic origin because microcrystals of *dia*magnetic CO myoglobin do not orient even in 8.5 T [4.57, 58]

4.2 Dia- and Paramagnetic Anisotropy of Macromolecules

If a molecule (without an unpaired electron) is placed in a uniform magnetic field, the intramolecular electronic screening currents generate a diamagnetic moment μ proportional to the magnetic field H

$$|\boldsymbol{\mu}| = \chi |\boldsymbol{H}|,$$

This diamagnetic response is a universal property of matter, only in less than 1 % of the presently known compounds it is overshadowed by a ferromagnetic behaviour [4.60, 61].

The diamagnetic susceptibility of a molecule is closely related to its structure and therefore is generally anisotropic. The diamagnetic moment $\boldsymbol{\mu}$ and the field $\boldsymbol{H}$ are not necessarily antiparallel, but three orthogonal principal axes of the susceptibilty tensor always exist along which $\boldsymbol{\mu}$ and $\boldsymbol{H}$ are antiparallel [4.62]. If χ_{xx}, χ_{yy} and χ_{zz} are the corresponding principal values of the susceptibility in these three directions, the differences $\chi_{zz} - \chi_{xx}$ and $\chi_{zz} - \chi_{yy}$ are the anisotropies $\Delta\chi$ of the molecular diamagnetic susceptibility. The average susceptibility is $(\chi_{xx} + \chi_{yy} + \chi_{zz})/3$ [4.62]. Various definitions and units of diamagnetic susceptibilities are used in the literature. We give some conversions for convenience. The *volumetric* susceptibility χ is defined as the magnetisation M for a sample per unit magnetic field strength H, $\chi = M/H$ in both the cgs Gaussian and the MKSA systems of units. It is dimensionless in both systems and one cgs Gaussian unit of it corresponds to 4π MKSA units. The mass susceptibility χ_g equals χ/m, m being the mass of the sample. The *specific* susceptibility χ_ϱ equals χ/ϱ, ϱ being the mass density, and the *molar* susceptibility χ_m equals $M\chi/\varrho$, M being the molecular weight. Further, χ_m has the dimensions

$$\left[\frac{cm^3}{mol}\right] = \left[\frac{erg}{Oe^2 mol}\right] = \left[\frac{erg}{G^2 mol}\right] \quad or \quad \left[\frac{m^3}{mol}\right] = \left[\frac{J}{(Am^{-1})^2}\frac{Am}{Vs}\frac{1}{mol}\right]$$

$$= \left[\frac{J}{T^2}\frac{Vs}{Am}\frac{1}{mol}\right].$$

The susceptibility per molecule, χ_0, equals χ_m/N_A (with N_A being Avogadro's number) and has dimensions

$$\left[\frac{erg}{G^2}\right] \quad or \quad \left[\frac{J}{T^2}\frac{Vs}{Am}\right].$$

Magnetic orientation involves ratios of magnetic energies to thermal energies per molecule which are

$$\left[\frac{\Delta\chi_0 H^2}{kT}\right] = \left[\frac{\Delta\chi_0 B^2}{kT}\right] \quad or \quad \left[\frac{\mu_0 \Delta\chi_0 H^2}{kT}\right] = \left[\frac{\Delta\chi_0 B^2}{\mu_0 kT}\right]$$

in the cgs Gaussian or MKSA systems, respectively ($\mu_0 = 4\pi \times 10^{-7}$ Vs/Am). The diamagnetic anisotropy of benzene, for example, amounts to $\Delta\chi = -6.71 \times 10^{-7}$ [cgs Gaussian] or -8.44×10^{-6} [MKSA], $\Delta\chi_\varrho = -7.64 \times 10^{-7}$ [cm³/G] or -9.60×10^{-9} [m³/kg], $\Delta\chi_m = -59.7 \times 10^{-6}$ [cm³/mol] or -7.50×10^{-10} [m³/mol] and $\Delta\chi_0 = -9.91 \times 10^{-29}$ [erg/G²] or -1.24×10^{-33} [(J/T²) (Vs/Am)]. The small differences in $\Delta\chi$ obtained for gaseous, liquid or crystalline benzene with various techniques have been discussed recently [4.63].

The diamagnetic anisotropies of many molecules have been determined directly by producing a single crystal of the molecular species and measuring the torque when the crystal is suspended in a uniform magnetic field [4.64–66]. Other estimates of the diamagnetic anisotropy were derived from measuring the *magnetic birefringence* in the liquid state (Cotton-Mouton effect), provided that the optical polarisability tensor of the molecular was known [4.66–69]. Finally the diamagnetic intramolecular screening currents lead to a reduction of the magnetic field acting on the various nuclei of the molecule. From the corresponding diamagnetic shift of the ^{13}C or ^{1}H lines in *nuclear magnetic resonance* experiments, additional *local* information about the diamagnetic shielding currents within the molecule has been acquired [4.66].

The most striking observation is the large diamagnetic susceptibility χ_{zz} of *aromatic molecules* if the magnetic field is applied perpendicular to the molecular $x-y$ plane. For benzene, for example, the principal molar susceptibilities (in 10^{-6} cm³/mol) are $\chi_{xx} = \chi_{yy} = -34.9$ and $\chi_{zz} = -94.6$, resulting in an anisotropy of $\Delta\chi = \chi_{zz} - \chi_{xx} = -59.7$. This large value of $\Delta\chi_m$ is caused by the diamagnetic "ring current" which is generated if the field is applied perpendicular to the plane of the conjugated benzene ring [4.70, 77].

For *non-aromatic molecules* (for which a non-local ring current can be excluded) the diamagnetic anisotropy of a molecule is, to a first approximation, the sum of the local anisotropies of the interatomic bonds composing the molecules. For example, one carbon atom has an isotropic susceptibility, but if two carbon atoms are bonded by a single, double or triple bond, this bond has been found to cause diamagnetic anisotropy (Table 4.1), which is of opposite sign for single $C-C$ bonds and for multiple bonds. While a single bond has its largest diamagnetic value for a field parallel to the bond axis and therefore orients perpendicular to the field, the reverse is true for double and triple bonds; they align parallel to the field and show largest diamagnetism for fields

Table 4.1. Observed diamagnetic anisotropies of certain bonds

$\overset{y}{\underset{\scriptstyle z}{\llcorner}}$	$\Delta\chi_m = \chi_{zz} - \chi_{xx}$ [10^{-6} cm³/mol]	Ref.
$C-C$	-1.3 ($\pm 25\%$)	[4.69]
$C=C$	$+8.2$	[4.72]
$C\equiv C$	$+37$	[4.73]
$C=O$	$+6.6$	[4.74]

perpendicular to the bond axis. As already pointed out by *Lonsdale* [4.64] introduction of double or triple bonds leads to a large *optical anisotropy* with the highest polarisability for the electric field parallel to the bond axis. This latter fact is important for magnetic birefringence experiments.

The diamagnetic anisotropies of other bonds such as the $C-H$, $C-Cl$ and $C-O$ bonds are less well known. Estimated values obtained mainly from NMR observations are $\Delta\chi_{C-C} \simeq 1 \times 10^{-6}$ [4.75] and $\Delta\chi_{C-H} \simeq \Delta\chi_{C-O} \simeq +3 \times 10^{-6}$ [4.76] [cm/mol]. For further information on many other measurements and theoretical calculations of the diamagnetic anisotropy we refer to the excellent review by *Bothner-By* and *Pople* [4.62]. Of particular interest for the experiments described in Sect. 4.3 are the estimates for the diamagnetic anisotropies of *nucleic acids* and *proteins*. According to calculations by *Veillard* et al. [4.77] the flat *base pairs* adenine-thymine and guanine-cytosine have almost the same diamagnetic anisotropy as a benzene ring with the largest diamagnetism for a field perpendicular to the molecular plane. Recent susceptibility measurements with a sensitive SQUID magnetometer on oriented films of DNA point to somewhat higher values [4.78]. Since in the DNA double helix the base pairs form a parallel stack, a free DNA molecule in a solution orients itself perpendicular to an applied magnetic field, Sect. 4.3.

Recently *Worcester* [4.79] suggested that since the *peptide bond* in proteins partially has the character of a double bond at resonance between the two bond structures, Fig. 4.2, it should have a diamagnetic anisotropy. Based on early experimental data of *Lonsdale* [4.64, 65] he gives $\Delta\chi = +8.8 \times 10^{-6}$ cm^3/mol for the peptide bond. More recently, *Pauling* [4.80] confirmed this anisotropy but arrived at the smaller value $\Delta\chi_m = 5.36 \times 10^{-6}$ cm^3/mol in better agreement with experimental findings [4.81, 82]. Thus the anisotropy of the peptide bond is apparently of the same sign and similar magnitude as for the $C=C$ and $C=O$ double bonds (Table 4.1). Consequently the plane of the peptide bond tends to orient parallel to an external field (as does the plane of the $C=C$ double bond). Since in an α helix the planes of the peptide bonds are parallel to the helical axis, it becomes clear why α helical proteins in solution can align their helical axis parallel to the magnetic field without the assistance of aromatic groups (Sect. 4.2). In addition, three out of the 20 amino acids occurring in proteins (as side groups of the polypeptide chain) are aromatic and therefore have $\Delta\chi$ values comparable to benzene: *Veillard* [4.77] reported the value $\Delta\chi_m = -95.3$ for tryptophane and -60 [10^{-6} cm^3/mol] for both tyrosine and phenylalanine.

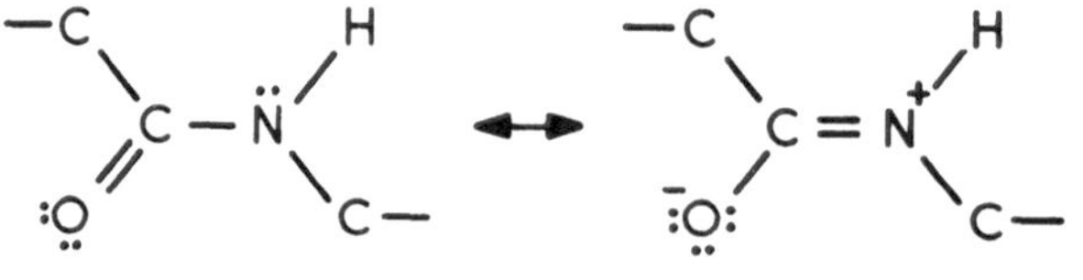

Fig. 4.2. Two bond structures at resonance result in a partial double bond character of the peptide bond

Hence, if coupled in their orientation with respect to the polypeptide chain, aromatic amino acids might contribute substantially to diamagnetic anisotropy in proteins.

If a molecule contains unpaired electrons its overall spin is different from zero, generally resulting in a paramagnetic moment μ. Its paramagnetic susceptibility (per molecule) is then given by *Curie's* law $\chi = \mu^2/3kT$. The term μ is proportional to the g factor which describes the coupling between spin and orbital momentum. For anisotropic groups, such as the iron-containing planar aromatic porphyrin group ("haeme" group), the g factor is usually anisotropic. For example, with strong spin-orbit coupling at low temperatures, g equals 2 normal to the haeme plane whereas $g \sim 6$ within the haeme plane, as deduced from ESR and measurements of the torque in a homogeneous magnetic field on single crystals of myoglobin [4.83]. (For details about the paramagnetic properties of iron-containing proteins we refer to a review article by *Kotani* [4.83].) Interestingly some spin-orbit coupling persists even at room temperature, resulting in a *paramagnetic* anisotropy $\chi_{zz} - \chi_{xx}$ per haeme group which is numerically (for ferri-myoglobin at 300 K) some 30 times higher than the diamagnetic anisotropy of benzene and has *equal* sign [4.83]). *Nakano* et al. [4.84] reported a $\Delta\chi_m$ value of $-1100 \times 10^{-6}\,\mathrm{cm^3/mol}$ for the haeme group in deoxyhaemoglobin. Hence such haeme groups orient substantially better in their paramagnetic than in their diamagnetic state: in both cases they tend to align their planes parallel to the magnetic field. Finally one has to mention the anisotropy caused by the crystal field on the ligands, which could lead to an effective orientation even if the g factor were isotropic.

4.3 Magnetic Orientation of Macromolecular Systems

The orientation of macromolecular systems in magnetic fields is usually caused by their overall anisotropic magnetic susceptibility. Various magnetic orientation effects in liquid crystals have been known for a long time [4.85]. The first experimental evidence for magnetic anisotropy of organised biological structures was found by direct measurements of the torque on mechanically suspended fibres of celluloses, silks, keratines and collagens in a small homogeneous magnetic field in 1942 [4.86]. A similar study on muscle followed in 1958 [4.87]. The magnetic orientability of biological particles in solution was first discovered in 1970 [4.7] by direct microscopic visualisation of the magnetically induced rotation of the outer segments of retinal rods. This work and the report by *Geacintov* et al. [4.88] that the fluorescence of chlorella cells became anisotropic in a strong magnetic field because of magnetic orientation of the chloroplasts [4.89] have triggered a rapid development in this field (Sect. 4.3.4). An interpretation of the early experiments in terms of diamagnetic anisotropy of the constituent subunits was given by *Hong* et al. [4.90a] in 1971. This initiated a discussion about the relative contributions of diamagnetic anisotropies of lipids, aromatic amino acids, α-helical parts of proteins and chromophores in these complex biological objects. Clarification resulted mainly from orientation studies on simpler systems such as α-helical polypeptides: they form liquid crystalline phases and therefore are substantially orientable already by *small* magnetic fields [4.91] (Sect. 4.3.2). Alcanes and phospholipids generally require stronger magnetic fields and more sophisticated optical techniques for orientation to be detected [4.6] (Sect. 4.3.4). It is now possible to investigate by *high* field magnetic birefringence the structure of smaller isolated particles like viruses in dilute solution [4.92] (Sect. 4.3.5). High magnetic fields have also become important in the study of intra- and intermolecular orientational correlations in liquids. For example, the rigidity of polymer chains has been studied in various dilute polymer solutions (Sect. 4.3.2a): long-range order due to intermolecular interactions was investigated in solutions of charged polymers ('polyelectrolytes'), in polymer melts (Sect. 4.3.2b) and in thermotropic liquid crystals near the isotropic-nematic transition temperature (Sect. 4.3.3). Some of this work was briefly reviewed recently [4.93].

4.3.1 Magnetic Birefringence, Cotton-Mouton Effect

The magnetic anisotropy of small molecules is weak and even in strong magnetic fields only modest alignment occurs. The deviation from random orientations of the molecules results in an optical birefringence (Cotton-Mouton effect) which can be measured with extreme sensitivity (Fig. 4.3). This is perhaps the most sensitive way to detect magnetic orientation in liquids and gases.

The theory of the Cotton-Moutton effect of *independent rigid molecules* was first developed by *Langevin* [4.95] and *Born* [4.96] and was extended later on [4.67, 68, 97, 98]. We summarise first the simple case of non-interacting rotationally symmetric molecules with anisotropic diamagnetic susceptibility ($\Delta\chi_0$) and optical polarisability ($\Delta\alpha_0$). Let $f(\theta)$ be the magnetic field dependent orientation distribution function of the molecules which is given by Boltzmann statistics, and θ is the angle between the magnetic field direction and the molecular z axis. Each molecule of the ensemble at angle θ contributes on the average $(3\cos^2\theta - 1)\Delta\alpha_0/2$ to the macroscopic optical anisotropy. Integration of the latter function weighted with $f(\theta)$ results in the effective optical anisotropy per molecule; for $\Delta\chi_0 H^2 \ll kT$ it turns out to be

$$\Delta\alpha_0 \cdot \frac{\Delta\chi_0 H^2}{15 kT} = \Delta\alpha_0 \cdot \Phi.$$

The term Φ is sometimes called degree of orientation. With the Lorentz-Lorenz formula (relating polarisability and refractive index) the magnetic birefringence Δn becomes in cgs Gaussian units[1]

---

1 in MKSA units (4.1) becomes

$$\Delta n = \frac{1}{18\,\varepsilon_0} \frac{(n_0^2 + 2)^2}{n_0} N\Delta\alpha_0 \frac{\mu_0 \Delta\chi_0 H^2}{15 kT}.$$

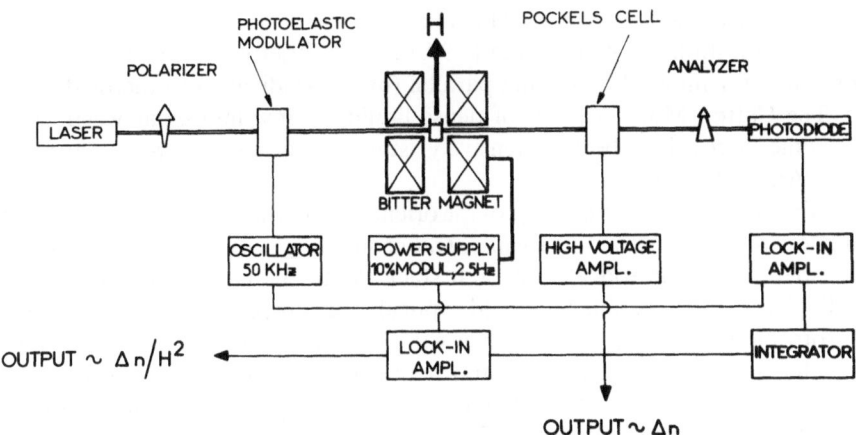

Fig. 4.3. Sensitive magnetic birefringence setup. Laser light (usually 632.8 nm wavelength) propagates horizontally through the sample in a vertical magnetic field (≤ 13.5 T) produced by a Bitter solenoid with 4 mm horizontal and 5 cm vertical bore. Polarizer and analyzer are crossed and at 45° with respect to H. A small 50 kHz modulation of the birefringence is produced by a photoelastic modulator resulting in a 100 kHz intensity modulation of the photodiode output. Any superimposed steady state (magnetic) birefringence results in an additional 50 kHz photodiode output which is phase-sensitively detected, dc converted and used as error signal in a feedback loop to compensate the steady state birefringence, by a Pockels cell. Hence in the compensated mode the voltage across the Pockels cell is a direct measure of the magnetic birefringence Δn. A low-frequency modulation of H induces a LF modulation of Δn which is directly proportional to the Cotton-Mouton constant and can be detected continuously by a second lock-in amplifier. At a sample length of 1 cm (sample volume ~ 1 ml) the resolution in Δn is close to 10^{-10} for transparent liquids. It is essentially limited by mechanical vibrations of the hydraulic cooling system of the *Bitter* coil [4.94]

$$\Delta n = \frac{2\pi}{9} \cdot \frac{(n_0^2+2)^2}{n_0} \, N\Delta\alpha_0 \, \frac{\Delta\chi_0 H^2}{15\,kT} = \Delta n_{\text{sat}}\Phi. \tag{4.1}$$

Here n_0 is the mean refractive index of the solution, $N = cA_0/M$ is the number of molecules per unit volume, c, A_0, M are concentration, Avogadro's number and molecular weight, respectively and Δn_{sat} denotes the birefringence of a completely oriented sample. The general validity of the Lorentz-Lorenz description of the local optical electric field is not clear. Field corrections are frequently ignored for macromolecular solutions and hence

$$\Delta n = \frac{2\pi}{n_0} \, N\Delta\alpha_0 \, \frac{\Delta\chi_0 H^2}{15\,kT} \tag{4.2}$$

is used instead of (4.1). The Cotton-Mouton constant CM is defined for sufficiently small magnetic fields by

$$\Delta n = \text{CM}\lambda H^2, \tag{4.3}$$

with λ being the wavelength of light. The variation of Δn or Φ in stronger fields and its saturation can be found in [4.97, 99, 100]. Briefly, Φ tends towards $+1$ (or -0.5) for molecules aligning parallel (or perpendicular) to the field.

The Cotton-Mouton effect of dense liquids of low molecular weight and of some gases has been extensively studied during the past 60 years [4.62, 63, 67, 98, 101].

We consider now correlations of the orientation of the N molecules. The most simple case is non-interacting *rodlike stacks* consisting of say N_0 molecules with z axes parallel to the rod axis in each stack. With the current assumption of additivity of the diamagnetic and optical anisotropies, these stacks have anisotropies $N_0 \Delta\chi_0$ and $N_0 \Delta\alpha_0$. Because their number density is N/N_0, the rhs of (4.1) has to be multiplied by N_0 and hence their magnetic birefringence and orientation is N_0 times higher at identical N. A non-parallel orientation of the molecules within the stack reduces the Cotton-Mouton constant. For example, a rotationally symmetric arrangement of molecules having an angle δ between their z axes and the symmetry axis of the stack reduces CM by

$$(\tfrac{3}{2}\cos^2\delta - \tfrac{1}{2})^2.$$

If the freely rotatable object is an anisotropic *cluster* (or domain) of a certain spatial extension – the so-called orientational correlation length – involving an effective number N_1 of parallel molecules, the rhs (4.1) has to be multiplied by N_1.

In the more complicated case of (isolated) *flexible polymer* chains products of the monomeric diamagnetic susceptibility tensor and the optical polarisability tensor have to be averaged over all possible conformations of the chain [4.102, 103]. A simpler approach was chosen by *Stuart* and *Peterlin* [4.104] in 1950. The actual chain was replaced by an equivalent chain built up of rigid segments, the "*Kuhn* statistical segments", each of which can be freely rotated relative to the neighbouring segments. The segmental length l_s is usually larger than the actual monomer length l_0 (but shorter than the length L of the fully extended chain). The l_s increases with the chain rigidity in such a way that for long chains ($L \gg l_s$) the mean overall extension (radius of gyration) of the real chain and the model chain are identical [4.103]. Stuart and Peterlin considered the *Kuhn*-segment as rods orienting independently in a magnetic field and hence their result corresponds to (4.1), if $\Delta\chi_0$, $\Delta\alpha_0$ and N are replaced by $\Delta\chi_0 \cdot l_s/l_0$, $\Delta\alpha_0 l_s/l_0$ and $N \cdot l_s/l_0$, respectively. The Cotton-Mouton constant of the polymer is l_s/l_0 times higher than the value for independent monomers. This rather qualitative model has been frequently applied to CM measurements of dilute polymer solutions [sometimes corrected for a non-rotationally symmetric orientation of (side) groups with respect to the segmental direction]. The resulting l_s values can be considered as an order-of-magnitude estimate for the orientational correlation length along the polymer chain.

Another description of a long chain molecule dates back to *Kratky* and *Porod* [4.103], appropriate for *semiflexible* chains. The chain bends in a characteristic homogeneous "worm-like" fashion and the parameter describing its rigidity is the persistence length P. It is defined as the chain's end-to-end vector averaged over all chain configurations and in the limit of vanishing segmental length [4.103], and is proportional to the bending elastic modulus of the chain [4.105]. The *Kuhn* and *Kratky-Porod* models are related by $2P \cong l_s$ [4.103]. *Wilson* [4.106] and *Weill* [4.93] have calculated the Cotton-Mouton constant for the "worm-like" model by the appropriate tensor averaging [4.102]. It turns out [4.93] that the rhs of (4.1) has to be multiplied by

$$\frac{l_0}{L} \sum_i \sum_j \left\langle \frac{(3 \cos^2 \theta_{ij} - 1)}{2} \right\rangle ,$$

θ_{ij} being the angle between the directions of the ith and jth chain segment, and $\langle \ \rangle$ denoting averaging over all configurations. *Wilson* and *Weill* end up with the identical result in the case of long chains ($P \ll L$): the polymeric Cotton-Mouton constant exceeds the monomeric value by $2P/3l_0$, and is independent of the polymeric molecular weight. Such a tensor averaging for *Kuhn*'s segmental chain and ($l_s \ll L$) results in $l_s/3l_0$ instead of l_s/l_0. In the limit of short chains ($P \gg L$), rod-like behaviour is found and CM is proportional to the molecular weight (L/l_0). *Weill* has given an analytic expression of the above double sum for any chain length:

$$\frac{l_0}{L} \sum \sum \langle \ \rangle = \frac{2P}{3l_0} \left\{ 1 - \frac{P}{3L} \left[1 - \exp\left(-\frac{3L}{P} \right) \right] \right\}. \tag{4.4}$$

Therefore, the persistence length of polymers can now be determined without knowledge of $\Delta\alpha_0$ and $\Delta\chi_0$ from CM measurements on two (or more) fractions of suitable and well-known L. One can sometimes use solutions of monomers and high molecular weight polymers for this purpose.

For very flexible polymers the worm-like model is less appropriate. Bond angle restrictions require more *specific models* such as the, so-called, rotational isomeric state model [4.103]. Cotton-Mouton constants can be calculated in analogy to the electro-optical Kerr constant (electric birefringence).

4.3.2 Polymers

a) Isolated Chains

Some experimental Cotton-Mouton constants of dilute polymer solutions are summarised in Table 4.2. A chemically simple polymer is *polyethylene* $(CH_2)_x$ (Fig. 4.4). The diamagnetic bond anisotropy of $C-C$ is very small (Sect. 4.2), but still much larger than the $C-H$ anisotropy [4.66, 69]. This combined with the high chain flexibility results in a very weak Cotton-Mouton constant in dilute solution [4.107, 108], which is in good agreement [4.98, 107, 108] with the rotational isomeric state model [4.106]. At concentrations $\geq 40\%$ in CCl_4 an

Table 4.2. Cotton-Mouton constants and persistence length of some polymers in dilute solutions[a]

Polymer	See Fig.	Solvent	Specific CM const. (10^{-13} G^{-2} cm^{-1} per mass fraction)	Ref.	Chain model	Persistence length [nm]
$(CH_2)_{16}H_2$	4.4a	Carbon tetrachloride	0.25	[4.69, 107, 108]	Rational isomeric state [4.103]	—
Polycarbonate	4.5	Chloroform	9.6	[4.107–109]	Worm-like [4.93, 106] Kuhn [4.104]	2.7 ($l_s/2$) = 1.7
Polystyrene	4.4b	Benzene teteachloroethylene	6.8 12.7	[4.110] [4.110]	Rational isomeric state [4.111]	— —
Poly-p-benzamide	4.7a	Conc. H_2SO_4	125	[4.107, 108]	Worm-like [4.93, 106]	75
Poly-1.4-phenylene-terephthalamide	4.7b	Conc. H_2SO_4	200	[4.107, 108]	Worm-like [4.93, 106]	120
Various mesogenic polyesters	4.8	1,1,2,2-Tetra-chloroethane	9–18	[4.112]	—	—
Poly(Tyr-Glu)	4.9	H_2O, $\mu = 1.1 \times 10^{-2}$ M H_2O, $\mu =$ 10^{-1} M	93 22	[4.113]	Worm-like [4.93, 106]	20 4
Polystyrene sulphonate	–	H_2O, $\mu = 2$ M	2.6	[4.114]	Worm-like [4.93, 106]	1.2
DNA	4.10	H_2O, $\mu = 10^{-3}$ H_2O, $\mu = 10^{-2}$ M H_2O, $\mu = 1$ M	460 290 280	[4.2, 6, 115]	Worm-like [4.93, 106]	113 71 69

[a] μ = ionic strength; all CM values measured at room temperature and $\lambda = 632.8$ nm

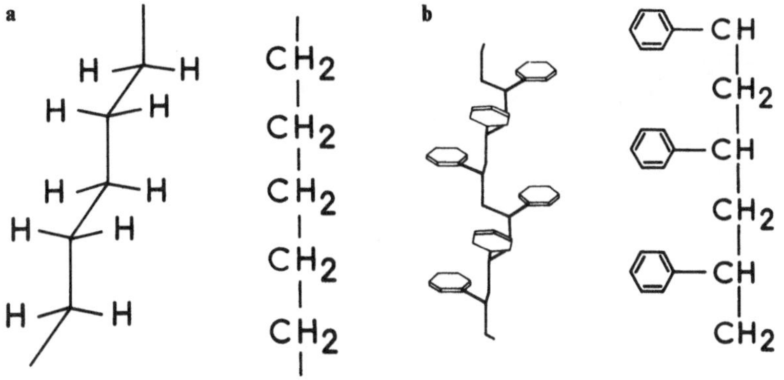

Fig. 4.4. (a) Polyethylene, (b) Polystyrene

onset of interchain correlations was observed (Sect. 4.3b), possibly related to a particular intermolecular correlation of the CCl_4 molecules [4.116].

Polycarbonates (Fig. 4.5) show a specific Cotton-Mouton constant almost 40 times higher than the alcanes [4.107–109], mainly due to the large $\Delta\chi$ and $\Delta\alpha$ values of the *in-chain* phenyl group. *Stamm* [4.107] and *Champion* et al. [4.109] derive different chain rigidities from identical experimental results (Table 4.2). Part of the problem is related to the reliable determination of effective diamagnetic and optic anisotropies with respect to the axis of the polymeric backbone. *Champion's* approach [4.109] is to combine magnetic and flow birefringence data with estimated $\Delta\alpha_0$ and $\Delta\chi_0$ values of the phenyl group; the resulting information about the local conformation of the backbone and the segmental length stems from the application of the *Stuart-Peterlin* theory. *Stamm's* CM data of polymeric polycarbonate and of a monomeric model compound result in $P = 2.7$ nm, when analysed according to the worm-like model see Eqs. (4.1, 4). This P value seems to be a lower limit because the above treatment certainly somewhat overestimates the effective $\Delta\alpha \cdot \Delta\chi$ value.

If the predominant anisotropic group, such as the phenyl group of *polystyrene* (Fig. 4.4b), is located in a side chain of the polymer, the analysis of Cotton-Mouton data in terms of polymeric backbone rigidity is less obvious. The degree of rotational mobility of the side group with respect to the main chain should be known from other sources. It should not be too high since otherwise the effective anisotropy along the backbone becomes small. The rotational freedom of the phenyl side group of polystyrene is known to be restricted [4.111, 117, 118]. The CM values of polystyrene in benzene (or tetrachlorethylene) (Table 4.2 and Fig. 4.6) exceed the monomeric Cotton-Mouton constant (i.e. essentially benzene) by a factor of 1.4 (or 2.6, respectively). This compares with a correlation parameter of 1.69 calculated by *Tonelli* et al. [4.111] based on the rotational isomeric state model and confirms the great flexibility of polystyrene in solution. In contrast, large CM values combined with strongly non-linear magnetic birefringence curves were reported for polystyrene in

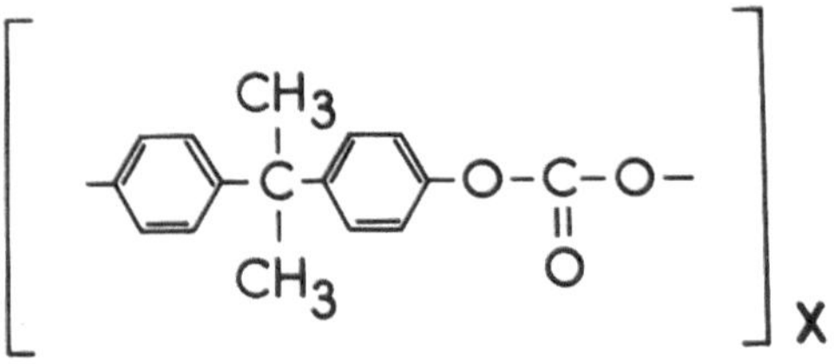

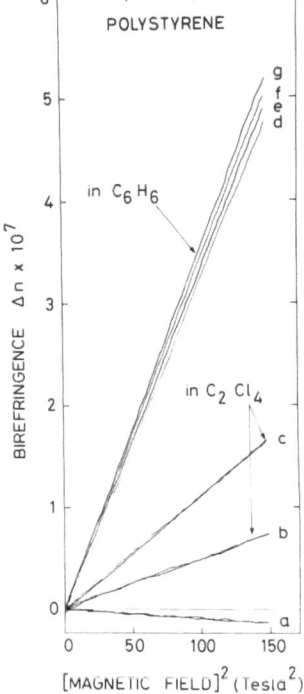

Fig. 4.5. Polycarbonate (Biphenyl -A-Pc)

Fig. 4.6. Cotton-Mouton effect of solutions of poly- ▶
styrene in *tetrachloroethylene* (*a*) $c=0$, (*b*) $c=7\%$,
(*c*) $c=15\%$ and in *benzene* (*d*) $c=0$, (*e*) $c=7\%$, (*f*)
$c=14\%$, (*g*) $c=21\%$, $T=22°$, $\lambda=632.8$ nm, optical path
$=3$ cm, [4.110]

cyclohexane and benzene, suggesting the existence of micron-sized aggregates in these samples [4.119]. As such anomalies have never been found by others [4.107, 108, 110, 120] we feel they are related to impurities.

Polybenzamides have been investigated by *Tsvetkov* et al. [4.121] and *Stamm* [4.107] in sulphuric acid, in 1-methylpyrolidon (MP) and in dimethyl-acetamide (DMAA) with LiCl. These polymers are of special interest because they have high mechanical and thermal strength and are liquid crystalline in concentrated solutions, resulting in various technological applications (for example, high modulus "Kevlar and Nomex fibres"). From low-field CM measurements, using the Stuart-Peterlin approach, Tsvetkov et al. suggested a persistence length of more than 500 nm ($l_s \geq 1000$ nm) for PBA (Fig. 4.7) in concentrated H_2SO_4. From precise high field CM measurements Stamm found (at concentrations below 2%) P values of 75.0 nm (120 nm) for PBA (PPTA, Fig. 4.8) applying the worm-like chain model. He also found [4.107] conformational changes of PBA in DMAA (at ionic strengths above 1.5% LiCl) and states with intermolecular order.

Interesting applications are expected from polymers carrying so-called mesogenic groups in the main chain which favour liquid crystalline order (Fig. 4.8). These new polymers have liquid crystalline phases in the melt at temperatures well below the onset of chemical decomposition (Sect. 4.3.2b).

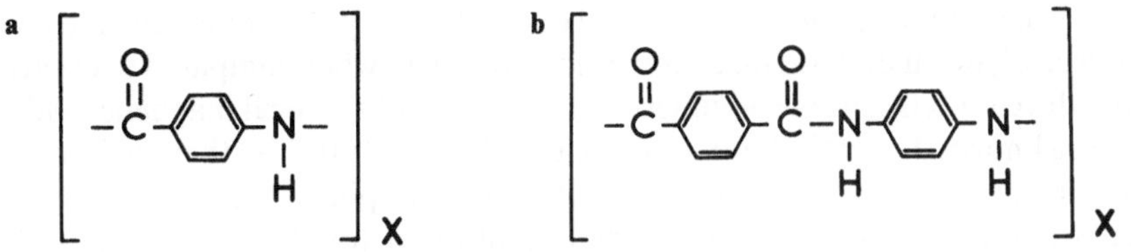

Fig. 4.7. (a) Poly-p-benzamide; **(b)** Poly-1,4-phenylene-terephtalamide

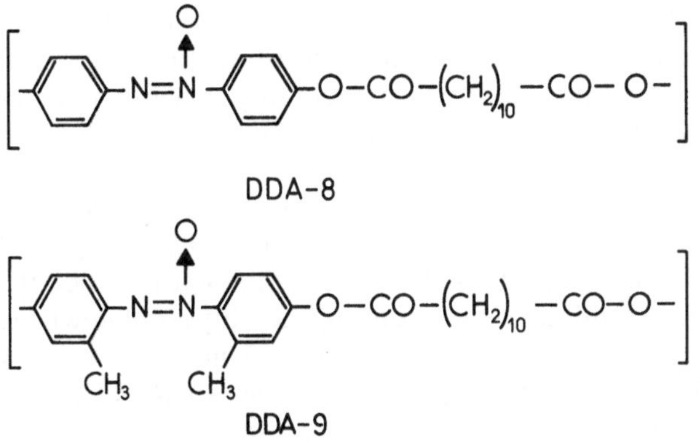

DDA-8

DDA-9

Fig. 4.8. Two polyesters with "mesogenic" groups and flexible spacers in the main chain. Both DDA-8 and DDA-9 [4.112] have liquid crystalline (or "meso"–) phases in the melt

Cotton-Mouton experiments have shown so far [4.112] that individual chains of these polymers in dilute solution have no tendency to form intrachain order (such as backfolding of mesogenic segments). They are highly flexible random coils in solution.

We now turn to polymers in aqueous solutions. In water, chemical dissociation of some polar groups usually occurs and an electrically charged macromolecule is obtained surrounded by small "counter" ions of opposite polarity. Such solutions which sometimes contain additional salt (i.e. additional ions of both polarities) are called *polyelectrolytic*. Almost all biopolymers are polyelectrolytes. At high charge density of the macroion and low counterion concentrations (i.e. weak screening) the chain conformation and the persistence length are determined by long-range electrostatic interactions between segments of the chain. At low charge densities and high counterion concentrations (i.e. strong screening) the chain can be considered neutral and its conformation is given by its local "steric" rigidity. This electrostatic effect has been known qualitatively for a long time. A theoretical description of the electrostatic persistence length of polyelectrolytes has been advanced recently [4.122–124 and references therein]. Various experimental techniques such as flow birefringence, dichroism [4.125, 126] and electro-optic relaxation [4.127, 128] have since been employed to check the predictions, showing the actuality of this basic problem.

We believe that *magnetic* birefringence is particularly well suited for studying polyelectrolytes since it has many intrinsic advantages when compared to electric field induced orientation techniques. For example, (a) the local magnetic field is very well known. It equals the external field within about 10^{-6} in the usual case of diamagnetic solutes and solvents. (b) It can be applied "steady state" and orientation can be studied in true thermal equilibrium. (c) There is only one well-defined orientation mechanism, described by (4.1), and (d) diamagnetic anisotropies are additive quantities to a good approximation (Sect. 4.2). (e) Very small amounts of sample are needed. In general, (a–e) are not true when electric or flow fields are applied to electrolytic solutions. However, the degree of molecular alignment obtained in strong magnetic fields is sometimes comparatively poor; even with sensitive birefringence techniques single-chain properties at sufficiently low concentrations sometimes can hardly be investigated. Obviously stronger steady magnetic fields could be helpful.

Cotton-Mouton experiments are available on the rather flexible polypeptide *polytyrosine glutamic acid* (Fig. 4.9) [4.113], on *polystyrene sulphonate* [4.114] and on *nucleic acids* (Fig. 4.10) [4.2, 6, 94, 129]. Poly (Tyr-Glu) carries the aromatic tyrosine in the side group; since the side group mobility is unknown the absolute persistence length cannot be deduced from CM measurements. A strong electrolytic stiffening is probably responsible for the increase of CM at low ionic strengths (Table 4.2).

The Cotton-Mouton constant of polystyrene sulphonate at high ionic strength is very weak and does not depend on ionic strength. This corresponds to a "steric" persistence length of about 1.2 nm [4.114]; its increase observed at

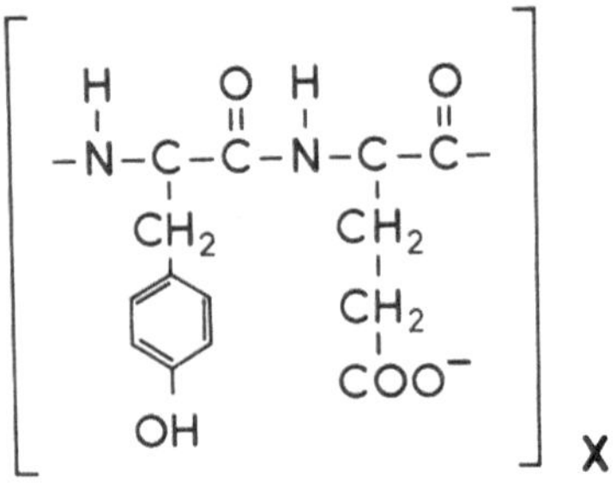

Fig. 4.9. Poly-tyrosine-glutamic acid

Fig. 4.10. Double-helical DNA [4.130]

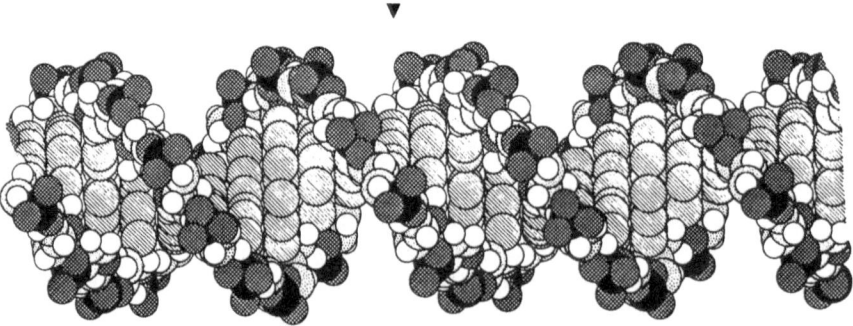

lower *ionic* strength is due to the electrostatic stiffening of individual chains and quantitatively agrees with the theoretical predictions [4.122, 124]; it becomes molecular-weight dependent because at low ionic strength Debye's electrostatic screening length is no longer small compared to the chain length L [4.114].

For DNA an increase of CM at ionic strength below some 10^{-2} M NaCl was reported earlier [4.6]. New measurements performed with improved experimental resolution and at lower concentrations enabled lower ionic strengths to be explored; it turned out [4.94] that the electrostatic persistence length deduced agrees with the *Odijk* theory over an ionic strength range of more than 3 orders of magnitude. At high ionic strength the Cotton-Mouton constant runs into a plateau given by the "steric" persistence length. The P values (Table 4.2), obtained by using the now well-established values $\Delta\alpha_0 = 12.8 \times 10^{-24}$ cm^3 per base pair [4.131] and the preliminary experimental $\Delta\chi_0 \approx 90$ cm^3/mol [4.78], agrees fairly well with results from other techniques [Ref. 4.132 and references therein]. Changes in rigidity with external parameters, for example during thermal denaturation, can be monitored directly by the Cotton-Mouton constant [4.2].

b) Interacting Chains

This section deals with ordering phenomena occurring at higher polymer concentrations. Knowledge about long-range intermolecular order and the forces involved are important for understanding not only macroscopic properties of liquid crystals, gels and polymer melts, but also biological processes like protein self-assembly and cell division. Intermolecular forces inducing parallel arrangements of molecules result in an increase of the magnetic orientability. Therefore strong magnetic fields have a broad variety of applications: they range from sensitive detection of the onset of weak orientational correlations to full macroscopic orientation of a sample with well-developed long-range order.

It is well known that charged *spherical* particles with typical diameters of 100 nm organise themselves in water above a critical concentration. They may form Wigner-type lattices with concentration-dependent lattice constants of many diameters because of their mutual electrostatic repulsion [4.133, 134]. Rod-like particles such as tobacco mosaic virus (length ~300 nm, diameter ~20 nm) at elevated concentrations show an almost parallel, liquid crystal-like, orientation extending over large domains [4.135–139]. *Onsager* in 1949 [4.140] was the first to predict such ordered phases of rods on the basis of entropy arguments. Considering only hard core interactions, he calculated two critical concentrations for the onset and full development, respectively, of the anisotropy. Both are inversely proportional to the rod length. Much theoretical work including work on semiflexible chains is available now [4.141, 142]. However, in the charged-rod water systems ordering has been observed already at substantially lower concentrations than predicted by hard core theories; hence more long-range (electrostatic and eventually electrodynamic) forces are probably involved [4.143–147]. In spite of the experimental and theoretical progress in this field a satisfactory quantitative description of both the electrostatic and the van der Waals potential between the macromolecules is still missing. For example, it is not yet clear how the long-range superstructure sometimes observed in such solutions can be related to the macromolecular structure; is there a general relation between the superstructure and the helical structure of the macromolecule at substantial dilution? Is the spatial extension of order connected to the macromolecular flexibility? In the following section we shall illustrate the potential value of high magnetic fields in this research area by summarising some experiments on concentrated polyelectrolytes.

I) Polyelectrolytic Solutions

Poly(Tyr-Glu) is a rather flexible polyelectrolyte at solvent conditions as mentioned above (Sect. 4.3.2a). Despite this, a transition into a phase of locally parallel ordered chains (involving interchain interactions) is suggested from the observation of a *sudden increase of the magnetic birefringence* at concentrations around 2.5 % [4.113]. Neutron scattering data were consistent with a mean radial intermolecular distance of about 6 nm in this case and the order seemed to extend only over a few hundred Kuhn segments. The observed sharp decrease of CM with ionic strength indicates the importance of electrostatic interaction for parallel ordering. Many currently studied polyelectrolytes, such as polystyrene-sulphonate, have flexibilities, charge densities and diamagnetic anisotropies comparable to Poly(Tyr-Glu). Unfortunately most of them cannot be oriented in magnetic fields easily and have an undefined flexibility which depends on the solvent parameters.

Therefore very long thin charged rod-like *bacteriophages* (Pf1, fd) were studied *as model systems* [4.3, 92, 148]. They have a highly ordered internal structure resulting in a large diamagnetic anisotropy (Sect. 4.3.5) and form large birefringent domains at concentrations above about 1 % (Fig. 4.11a); at the critical concentration the Cotton-Mouton constant suddenly increases by more than 3 orders of magnitude [4.3] and saturation of the magnetic birefringence is observed in fields of about 5 T (Fig. 4.12). The rods are aligned essentially parallel to the magnetic field and may maintain their macroscopic orientation for hours at zero field (Fig. 4.11b), so enabling investigation of the structure and dynamics of the oriented polyelectrolytic liquid crystal by small-angle neutron and light scattering [4.3, 148–150]. The result indicates parallelism of the rods; in the radial direction a liquid-like structure factor is found with a well-defined concentration-dependent $(c^{-1/2})$ average next-neighbour distance of up to 70 nm. The effective diameter of rods deduced from the experimentally excluded volume is about 3 times larger than their geometric diameter, reflecting the thickness of the ionic cloud; it is however still about two times smaller than the mean radial next-neighbour distance, suggesting substantial freedom of non-collective lateral and axial motions. In the axial direction distances are random, no smectic-like layer structure could be found [4.149, 150], the light scattering is dominated by single-particle interference. Quasielastic light-scattering experiments by *Oldenbourg* reveal [4.149, 150] a strongly anisotropic translation

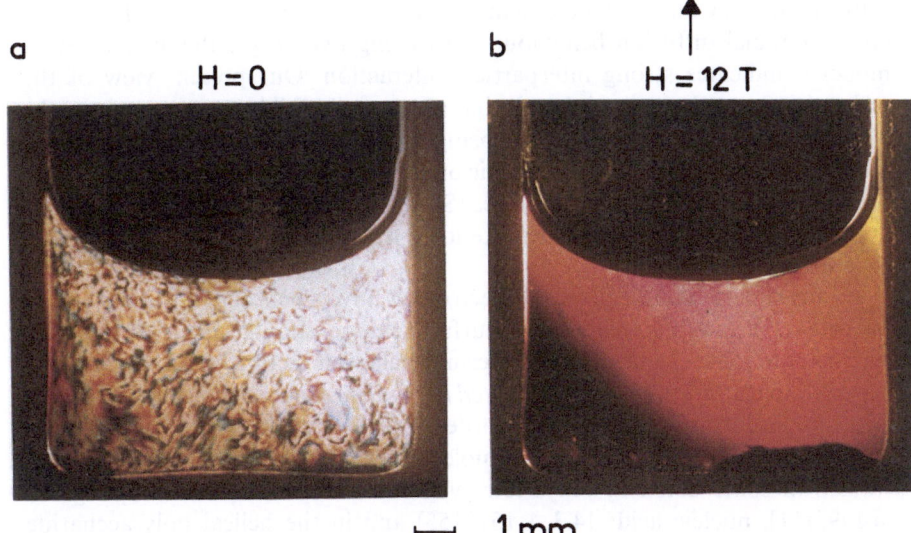

Fig. 4.11a, b. In solutions of the highly charged rod-like virus particle Pf1 (length ~ 1980 nm, diameter ~ 6 nm, pH 7.5, 10 mM Tris HCl, $T = 20\,°C$), a long-range parallel ordering of rods occurs above a critical concentration of ~ 0.6 % w/w. The preferential radial next-neighbour distances (~ 33 nm at $c = 2.4\,\%$ w/w) are maximised, corresponding to homogeneous space filling large birefringent domains which are (**b**) completely oriented in a magnetic field above ~ 5 T [4.148]. Photographs of 1 cm wide cells with 0.1 cm optical path between diagonal crossed polarisers were taken before (**a**) and after (**b**) exposure to a vertical magnetic field of 12 T ($c = 2.4\,\%$ w/w). The magnetic orientation is maintained in zero field for minutes to hours depending on the environment (temperature gradients, shaking). Similar ordering effects occur in solutions of a shorter bacteriophage fd (length 880 nm, diameter 6 nm) at concentrations above ~ 1.1 % w/w [4.3, 92, 148–150]

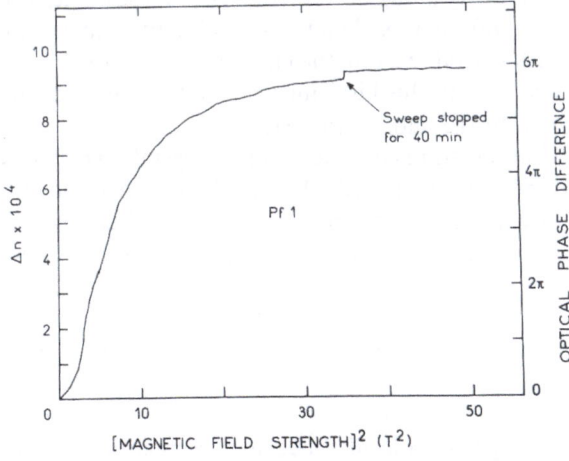

Fig. 4.12. Magnetic birefringence of the rod-like bacteriophage Pf1 in a liquid crystalline aqueous solution (at $c = 14.8$ mg/ml, $T = 20\,°C$, $\mu = 10$ mM Tris HCl, $\lambda = 632.8$ nm, optical path = 0.2 cm, field sweep velocity = 6.2×10^{-4} T/s). The arrow indicates a 40 min stop of the magnetic field sweep. After this time no further increase in Δn is observed

diffusion coefficient. Axial diffusion (along the magnetic field) is close to the free diffusion which would occur along the rod axis in isotropic solution, whereas the complex radial diffusion behaviour – involving two very different relaxation modes – indicates strong interparticle interaction. Our present view of the ordered phase of bacteriophages is that of a very "soft" lyotropic nematic, the large excluded volume of the rods being controlled by the counterion cloud. Observing full and stationary magnetic orientations of this system facilitates not only the production of oriented fibres (Sect. 4.4) but makes possible studies of anisotropic electrolytic properties like ionic fluctuations and ionic densities in ordered polyelectrolytes.

Many rigid or semiflexible polyelectrolytes have a helical conformation; the molecular surface as well as the surface charge should reflect the helical symmetry to some extent, and any intermolecular interaction of this symmetry is expected to cause a complicated *twisted long-range order* in solutions above the critical concentration. Long-range order with periodicities much larger than both intermolecular distances and molecular lengths has been observed and studied in some detail in polyelectrolytic solutions of the bacteriophage fd [4.149, 151], nucleic acids [4.3, 4, 151–155] and in the helical polysaccharide xanthan [4.156].

For fd and xanthan the *long periods* range from 10 µm to several 100 µm depending on concentration and ionic strength. They are easily measurable with laser light diffraction or can be seen in a polarising microscope; two examples are shown in Fig. 4.13. Combined measurements of optical rotation, birefringence and scattering of light and neutrons clearly demonstrate that the xanthan and fd rods are organised like the molecules in a neat cholesteric liquid crystal (Fig. 4.14) except for a much larger lateral interparticle distance.

Just as in classical cholesteric liquid crystals, molecules are parallel within planes, but have mean radial distances and interplane spacings d up to 10 times larger than the rod diameter. The twist angle between molecules in neighbouring planes defines a screw with periodicity along the direction $A - A$, the cholesteric axis. Since individual rods have different axial and radial refractive indices, the birefringence is modulated with period D along the cholesteric axis; the bright-dark layers seen in Fig. 4.13 are due to this birefringence modulation and thus make *visible* the planes normal to the cholesteric axis.

Various *magnetic field effects* occur in cholesteric liquid crystals. They have been extensively discussed in the literature [4.1, 85, 152, 157], so we mention only those relevant to the present experiments on polyelectrolytes. When orientational constraints due to wall effects can be neglected, the cholesteric axis for molecules with $\Delta\chi < 0$ or $\Delta\chi > 0$ tends to align parallel (a) or perpendicular (b) to H. In case (a) once all molecules are aligned normal to H by this process no further effects take place in stronger fields. The period D is magnetic field independent. In (b) even for complete alignment of all $A - A$ axes normal to H many molecules are still not oriented along H. Hence in stronger fields when the magnetic torque on the non-aligned molecules becomes comparable to the restoring elastic torque for planar twist (Sect. 4.3.3c) the cholesteric screw

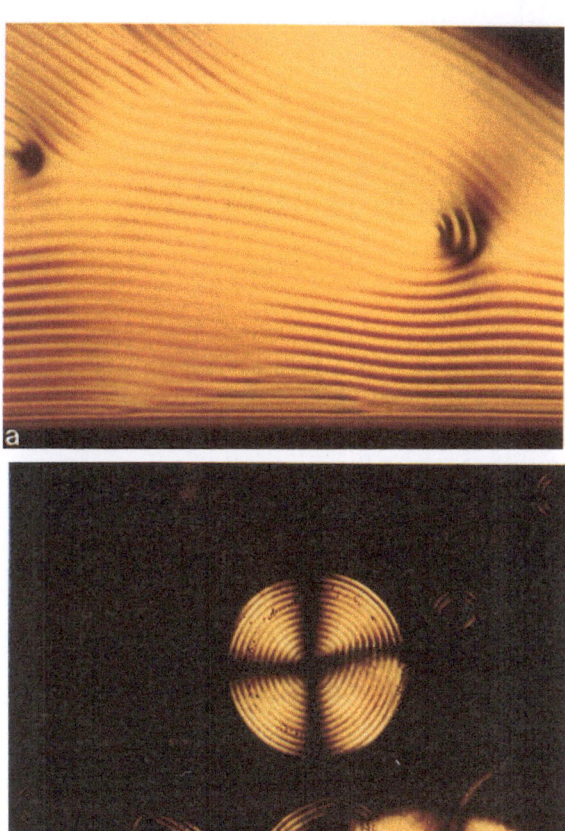

Fig. 4.13a, b. Microphotographs of ordered polyelectrolyte solutions in 0.1 cm thick optical cells between horizontal and vertical polarisers using white light. (**a**) Solution of the rod-like bacteriophage fd at 20 mg/ml in 10 mM Tris HCl, pH 7.5. The rods are ordered as in a cholesteric liquid crystal (Fig. 4.14) with lateral next-neighbour distances of ∼33 nm (∼5.5 diameters) on the average here. The cholesteric period (2D) is 61 μm [4.149]. (**b**) Helical semi-rigid xanthan molecules also form a cholesteric polyelectrolytic phase; at a concentration of 116 mg/ml the period (2D) is about 12 μm. Between the first and second critical concentration for the onset and full development of the cholesteric phase, respectively, a phase separation is observed and anisotropic droplets, surrounded by less concentrated isotropic solution, can be seen. They have a spherical shape apparently due to surface tension, resulting in a very distorted twisted structure similar to the "spherolites" in solutions of PBLG [4.139]. Note that a spherical structure with radial cholesteric axes everywhere is topologically impossible, in accordance with the observation of one radial line of defects in each droplet [4.156]

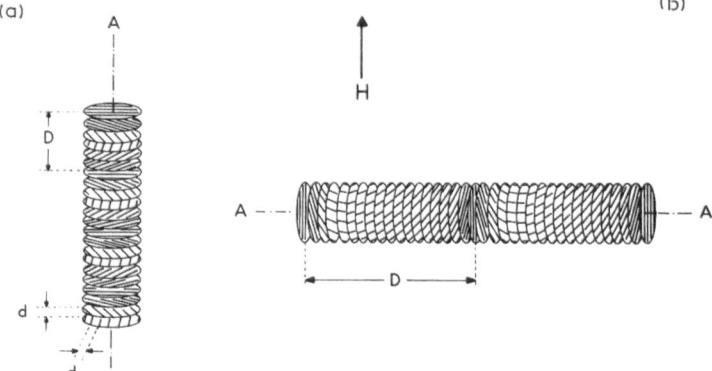

Fig. 4.14a, b. *Cholesteric phase* of rod-like polyelectrolytes in aqueous solution (see text for explanation)

rearranges, D increases and diverges at a critical field H_c, given [4.85] by

$$H_c = \frac{\pi^2}{2 D_0} \left(\frac{K_2 \cdot l_0}{\Delta\chi_0 L N} \right)^{1/2}$$

using the symbols introduced in Sect. 4.3.1; D_0 is the optical periodicity in zero field and K_2 is Frank's elastic constant for twist deformation, which can be measured through H_c.

In thin samples with dominating wall effects the cholesteric axes can be forced to maintain certain directions with respect to the magnetic field. Since magnetic orientation of the cholesteric as a whole is thus prevented, K_2 or other elastic constants can eventually be measured even for molecules with $\Delta\chi < 0$.

The magnetic field is thus useful in several ways: to determine the sign of the diamagnetic anisotropy of the molecules in the ordered phase; to deduce information about the molecular order by observing the magnetic alignment and deformation behaviour under various boundary conditions, to determine elastic constants and to keep a sample macroscopically aligned. For example, the cholesteric solutions of fd belong to case (b), Fig. 4.14 [4.149]: the cholesteric axes orient perpendicular to the magnetic field ($\Delta\chi > 0$) in agreement with the observed [4.3, 5, 92, 148] tendency of the individual virus to orient along H. Because of the large diamagnetic anisotropy [4.92] and the small twist elastic constant K_2 ($\sim 10^{-8}$ dyn), the critical field for untwisting turns out to be small, amounting only to ~ 0.8 T for the sample in Fig. 4.13a [4.149]. This is of practical importance because the fully oriented ("forced") nematic state above H_c can be easily studied with a superconducting magnet.

As mentioned earlier, DNA and other double and triple helical *nucleic acids* – unlike bacteriophages and normal liquid crystals – orient perpendicular to H ($\Delta\chi < 0$). Long-range order and magnetic field effects in concentrated polyelectrolytic solutions of these molecules have been investigated by *Jizuka* et al. [4.152–155] and by *Sénéchal* et al. [4.3, 4]. The first group studied viscous high molecular weight samples in relatively weak fields ($<2.5\,T$) using magnetic birefringence, linear and circular dichroism, light-scattering and x-ray diffraction. The other group prepared almost rod-like fragments not much longer than the persistence length and obtained fairly liquid samples with better developed order, avoiding the chain entanglements; they used orienting fields up to 18.5 T and measured birefringence, small-angle neutron scattering and light diffraction. The observed critical concentrations are typically around 10 % w/w, which is an order of magnitude higher than for the bacteriophages, because for the semiflexible nucleic acids the length scale controlling the Onsager transition, given by twice the persistence length [4.141], is about an order of magnitude smaller (Sect. 4.3.2a) than the length of the rod-like phage. The short-range order turns out to be similar to that of the phages described above, but with much smaller mean next-neighbour distances ($\sim 4.7\,nm$) and thinner ionic cloud according to the higher concentration. The long-range periodicity is much shorter and best measured by Bragg diffraction of laser light (Fig. 4.15). For weak boundary conditions, the periodicity aligns parallel to a strong magnetic field just as expected for case (a) in Fig. 4.14.

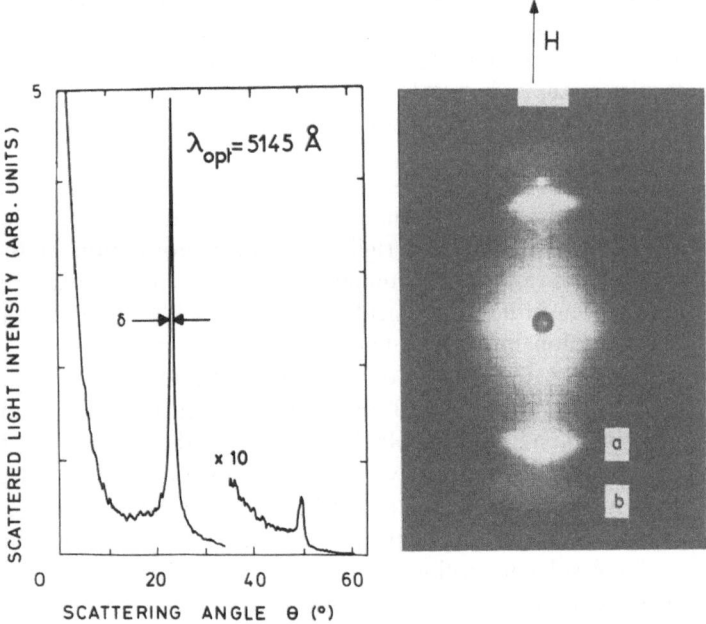

Fig. 4.15. A periodicity of long wavelength and very long coherence length apparently exists in aqueous solutions of semiflexible fragments of nucleic acids, amounting to 1.20 μm for Poly $A\cdot$ Poly U of mean molecular length $\sim 1700\,nm$, concentration $\sim 140\,mg/ml$, $4\times 10^{-3}\,M\,NaCl$, $T=20\,°C$, cell thickness 0.1 cm. It gives rise to the sharp diffraction of visible light and is oriented parallel to a strong magnetic field (13 T). (a) $\lambda=482.5\,nm$, (b) $\lambda=632.8\,nm$ [4.4]

Under strong boundary conditions, with magnetic field normal to the axis of the long period, *Jizuka* observed a magnetic field dependent increase of the long period, but a critical field could not be found. All the information quoted so far is consistent with a classical cholesteric lyotropic structure [4.154]. However, *Sénéchal* et al. [4.4] found that both the *Cotton-Mouton constant* and the high field *saturation birefringence are too small* to be accounted for by domain rotation and full high field alignment, respectively, of an ideal cholesteric. They suggested a cylindrical multishell model which is consistent with most of the data, but there is also agreement with a 3-dimensional twisted structure like *Robinson*'s spherolites [4.139].

In the examples discussed above, *repulsive* electrostatic long-range forces were important. It has been mentioned [4.4, 146, 147, 149, 150] that a small contribution of *attractive* forces due to electronic or ionic fluctuations cannot generally be neglected. In our view, the existence of such a force is the only reason for the twist between helical molecules at these large intermolecular distances.

II) Liquid Crystalline Polymer Solutions

Some of the concentrated polyelectrolytes treated in the last section are ("lyotropic") liquid crystals; they therefore essentially belong under this heading here. However, we want to distinguish between systems dominated by inter-molecular long-range electrostatic repulsion and more concentrated systems, usually in organic solvents, where more specific steric short-range interactions between the molecules or their chemical side groups also come into play. The macroscopic properties of these two classes of liquid crystalline polymers are often hardly distinguishable.

It is well known that concentrated solutions of some polypeptides can be nematic or cholesteric (with typical radial molecular separations of the order of 1–2 nm) when the solvent favours a rigid α-helical conformation of the peptide backbone. The long-range orientational correlation then implies strong magnetic orientability even in moderate fields. The presumably first observa-tion of magnetic orientation in such systems was in 1967; *Sobajima* [4.91] reported a strongly anisotropic motion of methylene chloride molecules as *solvent* for *poly-γ-benzyl-L-glutamate* (PBLG), deduced from high-resolution 1.4 T nuclear magnetic resonance. (No NMR signal from the polymer itself could be observed because of the strongly restricted segmental motion in the nematic phase, implying very broad polymer NMR lines.) Solvent molecules couple their orientation to that of the polypeptides. From the very long build up time (~100 min) of the solvent motional anisotropy after switching on the magnetic field, a slow reorientation of the α helices parallel to *H* was deduced. The orientation was confirmed by x-ray patterns of dry PBLG films prepared from the liquid crystalline state oriented in the magnetic field (Sect. 4.4). Sobajima's ideas were extended to other solvents and important information about the structure of the α helix was thus obtained, mainly by *Samulski* and co-workers [4.158–161].

An attempt to interpret the magnetic orientation in terms of diamagnetic anisotropies of the polymer constitutents was made in 1971/72: *Jizuka* and *Miyata* et al. [4.162–164] reported orientation perpendicular to *H* of the cholesteric helices of PBLG and magnetic untwisting. Similar but weaker effects were found in poly-γ-ethyl-*L*-glutamate (PELG) which contains no aromatic ring group; the authors concluded that besides the benzyl group the α-helical backbone is also diamagnetically anisotropic [4.165]. Interestingly, the observation of an overall orientation of PBLG *parallel* to the magnetic field excluded the existence of any side chain conformation in PBLG with the aromatic planes perpendicular or nearly perpendicular to the α-helical backbone. (Even any free rotation of the aromatic planes around an axis perpendicular to the backbone seems unlikely. In all these cases the intrinsic anisotropy of the backbone would be overcompensated by the mean side group anisotropy and an overall orientation of PBLG *perpendicular* to *H* should occur.) The observed *parallel* magnetic orientation is consistent with a recent model for the side chain conformation proposed by *Samulski* based on NMR data from magnetically oriented PBLG [4.166]. In addition these data give some evidence for interchain intercalation of the benzyl groups in liquid crystalline solutions of PBLG. *Jizuka* [4.167] and *Tohyama* and *Miyata* [4.82] gave $\Delta\chi$ values per α-helical peptide bond (Sect. 4.2), the latter deduced from direct measurements of the macroscopic magnetic torque on a previously oriented PBLG sample tilted 45° from the field direction. *Guha-Shridhar* et al. [4.168] measured the diamagnetic anisotropy of nematic PBLG with a magnetometer; their $\Delta\chi$ values led to the determination of the three principal elastic constants of lyotropic PBLG from optical observations of magnetic untwisting and from birefringence measurements mainly by *DuPré* and co-workers [4.168, 169]. From NMR of the weakly oriented solvent (H_2O), *Finer* and *Darke* [4.170] concluded that also the α-helical poly-L-lysine hydrobromide can be oriented in moderate magnetic fields.

The appearance of liquid crystalline phases in all of the polymer solutions discussed above is essentially due to the high rigidity of the polymeric backbone and intermolecular interactions ocurring at high concentrations. *Grossberg* et al. recently described the interesting idea that also flexible polymers at any dilution could be liquid crystalline: if the polymer contains rod-like so-called mesogenic groups (which as monomers would form a neat liquid crystal) and flexible spacers (Fig. 4.8), each individual polymer chain could collapse into a liquid crystalline globule at sufficiently small temperatures and favourable solvent conditions. Small globules forming at intermediate chain lengths should be very anisotropic, whereas large globules at high molecular weights should become isotropic [4.141]. A large number of such polymers has been synthesised in recent years and many display liquid crystallinity (*thermo*tropism) in the *melt* [4.171–173] (see the following section). Anisotropic globules in solution, clearly detectable by Cotton-Mouton experiments at different molecular weights, have not been found so far [4.112, 174].

III) Isotropic and Liquid Crystalline Polymer Melts

We now consider polymeric liquids *without solvent*. Much of the effort in polymer physics is related to studies of the *conformation of individual polymer chains* and their *correlation with neighbouring chains* in the liquid, amorphous and, more recently, liquid crystalline states.

As in the previous sections, strong magnetic fields are useful here mainly in two respects: to detect (small) orientational correlations of segments in the disordered melts via the Cotton-Mouton effect and, for highly correlated liquid crystalline systems, to produce strong orientation. In the latter case it should be possible to observe many of the magnetic field effects known for low molecular weight liquid crystals, but the field strength needed to produce the effects on a reasonable time scale could be substantially higher due to the generally elevated viscosities of the polymeric systems.

The local order of chains in the amorphous state is a debatable topic. Many contradictory models have been developed in the past; the one extreme *(Random-Coil model)* assumes a totally random configuration of the chain in the amorphous state as in the melt, just like the configuration in dilute solution; the other *(Bundle model)* predicts a short-range parallel order of the chains extending over some 1 to 10 nm [4.175]. Various techniques have been applied to solve this problem [4.108 and references therein]. Obviously magnetic orientation studies carried out directly in the amorphous state are of very limited significance because of the restricted freedom of any segments or bundles to rotate under field. There are, however, generally precursors of orientational correlations in the liquid state when cooled towards the transition temperature. These pretransitional effects are well known in liquid crystals above the isotropic to nematic transition (Sect. 4.3.3a). It can thus be expected that the existence of locally parallel chains in the amorphous state should be evident by a typical increase of the Cotton-Mouton constant when approaching the glass temperature T_G from above.

No such increase was found in melts of polystyrene [4.107, 176]. Above T_G the Cotton-Mouton constant stayed at an essentially temperature-independent small value which agrees well with the rotational isomeric state model of chains in the melt [4.107]. Below T_G CM is about 4 times smaller. This has been interpreted as a random coil conformation with high backbone flexibility and restricted but finite side group mobility in the amorphous state [4.107].

Extended Cotton-Mouton experiments have been carried out on melts of short *n-alcane chains* and *polyethylene* (Fig. 4.4a) [4.107, 108, 177]. A comparison of the results with experimental Cotton-Mouton constants from alcanes in solution demonstrates the existence of some short-range correlations in these melts. *Champion* et al. [4.177] pointed out difficulties in interpreting the data solely in terms of intermolecular correlations. However, CM shows a strong increase with decreasing temperature which coincides with the pretransitional $(T - T^*)^{-1}$ behaviour of liquid crystals. Since T^* turned out to be nearly independent of the molecular weight, the existence of pretransitional correlations of trans-segments has been proposed [4.107, 108].

During the last years, liquid crystalline phases were discovered in melts of a number of special polymers. As mentioned in Sect. 4.3.2a, b, Subsection II, their monomers contain a rigid *"mesogenic" group* located in the main chain (Fig. 4.8) or in a side group and are linked by flexible spacers. By an appropriate choice of mesogenic group and spacer the temperature range of the liquid-crystalline phase(s) could be made substantial (some 50–100 °C) and kept well below temperatures of chemical decomposition. Thus these polymers became interesting in several respects. For example, an academic question concerns the microscopic structural order which results from the competing effects of anisotropic interactions between mesogenic segments and the intrinsic tendency of the polymer chain to avoid extended configurations. Elastic properties should differ from those of low molecular weight liquid crystals. Potential industrial applications range from the possible production of strongly oriented high modulus polymer fibres to special optical coatings, for example obtained by freezing a cholesteric texture in a polymer film. Several reviews have appeared recently [4.171–173].

Magnetic orientation in nematic phases of various of these polymers has been studied in several laboratories since 1980 [4.174, 178–185]. For some of the polymers with mesogenic groups in the main chain strong orientation along H of at least the mesogenic groups was observed in fields around 1 T. This was clearly demonstrated by x-ray diffraction (Fig. 4.21 below) [4.178, 183, 185], NMR [4.182] and measurements of the diamagnetic anisotropy [4.180, 184]. Substantial new information about the structure of the nematic phases was thus obtained. There appears to be, however, a critical field below which no orientation occurs and, as do speed and final degree of alignment, it depends on the polymer's properties such as molecular weight, lenght of the spacer group and flexibility. Cholesteric polymers could not be aligned even at 12 T, underlining the importance of stronger fields for the study of a wider variety of these polymers. Interestingly, the experiments mentioned above revealed orientational order parameters of the mesogenic main-chain groups in the nematic state of about 0.6 to 0.8, which is some 50 % to 100 % higher than values for low molecular weight nematics. The property – promising for applications – can possibly be related to the particular isotropic-nematic transition behaviour found [4.182] in some of the polyesters. As outlined in Fig. 4.16, the polymer exhibits a pretransitional increase of the Cotton-Mouton constant $(T-T^*)^{-1}$ like classical liquid crystals such as PAA, but the difference between T^* and the real transition temperature T_c (clearing point) is much larger. This combined with a high value of the latent heat at T_c indicates that the polymeric transition is "much more first order"; following the Landau-de Gennes description of the transition it also leads to the prediction of a high-order parameter in the nematic phase. The values thus deduced agree with those derived from NMR [4.181, 182].

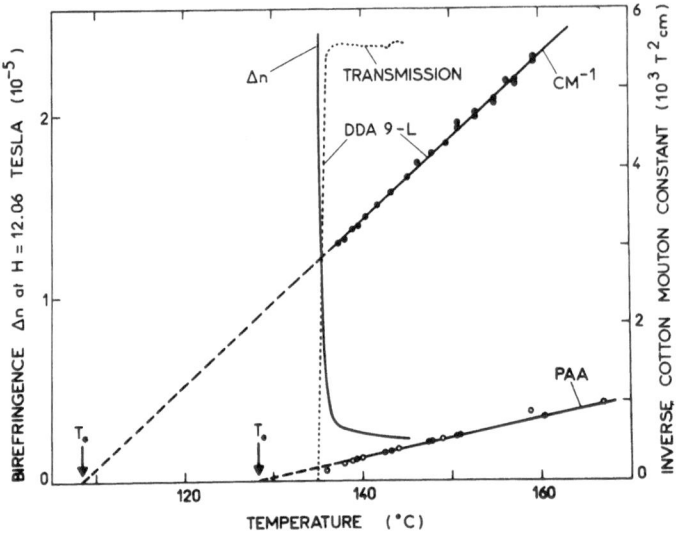

Fig. 4.16. The pre-transitional effects of a melt of the polyester DDA 9-L (Fig. 4.8) in comparison to a classical liquid crystal, *p*-azoxy-anisole $(CH_3-O-\langle O \rangle-N=N-\langle O \rangle-O-CH_3)$. Above the isotropic-nematic transition temperature T_c ($\simeq 132\,^\circ$C for PAA and $\simeq 135\,^\circ$C for DDA 9-L) the Cotton-Mouton constant (● data [4.182]; ○ data from [4.187]) of both samples is proportional to $(T-T^*)^{-1}$ over an extended temperature range. T^* is evaluated by linear extrapolation and T_c determined by the onset of strong turbidity, resulting in a sudden drop of optical transmission ($\cdots$) and by the divergence of the birefringence in a constant magnetic field (——). $T_c - T^*$ is much larger for the polymer ($\sim 27\,^\circ$C) than for PAA ($\sim 3.4^\circ$). Similarly the latent heat of the polymer at the transition exceeds that of PAA by about 5.4 times. Using the magnetic birefringence and calorimetric data the absolute values and temperature variation of the order parameter at *both* sides of the transition can be calculated using the *Landau-de Gennes* theory [4.85, 105]. The values thus predicted for the nematic phase of the polymer agree with the large experimental values cited in the text [4.182]

4.3.3 Classical Liquid Crystals

A large number of various anisometric – rod-like or disc-like – small molecules exhibit liquid phases with some orientational and translational long-range order at temperatures between the crystalline and isotropic liquid state; therefore they are called thermotropic liquid crystals. An example cited above (Fig. 4.16) is PAA. With molecular diamagnetic anisotropies typically like benzene and large correlation lengths in the liquid crystalline phases, they show many magnetic orientation effects, sometimes in fields far below 1 T which have been extensively described [4.1, 85, 157, 186]. Since this chapter deals mainly with *macro-molecules*, only some recent experiments involving high fields will be mentioned.

a) Pretransitional Effects in the Isotropic Phase

When the isotropic phase of a liquid crystal is cooled towards the nematic transition, increasing pretransitional orientational correlations occur very similar to the spin correlations in a paramagnetic substance above the Curie temperature. They result in an enhanced magnetic birefringence, observed by *Tsvetkov* already in 1944 [4.187]. *De Gennes* [4.85] pointed out that in Landau's mean field approximation, the Cotton-Mouton constant which is a measure of the local order parameter should, like the Curie-Weiss susceptibility, be proportional to $(T - T^*)^{-1}$, where T^* is the temperature of a *second*-order transition. This behaviour and small deviations from it occurring close ($\sim 2\,°C$) to the thermodynamically manifested first-order isotropic-nematic transition temperature T_c were accurately measured by *Stinson* and *Litster* in 1970 [4.188]; T^* is about 1° below T_c, therefore the isotropic-nematic transition is sometimes called a "weakly" first-order transition. Later birefringence measurements in high fields enabled the range of the pretransitional effects to be determined on the high-temperature side where CM is small [4.189, 190]. Also $T_c - T^*$ could be evaluated more precisely and the deviations of the order parameter from the $(T - T^*)^{-1}$ behaviour closer (a few tenths of a degree) to T_c [4.191] were evaluated. Even in the vicinity of T_c, CM exceeds the high temperature and solution values only by about 2 orders of magnitude. In terms of the number N_1 of correlated molecules (defined in Sect. 3.1) this means that the orientational correlation is still fairly short range ($N_1 \sim 10^2$). Thus the degree of orientation ($\sim N_1 \Delta \chi H^2 / kT$) remains small even in a field of 10 T. This is also reflected by the strictly linear relationship between Δn and H^2 found by *Poggi* and co-workers over all field strengths investigated (< 13.5 T), and by the high field value of Δn ($\sim 10^{-4}$) which is small compared to Δn values of completely oriented nematics ($\Delta n_{sat} \sim 10^{-1}$). In contrast to these results, *Keyes* and *Shane* [4.192] recently reported a linear H dependence of $(\Delta n / H^2)$ up to 10 T particulary pronounced near T_c. No explanation has yet been offered for this experimental disagreement.

It is obvious from simple thermodynamic considerations that T_c *itself should depend on the magnetic field strength:* since in, say, 10 T the isotropic and nematic phases are very weakly and strongly oriented, respectively, the relative shift of T_c is of the order $\Delta \chi H^2 / Q$, with $\Delta \chi$ being the diamagnetic anisotropy of the nematic phase and Q the latent heat [4.193]. This relation has been verified experimentally by *Rosenblatt* [4.193] in strong magnetic fields; the absolute shift amounts to only $5 \times 10^{-3}\,°C$ at 15 T.

b) Orientational Fluctuations in the Nematic Phase

In the nematic phase the orientational correlation lengths are macroscopic. Hence almost completely oriented samples form spontaneously in small containers with walls treated so as to impose a direction to those molecules in contact with the wall. The intrinsic (zero-field) birefringence of such samples is very large, typically of the order of $\Delta n \sim 10^{-1}$. There are, however, thermal fluctuations of the direction of the long axes of the molecules around the average (macroscopic) axis of alignment. Their amplitudes can be slightly reduced in a strong magnetic field. The corresponding small increase of birefringence Δ

(typically some 10^{-5} at 10 T) was theoretically predicted by *de Gennes* [4.85] and observed by *Poggi* and *Filippini* [4.194]. In this case Δ is proportional to the magnetic field, essentially because fluctuations with wavelengths longer than the magnetic correlation length ξ are quenched [4.85]. This ξ is a characteristic length resulting from a balance between magnetic and elastic torques, roughly corresponding to the shortest distance between a molecule clamped in a given direction and molecules parallel to the magnetic field. Depending on the geometry (i) it equals

$$\xi = (K_i/\Delta\chi)^{\frac{1}{2}} H^{-1},$$

with K_i the elastic constants for splay ($i=1$), twist ($i=2$) or bend ($i=3$), respectively. Since for small changes in ξ the birefringence increment Δ must be proportional to ξ^{-1}, it follows that $\Delta \sim H$. *Malraison* et al. [4.195, 196] extended *de Gennes'* theory [4.85], originally developed for a mean elastic constant, to the more general case of different splay, twist and bend elastic constants. They also measured Δ/H over the complete nematic temperature range. Including the temperature-dependent values of the elastic constants determined by other techniques they found good agreement between their theory and experiments, except for temperatures near the nematic isotropic transition. Because of the naturally limited sample length and the small Δ values, such experiments necessitate the combined application of strong fields and sensitive birefringence techniques.

The orientational order parameter of nematic liquid crystals has been estimated from measurements of the anisotropy of the diamagnetic susceptibility of a nematic glass produced by quenching a liquid sample in a strong magnetic field [4.197].

c) Untwisting in the Cholesteric Phase

As mentioned above, the coherence length ξ is proportional to H^{-1}. Transitions of *nematic* into aligned phases usually occur at a critical magnetic field H_c above which $\xi(H)$ is smaller than a typical dimension of the sample, for example the distance between the confining glass plates which can be around 100 μ. With common values for K_i and $\Delta\chi$, H_c ranges well below or around 1 T. In *cholesterics*, however, the equivalent typical dimension is the pitch. Because the pitch in normal cholesterics is generally comparable to optical wavelengths or even smaller, the critical fields for untwisting are somewhat above presently available field strengths. Therefore magnetic untwisting has been observed so far only in special cholesterics with large pitches, such as the α-helical polypeptides, concentrated solutions of virus (Sect. 4.3.2b, Subsection II) or solutions of cholesterics in nematics [4.1, 85, 115]. Certainly, also some of the other magnetic field effects often discussed could become observable in very strong fields (~ 30 T) for a number of cholesteric materials, depending on the sign of $\Delta\chi$, the size of $\Delta\chi/K_i$ or the influence of the wall.

4.3.4 Membranes and Micelles

a) Background

We now turn to so-called amphiphilic molecules, which consist of a polar head group soluble in water (hydrophilic) and an aliphatic tail (hydrophobic). Therefore, when suspended in water at low concentration, they form organised aggregates such as double layers (membranes), vesicles or micelles in order to maximize both exposure of head groups to water and mutual contact of tails. Similarly inverted structures might occur in oil. In oil/water mixtures the amphiphiles tend to cover the interface, thus acting as detergents. At higher concentrations the aggregates can form various ordered or sometimes liquid crystalline phases, depending on chemical composition, solvents, presence of ions or of other molecules like alcohols, temperature, etc. In contrast to the polymeric systems, both the number of molecules contained in these objects (corresponding to their molecular weight) as well as their shape are often variable and frequently not known. This and their multicomponent nature usually make structural investigations of these systems sometimes more complex than for lyotropic polymers or thermotropic liquid crystals. Organised biological membranes containing proteins in the amphiphilic layer are treated separately in Sect. 4.3.4c.

b) Artificial Multilayers and Micelles

I) Expected Magnetic Field Effects

The number of correlated molecules within such systems can be enormous and substantial magnetic orientation effects are expected, although quite often groups with large $\Delta\chi$ values like aromatics are missing. A major contribution to $\Delta\chi$ might originate from the more or less ordered hydrocarbons in the hydrophobic layer. Some values are compared in Table 4.3. The uppermost value results from a simple estimate of the difference between normal and planar χ values for a sheet of normally oriented all-trans $(CH_2)_{18}$ chains using the $C-C$

Table 4.3. Diamagnetic anisotropy[a] of oriented hydrocarbon systems and bilayer membranes

C_{18}	-6.5 ($\pm 25\%$)	Estimated from $C-C$ bond anisotropy
Partially crystalline polyethylene $(CH_2)_x$	-1	[4.198]
Crystalline stearic acid $CH_3(CH_2)_{16}COOH$	-9.7 ($\pm 20\%$)	[4.64]
Crystalline dipalmithoyl--L-α-lecithin	-9	[4.198]
Egg lecithin membranes	-0.27	[4.199]

[a] Volumetric susceptibilities in 10^{-8} cgs units. The densities of all systems mentioned are very close to 1 g/cm^3

bond anisotropies (neglecting $C-H$) cited in Sect. 4.2. It is the same order of magnitude as experimental values from stearic acid crystals. These values should be considered as upper limits for a pure hydrocarbon layer. Orientational disorder lowers $\Delta\chi$ substantially, as can be seen by comparison with partially crystalline polyethylene containing disordered amorphous regions. It is seen from the sign of $\Delta\chi$ that individual extended hydrocarbon molecules tend to align perpendicular to H and hence the plane of a layer or membrane aligns parallel to H. Dipalmitoyl lecithin (DPL), a molecule consisting of two hydrocarbon chains and a polar head group, aggregates in water into bilayer membranes which are widely used as models for biological membranes. *In crystals* of DPL, bilayers are formed with parallel hydrocarbon molecules; the elongated head groups are roughly perpendicular to the hydrocabon long axis and all point in a given direction in the bilayer plane. Thus the crystals are biaxial. From the observed *biaxial* orientation behaviour of such crystals in a magnetic field, *Sakurai* et al. [4.198] estimated the normal-planar anisotropy of DPL bilayers (Table 4.3) and deduced that the head group has a remarkable diamagnetic anisotropy. They did not evaluate the head group anisotropy quantitatively, however. Interestingly, the normal-tangential anisotropy of an egg lecithin membrane, as derived from the slow alignment along H of large elongated vesicles, turned out to be much smaller than the above values derived from crystals [4.199]. This can be accounted for primarily by the substantial molecular disorder in the egg lecithin membranes, which consist of a mixture of various different amphiphilic molecules. The egg lecithin anisotropy is more than two orders of magnitude smaller than the usual anisotropy in liquid crystals ($\Delta\chi \sim$ some 10^{-7}).

Nevertheless strong orientation of such hydrocarbon systems must occur in high fields. Using, for example, as typical values $\Delta\chi = 2 \times 10^{-8}$, a layer thickness d of 5 nm, $H = 20\,T$, $T = 300\,K$, orientation would become substantial (i.e. $\Delta\chi H^2 aL^2/kT > 1$) for flat rigid layers with area L^2 larger than 100×100 nm^2, or for correspondingly smaller multilayer stacks. Ordered rigid objects larger than this are obviously quite common in many emulsions or biological systems. In addition, the presence of aromatic (head) groups or protein might substantially enhance $\Delta\chi$ numerically and even change its sign. Magnetic orientation and birefringence of rigid non-interacting objects in suspension such as disk-, tube- or cigar-shaped vesicles can be described straightforwardly by the equations in Sect. 4.3.1. All one needs to do is to calculate the effective anisotropies of each object by geometrical summing over molecular anisotropies or over the effective values per unit area. In this way a magnetic birefringence experiment yields *information about the anisotropic shape of the object*, if the molecular quantities are known, or vice versa. In case the refractive index inside the object differs from the solvent index (for example, in oil-detergent-water systems), shape birefringence might become important compared with the intrinsic birefringence of the anisotropic shell [4.200].

Spherical vesicles or micelles obviously do not orient. However, the possibility of magnetic deformation into elongated ellipsoids has been theoretically discussed by *Helfrich* [4.201, 202]. From a balance of the magnetic and elastic torque on the surface elements, he derived that the relative deformation should be $\Delta r/r = \Delta \chi H^2 a r^2/12\kappa$ for small deformations; r is the unperturbed radius of the object and κ the bending elastic modulus of the shell. As defined by the elastic energy $\varepsilon = \kappa/2r^2$ per surface element for an isotropic elastic medium, κ is related to the bulk elastic constant K by $\kappa = Ka$. The magnetic birefringence due to deformation is of the order of

$$\Delta n \sim \Delta n_{\text{sat}} \cdot \frac{\Delta r}{r}$$

with typically $\Delta \chi = 2 \times 10^{-8}$, $H = 20$ T, $a = 5$ nm, $r = 25$ nm, $\Delta n_{\text{sat}} \sim 10^{-4}$ at a volume fraction of 10^{-2} and $\kappa = 10^{-13}$ erg for liquid crystals we estimate $\Delta r/r \sim 2 \times 10^{-3}$ and $\Delta n \sim 2 \times 10^{-7}$. Thus, though the deformation is small, it should be detectable in strong magnetic fields and *the bending elasticity of the microscopic objects can be directly determined*, if κ is not too large.

So far we have treated rigid particles and have omitted thermal fluctuations of shape. These are important when the elastic energy εr^2 in the sphere becomes smaller than kT, hence for $\kappa \gtrsim kT$. This situation possibly occurs for example in thin layers of detergent molecules in the presence of alcohols which seem to reduce the longe-range order [4.200]. In analogy to the persistence length in polymers (Sect. 4.3.1), *de Gennes* [4.200] introduced the idea of a persistence area ξ^2. In this description the flexible layer would be replaced by rigid plates of length ξ linked by ideally flexible joints. Here ξ should be of the order $\xi = e \exp(2\pi\kappa/kT)$, with e being a molecular length scale, for example the diameter of an amphiphilic molecule [4.200]. Therefore the number of molecules contained in a persistence area is about ξ^2/e^2 and magnetic orientation and the Cotton-Mouton constant are expected to be proportional to ξ^2/e^2; in fact $\xi^2/e^2 \simeq N$, as defined in Sect. 4.3.1 [4.203]. This concept, which has not been verified experimentally yet, is interesting because of the sensitive (exponential) dependence of magnetic orientability on K. It can be applied to any semiflexible layer system, i.e. if the extension of the objects is much larger than ξ. In this way κ could be determined also for very flexible vesicles or micelles, for which $\xi \ll r$. The region of cross-over ($\xi \sim r$) between flexible and rigid ($\xi \gg r$) vesicles is more complex and the magnetic orientation and deformation behaviour has not yet been worked out.

In the preceding section we considered the membrane as a layer with maximal symmetry, the symmetry axis being identical to the normal of the plane. However, bilayer membranes often have a complicated internal structure; phases are known involving tilting of the molecular long axes with respect to the normal (Fig. 4.17) or an in-plane orientation of the head groups. These phases are biaxial at least on a local scale and a large membrane can be considered as a planar ensemble of biaxial domains. Thus even if magnetic alignment of the membrane as a whole is prevented, for example in a multilayer sandwich-type sample confined between planar glass plates, magnetic orientation of the

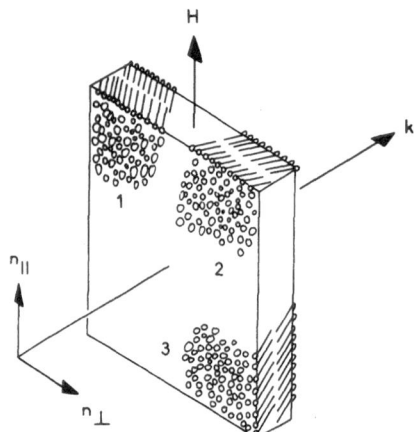

Fig. 4.17. In a membrane containing domains of tilted molecules, magnetic orientation could occur due to in-plane rotation of domains, e.g. if the membrane plane is fixed by wall effects parallel to H, domains 1 and 2, (3), have favourable (unfavourable) orientation; hence domain 3 rotates around k by 90° in a magnetic field sufficiently strong to bring the molecular long axes normal to H. This results in a birefringence $\Delta n = n_\parallel - n_\perp$. The orientation mechanism was first proposed by *Gaffney* and *McConnel* [4.204]. In-plane anisotropy can also result from a cooperative tilt of anisotropic head groups [4.205]

domains could occur as depicted in Fig. 4.17. The correlation lengths (defining the domain size) for the tilt of tails and heads, respectively, need not necessarily be related; they might also differ from ξ.

Thus quite a number of interesting magnetic field effects can be expected in emulsions of amphiphilic molecules.

II) Experimental Observations

We treat magnetic orientation studies of detergent containing systems and artificial membranes (phospholipid-layers) in two separate sections, because the latter are related to biological membranes.

In detergent-containing systems, magnetic orientation was first realized by *Lawson* and *Flautt* in 1967 [4.206]. Mixtures of typically 50 % water (D_2O), 42 % soap (sodium decyl sulphonate, SDS), 4% alcohol and 4% salt (Na_2SO_4) showed deuterium NMR spectra much like those in a magnetically oriented nematic liquid crystal. Since many such systems at elevated concentrations display some liquid-crystal like order, the ease of magnetic orientation of these phases in low fields has been frequently used since [4.207–211 and references therein]. This was most important for structural aspects, for example to decide whether the anisotropic objects forming the liquid crystalline phase were either disc-like or rod-like [4.207, 209]. Unfortunately a somewhat confusing classification and nomenclature was introduced: systems with alignment of the rotational symmetry axis of the aggregates parallel (or perpendicular) to H have been called "type I" (or "type II") or "positive" ("negative") diamagnetic anisotropy. Type I (type II) was usually associated with rod-like (disc-like) aggregates, which is obviously true only for layers with negative normal-tangential anisotropy such as the hydrocarbon-layers in Table 4.3. The reverse situation of positive normal-tangential anisotropy can easily occur when the molecules contain oriented aromatic groups. Another aspect concerned the possibility of using these phases as oriented matrices *for aligning a wide variety of small host molecules* as to whether they were soluble in an aqueous or aliphatic milieu. Periodic textures were observed to occur in a magnetic field and interpreted by an orientation-induced flow process [4.210]. It is noteworthy that addition of small polymer-coated ferromagnetic particles ("ferrofluids") to the emulsion seems to enable magnetic orientation [4.212] and deformation [4.213] already in very weak fields.

A substantial amount of intermolecular order might persist in detergent-containing systems even under conditions where turbidity is essentially absent and no liquid crystallinity exists. For example, soap molecules in pure water with or without salt form micelles which can adopt various sizes and shapes; oil-water-soap emulsions often turn transparent when a certain amount of alcohol is added, and are then called microemulsions. Structural investigations of such phases are important for the development of specific applications of soap. *Porte* and co-workers [4.214–217] studied the polymorphism and interactions of micelles of different soaps in dilute aqueous solution by high field magnetic birefringence. The soap molecules carried aromatic groups in the head region to enhance both diamagnetic and optic anisotropy. These measurements clearly demonstrated the existence of four distinct concentration regimes for sodium octyl benzene sulphonate in water. They have been interpreted in terms of *magnetic orientation of* free monomers and *more or less anisotropic micelles* [4.214, 215]. Possible contributions from other birefringence-producing mechanisms such as those outlined in (Sect. 4.3.4b, Subsection I) were not discussed. The experiments on cetylpyridium with NaBr in water are consistent with a worm-like model of fairly elongated micelles. Through a combination with light-scattering experiments a persistence length of about 20 nm at a micellar diameter of 6 nm was derived [4.216, 217].

Series of magnetic birefringence measurements were also performed on microemulsions [4.218]. Over a wide range in the phase diagram (of Na octylbenzene sulphonate, pentanol, decane and water) the Cotton-Mouton constant turned out to be small and consistent with an orientation of weakly anisotropic micelles, or of persistence areas, or with Helfrich's magnetic deformation mechanism. An extension of these measurements to monodisperse micelles at a series of calibrated radii, and eventually an independent determination of the yet unknown value of κ, should yield a better understanding of the behaviour of microemulsions in strong magnetic fields.

Large numbers of synthetic and natural lecithins or phospholipids in aqueous suspension spontaneously form bilayer structures (Fig. 4.17). They are widely studied as simple model structures (without proteins) for biological membranes. Many experiments are actually carried out on sandwich-like multibilayer stacks containing approximately equal amounts of lipid and water. Coplanar orientation of the bilayers is obtained between parallel glass plates; if more dilute emulsions are homogenised by ultrasound the bilayers form closed vesicles and samples of monobilayer vesicles with fairly narrow distribution of diameters can be prepared. Diameters might range from some 20 to 100 nm.

Under other conditions larger multibilayer "onion-like" vesicles are the most stable structure.

It was on oriented stacks of phospholipid bilayers that *Gaffney* and *McConnell* [4.204] in 1974 first found a direct magnetic field effect. They observed an anomalous spin label ESR signal for the planar orientation of *H* (Fig. 4.17). The ESR signal could not be fitted by any orientational distribution of the lipid molecules with rotational symmetry around an axis normal to the membrane. They proposed the mechanism of *in-plane rotation of tilted domains*, outlined in Fig. 4.17, which was consistent with the ESR results. Later, a small magnetic birefringence signal was observed [4.129] on multilayer stacks of egg lecithin and dimyristoyl lecithin, when the light beam pointed normal to the stack (Fig. 4.17). These data also agree with the above mechanism. Interpreted by a rotation of the tilted hydrocarbon chains and neglecting any anisotropy of the head, they would result in a mean domain size of the order of 100×100 nm^2. However, a recent ESR study [4.219] on spin labels attached to the head group of dipalmitoyl lecithin provided evidence of a *cooperative rotation of the head groups* in a magnetic field. Thus the magnetic birefringence data should be reinterpreted including head group orientation and anisotropies; the latter could be estimated from bond anisotropies.

In Sect. 4.3.4b, Subsection I, was outlined that suspended vesicles should have more freedom than fixed membrane stacks to reorient in a strong magnetic field. The possibility of measuring elastic constants, orientation correlation lengths and small anisomeries as well as their relevance for biomagnetic effects make such studies an interesting topic in spite of their complexity. The actual knowledge is fairly scarce; we are aware of only two experiments.

Boroske and *Helfrich* [4.199] observed with a phase contrast microscope that large cigar-shaped (~ 30 μm length and ~ 10 μm diameter) vesicles of egg lecithin slowly oriented parallel to a 1.5 T field. They deduced from the temporal sequence of the alignment the normal/tangential anisotropy of the volume susceptibility, which turned out to be rather small (Table 4.3). This can be explained by tilting, disordered lipid conformations or reorientations of tilted domains within the membrane. A magnetic deformation could not be found probably because of the use of *multi*layer membranes (large κ) and the difficulty of detecting small deformations with a microscope. High field birefringence experiments [4.6, 129, 220] were carried out on aqueous emulsions of dipalmitoyl lecithin, containing small (~ 60 nm) almost spherical bilayer vesicles and some larger onion-like multilayer vesicles of irregular shape. An electron microphotograph of such a sample (after drying on a substrate and staining) is shown in Fig. 4.18. Partial saturation of Δn was observed to occur in high magnetic fields, indicating that some of the larger aggregates can be completely oriented. Figure 4.18 shows the derivative of Δn with respect to H^2 measured at a fixed intermediate field strength by superimposing a small field modulation. It contains contributions from the small and large vesicles. Most pronounced changes can be seen at the pretransition and main transition, where the lipid chains transform with increasing temperature from an extended tilted to an

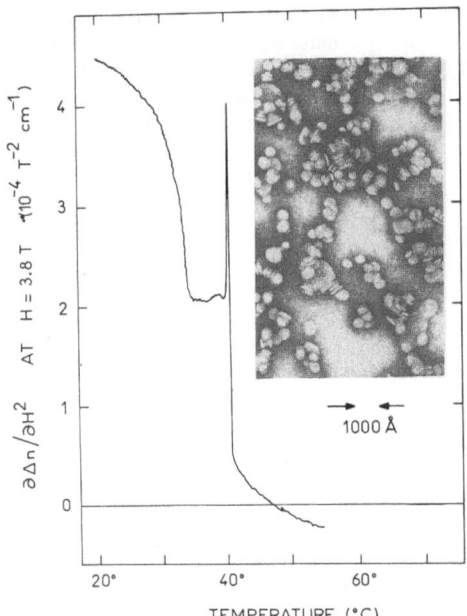

Fig. 4.18. Temperature dependence of the magnetic birefringence of suspensions of dipalmitoyl lecithin vesicles [$c = 20$ mg/ml, 5 mM La(NO$_3$)$_3$, $\lambda = 632.8$ nm, continuous variation of temperature at a rate of 0.5 K/min]. The magnetic birefringence suddenly changes at both the pretransition at 35 °C and the main transition at 41 °C of the lipid chains in the bilayer, so demonstrating the high sensitivity of this technique to detect transitions between phases of different lipid order [4.220].

extended non-tilted and to a disordered configuration. The magnetic anisotropy is related to the orientational order of the lipids. Since these transitions should also involve changes in orientational correlations, head group orientation, vesicles shape and κ, a quantitative interpretation of the detailed features of the data has not been made yet.

c) Organised Biological Membranes

Biological membranes consist of a phospholipid bilayer with various proteins embedded in or attached to it. The proteins control physicochemical and functional-biological properties of the membrane. Some proteins are known to cause structural changes within the neighbouring phospholipid bilayer, thereby influencing the bilayer's diamagnetic anisotropy; they can also contribute to the overall diamagnetic anisotropy of the biological membrane. In many biological systems large multilamellar arrays of biological membranes exist. It is therefore not surprising that already about 15 years ago full magnetic orientation in fields below 2 T was observed on outer *segments of retinal rods* [4.7] and *chloroplast*-containing cells of chlorella [4.88].

I) Outer Segments of Retinal Rods

The outer parts of retinal rods are cylindrical segments typically 50 μm long and 10 μm in diameter built up of a multilamellar stack of disc-shaped membranes. Membrane bilayers are perpendicular to the rod axis and consist to almost equal

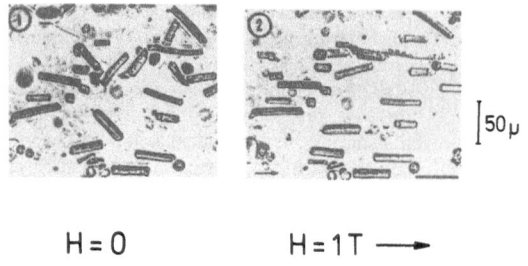

$]50\mu$

H = 0   H = 1T ⟶

Fig. 4.19. Magnetic orientation of outer segments of retinal rods in suspension can be observed already at modest fields $\lesssim 1.0\,T$. Their large diamagnetic anisotropy is mainly due to the highly ordered arrangement of α-helical proteins in the membrane stacks [4.7]

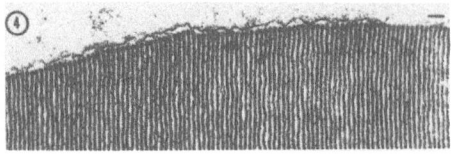

$]0.1\mu$

amounts of lipid and protein; about 85 % of the protein mass is the visual pigment rhodopsin. The early optical observation [4.7] that in suspension these segments aligned parallel to a magnetic field of 1 T (Fig. 4.19) has stimulated a number of studies, resulting in important information about the structural order of the disc membranes. From the time course of the orientation after switching on the magnetic field, *Chagneux* and *Chalazonitis* determined the diamagnetic anisotropy $\Delta\chi_R$ of the segments [4.221]: $\Delta\chi_R$ is completely suppressed by CO_2 gas [4.222] and decreases by about 20 % upon bleaching with intense light [4.223].

A better understanding of the large segmental anisotropy $\Delta\chi_R$ in terms of phospholipid bilayer and protein (aromatic amino acids, retinal chromophore and α-helical polypeptide chain) was initiated by the theoretical work of *Hong* et al. [4.90]. More recent contributions are from *Chabre* and co-workers [4.224, 225] as well as *Becker* et al. [4.226]. The *phospholipid membrane* cannot be the major contributor to $\Delta\chi_R$ since this would imply an orientation of the segments *perpendicular* to H, as pointed out in Sect. 4.3.4a. Nevertheless, its contribution is not negligibly small [4.227]. A contribution of the aromatic *retinal chromophore* of rhodopsin can presumably be neglected [4.224, 226]; as indicated by linear dichroism measurements [4.226] this group strongly reorients on bleaching, which would necessitate a bleach *enhancement* of $\Delta\chi_R$ in contrast to the measured decrease of $\Delta\chi_R$ [5.223, 224]. The magnetic linear dichroism data [4.224, 226] also show that on the average, only one out of 61 *aromatic amino acid residues* per rhodopsin molecule has a non-random orientation; it is aligned with the aromatic plane parallel to the segmental long axis. Therefore the aromatic groups all together account for only at most 15 % of $\Delta\chi_R$ [4.224]. *Chabre* pointed out that the major contribution to $\Delta\chi_R$ may originate from the *α-helical (≈ 50 %) part* of the rhodopsin molecules; the α helices seem to be oriented normal to the disc membrane plane [4.224]. This was confirmed by the observation of a high linear infrared dichroism from the $C=O$-bonds in the α helix of magnetically oriented segments [4.225]. Neutron small-angle scattering from oriented pellets, produced by sedimenting retinal segments in 1.7 T, has been used to determine which fraction of the rhodopsin molecule is embedded in the membrane and how it changes with bleaching [4.228]. Well-oriented samples for photochemical studies at low temperatures (77 K) could successfully be prepared only by quenching in a strong (17 T) magnetic field [4.229]. In this way disorienting effects due mainly to convection during cooling were overcome.

II) Systems Containing Chloroplasts

An important and rapidly developing field of biological and biophysical research concerns the molecular processes involved in the photosynthesis of plants. One particular structural aspect is the orientational order of the photosensitive molecules (chlorophyll) within the innercellular light-sensitive organelles, the chloroplasts. In 1971 *Geacintov* et al. [4.89] observed that the chlorophyll "*a*" fluorescence in a monocellular green alga (which contained only one chloroplast) was anisotropic in magnetic fields of 1.6 T. This is due to a strong alignment of the multilamellar disc-like chloroplast normal to *H*. A large number of mainly optical studies followed this observation, reviewed by *Hong* in 1976 [4.230] and theoretically described to some extent by *Knox* and *Davidovich* in 1978 [4.231]. We therefore restrict ourselves to a few remarks [4.230–243].

An attempt to explain the observed magnetic orientation of chloroplasts in terms of diamagnetic anisotropy of constituent molecules was made by *Geacintov* et al. From linear dichroism of rhodamin 6*G* attached to phospholipids they deduced that the membrane stacks orient normal to *H*; therefore the phospholipids contribute only weakly, if at all, to $\Delta\chi$. Besides some possible minor contribution from other lipids, carotenoid lipids, it is thus the proteins that determine $\Delta\chi$ [4.232]. The orientation relative to the membrane of various groups in the chlorophyll molecule (and their optical transition moments) has since been studied with many optical techniques mainly by *Breton* et al., though a satisfactory calculation of $\Delta\chi$ has not appeared yet. This might be partially due to uncertainties in the chloroplast's overall shape and to the variety of contributing molecular groups. Aromatic porphyrin rings of chlorophyll as well as aromatic tyrosines of the structural proteins tend to be perpendicular to the membrane plane, whereas α helices and carotenoid molecules are nearly in plane; other groups with unknown orientations possibly contribute to $\Delta\chi$ as well. Nevertheless, magnetic fields in the tesla range become a useful and easy tool for producing full alignment of chloroplast systems. Various studies of structural and optical aspects of photosynthesis have benefitted from this.

Neutron small-angle scattering by *Neugebauer* et al. [4.244] has shown that also bacterial chromatophores (planar pieces of purple membranes of the *Halobacterium halobium*) substantially orient in modest fields. As in retinal rods and chloroplasts, the membrane aligns normal to *H*. It was suggested by *Worcester* [4.79] that the ordered α helices of bacteriorhodopsin are responsible for the diamagnetic anisotropy of this system.

4.3.5 Other Biological Particles

Substantial progress in molecular biology is due to the knowledge of the atomic structure of biological particles like proteins as obtained mainly from x-ray work on crystalline samples. Unfortunately various bioparticles do not crystallise and different techniques (of comparably low structural resolution) have to be applied to gain information about their structure in a solution. Some techniques, for example NMR, ESR, Raman scattering or infrared absorption, are sensitive to local properties of single groups, others, like elastic light scattering, sedimentation and viscosity measurements, to the overall size and shape. In this section we wish to outline that magnetic birefringence and dichroism experiments, feasible even on small particles of modest $\Delta\chi$ in dilute solutions by using strong magnetic fields, can also yield *structural* information.

A *magnetic birefringence* experiment gives immediately the product $\Delta\chi \cdot \Delta\alpha$ for an isolated rigid particle (Sect. 4.3.1). Since measurements are usually carried out at wavelengths far from optical absorption bands, the birefringence originating from various anisotropic groups within the particle thus is determined by the anisotropies ($\Delta\alpha_i$) of these groups (species i), their relative orientation and their total number N_i per particle. To a good approximation it can be calculated by geometric summing

$$\Delta\alpha \cong \sum \Delta\alpha_i N_i f_i.$$

Here f_i can be considered as the average orientation factor for the groups of species i with respect to the particle's symmetry axis and $0 < f_i < 1$. An analogous relation holds for

$$\Delta_\chi \cong \sum \Delta\chi_i N_i f_i.$$

In both relations, contributions from shape anisotropies resulting from the particle's anisomery and differences between the mean optical (and magnetic) susceptibilities and those of the solvent have been neglected. This is generally justified for $\Delta\chi$ because of the negligibly small difference between external and internal magnetic fields in diamagnetic solutions, but shape birefringence can be important for larger macromolecules and virus particles. We are mainly interested in determining the f_i since they contain information about the structure of the particles. The N_i values are usually known from chemical analysis and $\Delta\alpha_i$ and $\Delta\chi_i$ values are available for various groups (Sect. 4.2); thus from CM one can deduce a relation between the f_i values. This relation simplifies when $\Delta\alpha$ is known from other experiments such as flow birefringence, electric birefringence or depolarised light scattering.

A measurement of the *linear magnetic dichroism* $\varrho = (A_\parallel - A_\perp)/A$ consists in determining $A_\parallel$, $A_\perp$ and A, the optical absorbances for light propagating perpendicular to H and polarised parallel, perpendicular to H and unpolarised, respectively. The term ϱ is related to the magnetic orientation function ϕ defined in Sect. 4.3.1 and to f_i. For example, if the groups (i) have non-overlapping absorbance bands and (e.g. the base pairs of DNA) the corresponding optical transition moment lies in the plane normal to the rotational symmetry axis of the diamagnetic tensor, ϱ is simply given by $\varrho = -3f_i\phi/2$ [4.245]. Hence in this case the relative magnitudes of the f_i's can be determined for all absorbing groups spectrometrically accessible in the experiment. Relative f_i values can also be obtained for overlapping dichroic bands though data analysis is less straightforward. Knowledge of $\Delta\alpha$ results in a determination of ϕ from magnetic birefringence and then absolute values of f_i can be determined from dichroism. Therefore if the number of groups which contribute to the optical and diamagnetic anisotropies in a bioparticle are limited to say 3 or 4, a combination of magnetic dichroism and birefringence may yield information about orientational order (i.e. f_i) of these groups; f_i values can be calculated for structural models and compared with experiment.

Following these lines the structures of several nucleic acid- and/or protein-containing bioparticles in solution have been studied by magnetic orientation. Cotton-Mouton constants were measured [4.92] on dilute solutions of the *rod-like bacteriophages* Pf1 and fd, which both consist of a cylindrical protein shell enveloping 6.5 % and 12.9 % (w/w) DNA, respectively. The data were interpreted in terms of the diamagnetic anisotropies of the α helices and aromatic amino acids in the coat protein and of the DNA bases, as $\Delta\alpha$ of a single virus in solution was determined separately from the saturation birefringence in high magnetic fields of a fully oriented polyelectrolytic liquid crystal (Fig. 4.12). It turned out that the contribution of DNA to $\Delta\chi$ is weak and that both the axes of the α helices and the planes of the aromatic amino acids are oriented parallel to the phage major axis. By using magnetic birefringence a temperature-induced conformation change occurring around 10 °C was observed for Pf1 in solution for the first time. Independently, f_i values for DNA and aromatic amino acids were estimated from the magnetic linear dichroism of completely oriented virus solutions (Fig. 4.20). The results from both techniques agree.

A similar high field CM study was carried out on the buoy-shaped bacteriophages T4 and T7 which contain about 50 % DNA as well as on their empty protein shells (ghosts) [4.246a]. It was found that the diamagnetic anisotropy of the protein shell is negligibly small and that in T4 about $f_i = 35$ % of the DNA packed inside the protein shell (in an unknown fashion) is oriented parallel to the phage's head-tail axis. Diamagnetic anisotropy was discovered recently [4.246b] even in some so-called spherical viruses, suggesting that anisotropic packing of nucleic acids might be quite common in compact biological structures.

Dilute aqueous solutions of *chromatin* (the DNA protein complex occurring in the eucaryontic nuclei) and of chromatin fragments (so-called *nucleosomes*) at low ionic strengths ($\leq 10^{-2}$ M NaCl) both showed about 3 times smaller CM constant than free DNA. This reduction in anisotropy quantitatively agrees with current structural models of chromatin, in which the DNA loops around the globular protein core, thus forming bead-like nucleosomes linked by free DNA.

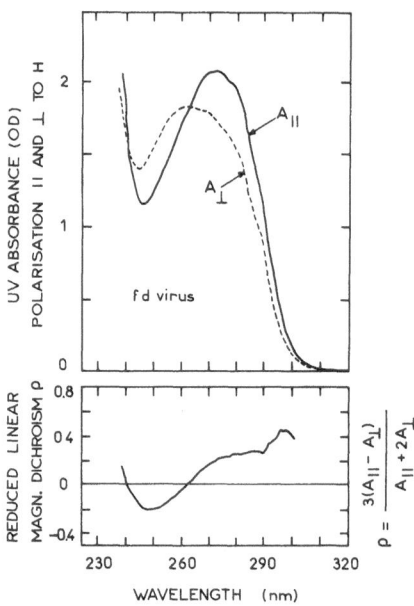

Fig. 4.20. The uv absorbance of magnetically oriented concentrated solutions of the cylindrical bacteriophage fd is strongly anisotropic (the magnetic linear dichroism saturates). The solution (at $c = 9.9$ mg/ml, 10^{-3} M Tris-HCl, pH 7.5) was oriented in a standard spectrometer cell (optical path 0.5 mm) by a vertical magnetic field of 12.2 T for 10 min, and appeared like the samples shown in Fig. 4.11b. The uv spectra were subsequently recorded for light polarised parallel and perpendicular to H; no disorientation could be observed within the first 30 min. Absorbance in the 240–260 nm and 270–300 nm regions mainly originates from the phage's DNA and aromatic amino acids, respectively. Both DNA bases and aromatic amino acids have in-plane, electronic transitional moments and the phage orients along H; therefore the negative (positive) dichroic bands observed for DNA (aromatic amino acids) indicate that the planes of DNA bases (aromatic amino aicds) are on the average more perpendicular (parallel) to the long axis of fd (Maret, unpublished result)

The results also suggest that the chain of beads is very flexible [4.6, 129]. It has been erroneously concluded from strong polarised and depolarised light-scattering signals that t-RNA, its shape being approximated by a prolate ellipsoid of revolution, aligns *along* a magnetic field [4.247]. A detailed analysis of the reported data [4.247] revealed that t-RNA orients *perpendicular* to H and that the dominant part of the aromatic base planes are preferentially perpendicular to the particles long axis [4.248]. This agrees with the observed negative birefringence and positive Cotton-Mouton constant of t-RNA. Probably the anisotropic light scattering originated from large anisometric aggregates.

Experimental Cotton-Mouton constants of dilute solutions of *fibrinogen*, an elongated protein involved in blood coagulation (Sect. 4.4), have been analysed in comparison with the rod-like bacteriophages fd and Pf1. From this it seems likely that fibrinogen contains about 30 % α helices oriented along the molecular long axis [4.249].

Murayama's microscope observation [4.250] clearly demonstrated that sickle cells, which are anomalously deformed *erythrocytes*, align with their long axis perpendicular to a field of only 0.35 T. *Costa Ribero* et al. [4.251] confirmed this recently and obtained a diamagnetic anisotropy of 2.7×10^{-18} erg/G^2 per sickle cell. They pointed out that this large anisotropy can be quantitatively accounted for by the paramagnetic anisotropy of the haeme group (Sect. 4.2) of the protein haemoglobin confined in the cell. Haemoglobin molecules in sickle cells are condensed into parallel fibres which are oriented along the cell's long axes. Within the fibres the average orientation of the haeme planes is almost normal to the fibres; thus alignment of the cell axes normal to H corresponds to strong orientation of the haeme planes along H, as expected (Sect. 4.2). There seems to be no need to include contributions of the cell membrane to $\Delta\chi$.

Leitmannova et al. [4.252] reported that abnormally shaped erythrocytes occur at higher density in a sample after exposure (for typically 10 min) to a 1.0 T magnetic field. In addition, about 10 % of erythrocytes were damaged ("haemolysed") in 1 T. Since this amount increased more than linearly with H, one could expect full haemolysis to take place in fields above 8–10 T. However, others [4.253] have found that less than 1 % of erythrocytes, if any, are haemolysed after 10 min exposure of blood to 20 T.

Cotton-Mouton constants and partial saturation of the magnetic birefringence in fields above 6 T have been reported [4.254] for suspensions of a peptidoglycanteichoic acid complex, the major component of a *bacterial cell wall*. The dominant contribution to $\Delta\chi$ seems to come from the ordered crosslinked polysaccharide chains within the cell wall. A quantitative analysis of these data appears to be difficult because $\Delta\alpha$ is not known, the particles are much larger than optical wavelengths (Sect. 4.3.1) and presumably polydisperse; since they have a closed sac-like shape, contributions to the birefringence could also originate from magnetic deformation.

4.4 Polymerisation Reactions and Locking of Alignment in High Magnetic Fields

A long-range orientational order makes possible strong magnetic alignment even of weakly diamagnetically anisotropic molecules in modest fields. For various applications, it is of interest to stabilise the aligned phase (e.g. by freezing, crystallising, drying, polymerising, cross-linking or other ways of gel formation). *Sobajima* [4.91] was the first (1967) to produce strongly oriented films of partially crystallised α helices by *slowly drying* a lyotropic solution of poly-γ-benzyl-L-glutamate (PBLG) in 0.7 T. These and other magnetically oriented and dried films or fibres, including poly-γ-ethyl-glutamate, polynucleotides and rod-like viruses, have become useful for structural studies mainly by x-rays, but also by optical, NMR and other techniques [4.5, 91, 152, 153, 158, 159, 161, 167, 178, 183, 185, 255]. As an example, Fig. 4.21 shows the oriented x-ray pattern from the two polymers DDA-8 and DDA-9 (cf. Fig. 4.8) with nematogenic groups in the main chain. One sample (a) was slowly *cooled* from the nematic melt below the crystallisation temperature in a vertical 16 T field. Strongly oriented crystallites result. The other (b) was rapidly *quenched* and the magnetically oriented nematic structure was preserved at room temperature.

Finer and *Darke* [4.170] obtained an oriented gel of poly-L-lysine-hydrobromide by cooling a concentrated aqueous solution through the liquid-gel transition temperature in 1.45 T. *Avirom* [4.256] stabilised the magnetic

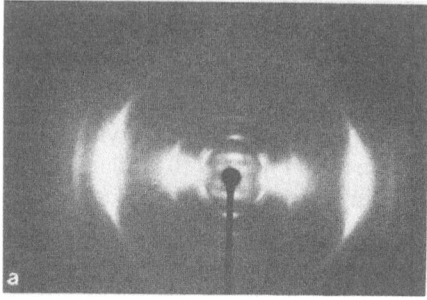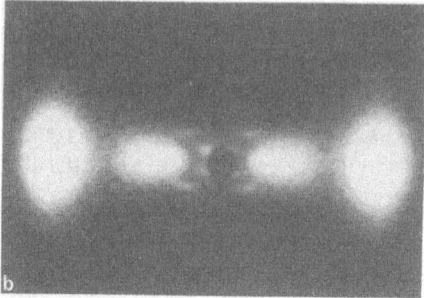

Fig. 4.21a, b. Oriented partially crystalline (**a**) or quenched nematic (**b**) phases can be produced when melts of polymers with liquid crystalline (nematogenic) groups in the main chain are slowly (**a**) or rapidly (**b**) cooled in a strong magnetic field from the nematic phase into the solid state. The x-ray patterns were taken from samples contained in 1 mm diameter capillaries with the x-ray beam normal to the magnetic field. (**a**) DDA-8 oriented by a vertical field of 16 T at 235 °C, cooling rate 1°/min to room temperature. (**b**) DDA-9 quenched in 12 T at 100°/min. The most outward equatorial reflection corresponds to a spacing of about 0.4 nm [4.183, 185]

orientation in PBLG by intermolecular chemical *cross-linking* with amide bonds between benzyl-alcohol groups. In 1973 *Liebert* and *Strzelecki* [4.257] introduced *polymerisation* under magnetic fields as a tool to produce oriented bulk polymers. A mixture of two types of nematic monomers (p-acryloyloxy-benzylidene p-cyano-aniline and di-p-acryloyloxybenzylidene p-diamino-benzene) copolymerised in 0.8 T resulted in an oriented crystalline phase which was almost free of defects. *Perplies* et al. [4.8] polymerised a methacrylic Schiff base in strong magnetic fields (7 T) and obtained an oriented polymer matrix, but only from the nematic and not from the isotropic phase of monomers. These authors also showed, in agreement with *Clough* et al., *Cser* and *Lorkowski* and *Reuther* [4.258–260], that a magnetically oriented nematic monomer can yield a polymer with smectic ordering. *Torbet* et al. [4.249] investigated for the first time a *biological polymerisation* process in magnetic fields; they found a highly oriented filamentous fibrin network when polymerising – as during blood coagulation – the blood plasma protein fibrinogen in fields up to 20 T. From neutron-diffraction patterns of these oriented gels (Fig. 4.22) new features of the structure and packing of the fibrin protofibrils inside the fibrous network could be obtained; structural information about fibrinogen and fibrin can be expected from the study of such magnetically oriented samples.

In addition, *Perplies* et al. and *Torbet* et al. [4.8, 9] have shown that the application of a magnetic field is needed – at least in some cases – only during primary stages of polymerisation. As indicated in Fig. 4.23, the orientation persists to increase after switching off the field, once an oriented polymer matrix has started to form. From such measurements it seems possible to gain new insight into the complex microscopic processes involved in the formation of microfibrils and gels during polymerisation or polycondensation. At present, however, the influence of molecular orientation on the reaction kinetics is not

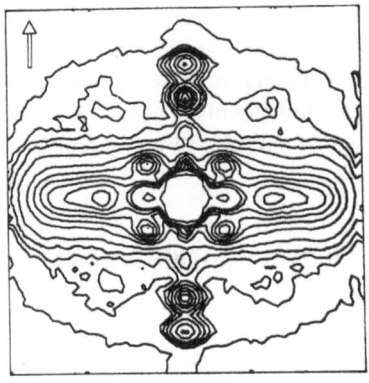

Fig. 4.22. Neutron diffraction pattern from fibrin gels ($c = 10$ mg/ml in H_2O buffer) oriented in a vertical magnetic field of 18 T. The 22.5 nm period meridional reflections presumably arise from a half staggered arrangement of monomers along the fibre axes and the 19 nm equatorial peak originates from the lateral protofibril packing [4.9]

well understood. For example, *Perplies* et al. [4.8] observed a 40 % *increase* in the polymerisation rate of bulk nematic monomers oriented by 7 T, whereas *Wojtczak* [4.261] found a strong *inhibition* of acetic aldehyde polymerisation in aqueous solution by fields of only 0.6 T. It was reported recently [4.262] that the gelation temperature of aqueous solutions of the polysaccharide agarose increases by 2 °C in 1 T. The significance of this observation seems unclear when compared with the very small shift of the nematic-isotropic transition temperature (Sect. 4.3.3a).

It is evident that magnetic fields can be used to bring molecules into a stable strongly oriented form well suited for structural analysis. The most important

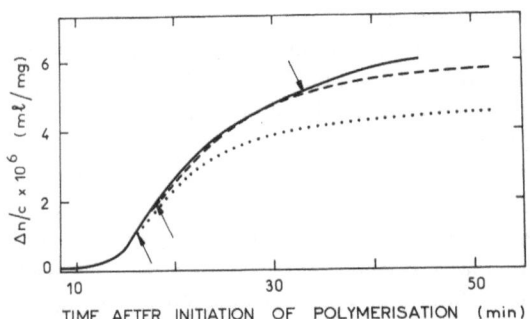

TIME AFTER INITIATION OF POLYMERISATION (min)

Fig. 4.23. Specific birefringence $\Delta n/c$ during polymerisation of fibrin in a strong magnetic field (11 T), initiated at $t = 0$ by adding the activating enzyme thrombin to a 5 mg/ml solution of fibrinogen. As polymerisation proceeds the growing fibrin fibres become sufficiently large and anisotropic to be strongly orientable prior to substantial network formation. Arrows indicate instances of switching off the magnetic field. (The lowest, middle, and upper arrow belong to $\cdots$, $---$, ——, respectively). In all three cases ($\cdots$, $---$, ——) the orientation futher increases even in zero field, indicating that the orientation of fibres becomes fixed in the forming network and that the final reaction step is fibre growth. The maximum value of ($\Delta n/c$) corresponds to nearly complete alignment of the polymer filaments, as seen independently by scanning electron microscopy and neutron diffraction [4.9, 249]

requirement for a successful application of this technique to produce alignment seems to be the existence of a highly ordered *liquid* (say liquid crystalline or fibrous) state prior to solidification. In contrast to usual mechanical orientation techniques, no elastic gel state with polymeric entanglements is needed and the orienting field can be applied steady state to small samples of any shape.

4.5 Separation of Macromolecules in Magnetic Field Gradients

In homogeneous magnetic fields, an orientational but no translational force acts on an *anisotropic* particle. In inhomogeneous fields, however, a translational force exists even for *isotropic* particles; it depends on the variation of H across the particle, i.e. $\mathrm{grad}(H^2)$, on the excess susceptibility of the particle as compared to the susceptibility of the surrounding medium and on its volume. In weakly diamagnetic media, ferromagnetic and paramagnetic particles are attracted into the high field region and diamagnetic particles are weakly repelled. These forces have been well described [4.263–4.266 and references therein] and various devices combining strong homogeneous magnetic fields with high gradients near ferromagnetic edges or wires have been developed and succesfully applied to separate ferro- or paramagnetic particles as small as a few microns in diameter. Promising applications range from filtration of low-grade iron ores and desulphurisation of coal to purification of waste water. This work has been reviewed by *Oberteuffer* and his colleagues [4.263, 267].

Separation is efficient when the magnetic force overcomes thermal agitation and gravitational forces. It builds up on a time scale controlled by the viscosity of the solvent. To separate diamagnetic particles or macromolecules of given size, shape and weight, the product $H \cdot \mathrm{grad} H$ should be maximised. Near a ferromagnetic wire or edge in a strong external field $H_0 (> M_s)$ the field drop – excess over H_0 – approaches a limit, which is just the saturation magnetisation of the ferromagnetic material (M_s), and spatially extends over a distance comparable to the surface radius of curvature R, hence $\mathrm{grad} H$ approximately equals M_s/R. Since R should not be smaller than the dimension of the particle or macromolecule to be separated, M_s/R is limited and higher separation efficiency can be obtained only by applying a stronger external field H_0, $H \cdot \mathrm{grad} H$ approaching $H_0 \cdot M_s/R$. Following these lines, *Gill* et al. [4.265] reported in 1960 the diamagnetic susceptibility of polystyrene latex and red blood cells from visual observation of the drift velocity (of typically 1 µm/s) in inhomogeneous fields ($H \cdot \mathrm{grad} H \simeq$ some $\mathrm{T}^2 \mathrm{cm}^{-1}$). *Melville* et al. [4.31, 268] separated deoxygenated, paramagnetic red blood cells without damage from normal blood flowing at velocities around 1 cm/min throughout a randomly coiled steel wire within a magnetic field; the $H \cdot \mathrm{grad} H$ value near the wire was estimated to be about 130 $\mathrm{T}^2 \mathrm{cm}^{-1}$. *Owen* [4.32] used a similar experimental set-up and similar conditions to separate paramagnetic methaemoglobin erythrocytes from diamagnetic oxygenated erythrocytes or leucocytes. Magnetic migration of ery-

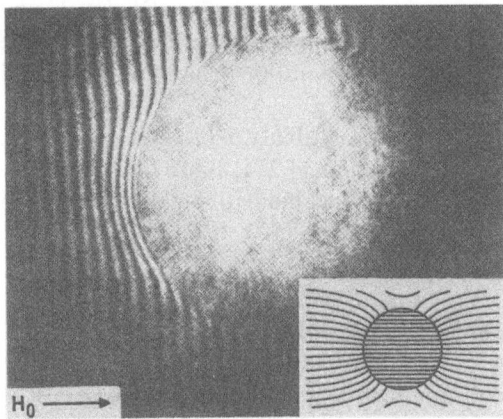

Fig. 4.24. Interferometric observation of the magnetosedimentation of the globular protein bovine γ-globulin in dilute aqueous solution. A cylindrical cell containing the solution and a coaxial steel cylinder (grey central region) is placed in one branch of a Jamin-type interferometer and is photographed end on. This optical set-up implies that at zero magnetic field, the uniform protein concentration (uniform refractive index) results in straight optical interference fringes (not shown); if a horizontal magnetic field H_0 of 1.2 T is applied perpendicular to the ferromagnetic cylinder, the field gradient near its surface as sketched in the insert creates a concentration gradient (refractive index gradient) which causes the interference fringes to bend [4.30]

throcytes was also noted by *Porath-Furedi* and *Yanai* [4.269]. By an optical interference technique *Simonsen* and *Gill* [4.30] were able to demonstrate a concentration gradient even of smaller proteins, the 169,000 molecular weight bovine gamma-globulin, in aqueous solution (containing a paramagnetic salt) building up within typically 10^2 min in $H \cdot \mathrm{grad} H \simeq 100\, \mathrm{T}^2\, \mathrm{cm}^{-1}$ close to a ferromagnetic rod (Fig. 4.24). By this method it was possible to determine the diamagnetic susceptibility of macromolecules in solution even at substantial dilution.

Selective removal of small diamagnetic (bio)-particles from suspension can also be achieved in fairly low fields and gradients if a complex with ferromagnetic particles is formed. This was first demonstrated by *Bitton* and *Mitchell* [4.270] on bacteriophage T7 bound onto magnetic particles, using a steel wool filter and only 0.1 T. More recently so-called ferrofluids have become important in this field. They are suspensions of ultramicroscopic ($\lesssim 100\,\mathrm{nm}$) ferromagnetic particles coated by polyelectrolytes which prevent interparticle agglutination. Coating the ferromagnetic particles by polymers which were specifically substituted with certain enzymes enabled *Mosbach* and *Andersson* [4.271] to remove the latter magnetically, thus avoiding the usual centrifugation and column chromatography steps. Similarly, when coated by a special polymer carrying a specific cell-surface ligand, magnetic ferrofluids were used to separate whole neuroplastoma cells in 0.1 $\mathrm{T}^2\,\mathrm{cm}^{-1}$ with better than 99 % efficiency under sterile conditions within only a few minutes [4.272]. Also drugs were concentrated magnetically at specific sites in organisms [4.273]. Obviously the chemical variety of coating polymers on ferromagnetic particles opens a wide field of applications for high gradient magnetic separation in biology and related fields.

Another way of enhancing the magnetic susceptibility of biological materials is to add strongly paramagnetic ions such as erbium [4.274]. Since Er^{3+} ions have a magnetic moment of 9.5 Bohr magnetons and bind to negatively charged surfaces, magnetic separation becomes more efficient.

4.6 Influence of Magnetic Fields on Living Cells

In this section we mention some newer findings demonstrating the interaction of magnetic fields (including weak fields) with more complex living systems. The microscopic processes involved are scarcely understood in many cases. We add some speculative remarks on how some biomagnetic effects could be related to the *magnetic properties* of the constituent molecules.

The work in "magnetobiology" (effects of external magnetic fields on biological matter) has been reviewed [4.34–36, 275–279]. The rapidly growing field of "*biomagnetism*" (weak magnetic fields created by living matter), which is beyond the scope of this article, has been reviewed by *Williamson* et al. and *Geselowitz* [4.280, 281].

4.6.1 Magnetic Field Detection by Animals

During the last decade strong evidence has been accumulated demonstrating that various animals ranging from simple amoebia and bacteria to sharks and homing pigeons are able to detect weak magnetic fields such as the earth's field (survey in [4.38–42]). Pigeons and bees seem to be sensitive to field variations some thousand times below the earth's field level. At least in some cases the animals are known to use this sense for orientation and navigation purposes. *Volvox* – an evolutionary very simple organism and a direct descendant of one of the earliest forms of life – can distinguish between magnetic fields of different intensities and directions [4.282]. This suggests that sensitivity to magnetic fields might be a rather general and important feature of living matter.

A first hint that ferromagnetic particles might be involved in a magnetic field "sensor" mechanism arose in *Blakemore* and colleagues' work on *bacteria* [Refs. 4.34, p. 354; 4.39, 43, 283]. They observed that some classes of bacteria migrate spontaneously towards geomagnetic north. This property seems to be essential for survival of the bacteria, since by swimming north along the earth's field lines, which are inclined towards the ground in the northern hemisphere, the bacteria always return to their optimal environment in the mud. These so-called magnetotactic bacteria synthesise almost cubic, ~ 50–100 nm sized, ferromagnetic monodomain magnetite particles. Each bacteria contains about 20 such particles aligned in a chain so that the total magnetisation points along the bacterial major axis. This compass needle like device is just large enough to allow alignment in the earth's field against thermal agitation [4.284]. It turned out that there is no north-south ambiguity in this device: bacteria in the southern hemisphere do swim towards south because their chains are magnetised in the opposite direction [4.285]. Magnetite is synthesised in the bacteria with headwards or tailwards magnetisation at about equal probability, but only the species with correct magnetisation polarity seems able to survive in a given hemisphere, except near the equator. Other simpler animals responding to the earth's magnetic fields are snails [4.286] and salamanders [4.287]. Amongst insects, flies (*Drosophila*) [4.288], *sandhoppers* (*Tenebrio* and *Talitrius*) [4.289] and more intensely *bees* and *hornets* [4.44, 277, 290, 291] have been successfully studied. Bees exhibit a certain "Missweisung" (deviation from the correct orientation) of their dancing figures and construct their combs in certain directions, both being related to the earth's magnetic field. They eventually adjust their circadian rhythm to small diurnal variations of the earth's field. In their abdomen, ~ 100 nm particles of magnetite have been found [4.44] and the amplitude and direction of the overall magnetisation have been measured directly by a SQUID magnetometer.

Magnetite was also found recently in *dolphins* [4.292] and *rays* [4.293]. It would be premature, however, to conclude from this that the magnetic sensibility of animals can generally be explained in terms of ferromagnetic "compass needles". Beautiful training experiments by *Kalmijn* [Refs. 4.40, p. 347; 4.294] on *elasmobranch fish* (sharks, skates and rays) demonstrated their ability on the one hand to orient magnetically, and on the other hand to detect electric fields as small as 10 nV/cm by their organ ampullae of Lorenzini located in the skin; since electric fields of such weak strength are in fact induced in the fish by swimming through the earth's magnetic field, an electrodynamic compass system might exist as well. *Jungermann* et al. [4.295] discussed the labyrinth of the inner ear as a possible location for the induction loop.

Various *migrating birds* like robins [4.296], warblers [4.297], indigo buntings [4.298] and blackcaps [4.299] have a magnetic compass. *Wilschko* and *Wilschko* showed that robins do not use the polarity of the magnetic field to determine north, but only its inclination with respect to gravity; there is a maximum in sensitivity at magnetic fields comparable to the earth's magnetic field. An increasing number of studies has been published (surveys in [4.42, 300, 301a, b]) in the past years on the magnetic sense of *homing pigeons*, mainly initiated by work of *Keeton* [4.302] and *Walcott* and *Green* [4.303]. They showed that superimposing an artificial magnetic field on the earth's magnetic field by fixing small bar magnets or Helmholtz coils to the pigeons' head caused a substantial loss in their orientation capability. Pigeons integrate some information about the earth's field during transport from the loft to the release site [4.304] and possibly possess a "geomagnetic map" [4.301] in addition to the magnetic compass system. Monodomain elongated magnetic particles (~ 100 nm) have been discovered [4.46] in tissues of pigeons too, apparently located between the pigeons' brain and skull [4.46, 305] and in the neck musculature [4.300].

Studies on humans have been controversial so far. *Baker* [4.306, 307] reported the ability of blindfolded humans to indicate the homeward direction when carried away from home; this ability apparently disappeared when the earth's magnetic field was strongly perturbed in the head region using Helmholtz coils mounted around the head. Attempts to reproduce this observation were not successful [4.308]. *Kirschvink* found magnetic remanence in human adrenal tissue [4.309].

Besides the "ferromagnetic" and "inductive" compass mechanisms [Refs. 4.34, p. 287; 4.310] other types of earth's magnetic field detectors in animals have been proposed. *Gunter* et al. [4.311] suggested that the animal detects a movement of large particles or organelles inside a cell due to the drag caused on them by the ionic currents if the cell is exposed to crossed internal electric and external magnetic fields. This electrophoretic process in a magnetic field leads to a (linear) increase of the particles' speed with H, and one might speculate that exposure of the animal to very strong magnetic fields would cause severe deviations. Other models of "magnetoreception" involving diamagnetic or paramagnetic properties of living material or optical pumping into magnetically split triplet states have recently been discussed and criticised by *Leask* [Ref. 4.40a, p. 318].

In birds the "induction" model does not seem to be appropriate mainly because of the finding that even a magnetic field produced by coils which *moved* with the bird (and thereby produced no induction) did perturb the bird. A mechanism involving the magnetic particles seems to be very attractive in the sense that it disagrees with neither experimental nor theoretical findings [4.312–315]. The physics of small monodomain or superparamagnetic particles in a magnetic field is well understood [4.314], however, little is known about the coupling mechanism of these particles to the nervous system. *Presti* and *Pettigrew* [4.300] suggested that some sensory fibres in the muscles particularly sensitive to stretch might be involved. *Kirschvink* and *Gould* [4.314] proposed four types (1–4) of "magnetoreceptors". The first (1) implies freely rotatable monodomain magnetite particles embedded in a membrane. The particle is coated with some electrically insulating material except for the regions at the magnetic poles. Since magnetite is a good conductor an orientation of the particle with the magnetic field normal to the membrane would open a conducting channel across the membrane. In (2) a superparamagnetic magnetic particle is locked into the membrane. In this case the magnetisation aligns without rotation of the particle in an external field; the transmembrane conduction change is performed by orientation of nearby membrane proteins ("satellites") which have ion pores and carry small magnetic rods. In (3) the torque in a magnetic field on a chain of monodomain particles would be detected by some mechanoreceptor, for example a hair cell. Detector (4) involves an array of regularly spaced superparamagnetic particles embedded in an elastic medium. Magnetic interparticle interaction stresses the elastic medium, the stress

depending on the direction of an external field. Other mechanisms or combinations of (1–4) could be imagined. Obviously a direct, perhaps electron-
microscopic observation of such an organelle would best elucidate the actual
situation. Another approach is to expose the animal to a strong external field
with and without field gradients and then test its navigational performance.

Such experiments on pigeons were actually performed in our laboratory
[4.316]. Submitted to steady fields of up to 10 T for 2 min, the pigeons did not
show any sign of harm during exposure. However, their average vanishing
direction, when released some 50 km from the loft, deviated (by some 40°–50° to
the left) from the homing direction taken by the unexposed control group of
pigeons [4.316]. Vanishing bearings from experimental and control pigeons both
showed the normal amount of scattering around the vanishing direction. These
results indicate that some modifications were caused by the strong field and
lasted for at least several days. If effects of field gradients can be neglected, the
data would not be consistent with detector mechanisms (1) and (2) because the
free rotatability of the magnetite particles should not be permanently affected by
a strong field. Torque and stress in (3) and (4) might have gone beyond a certain
yield limit during exposure and permanent damage occurred; this does not seem
very probable because of the observed normal behaviour of the pigeons during
exposure and the lack of increased scattering in their vanishing bearing. In the
framework of (1–4) we would thus be left with the possibility that some change in
the direction of magnetisation in the torque detector (3) occurred. Further
studies of this kind should certainly include strongly inhomogeneous fields and
perhaps ac fields, the latter to search for the possibility of demagnetisation.

4.6.2 Growth in Magnetic Fields

Audus [4.47, 317] was the first to observe an influence of the magnetic field on the
direction of growth in plants (magnetotropism). Compensating for gravitropism
(by slow rotation of the experimental set-up around the horizontal direction of
initial growth) he found that both cress *roots* and oat *shoots*, initially aligned
along the pole shoes perpendicular to a 0.4 T field, tended to grow into regions of
weaker magnetic fields by bending. One might think that the negative
magnetotropism could be due to the overall force acting on the diamagnetic
body of the root or shoot in the magnetic field gradient of about 0.6 T/cm
(Sect. 4.5). However, when the plants grew along the planar surface of an agar
gel, which had been oriented parallel to the magnetic field but normal to the field
gradient and was sufficiently rigid to withstand this force, magnetotropism could
still be observed. Hence this latter force could not be the physical origin of
Audus' observation. He also demonstrated that the mechanisms involved in
magnetotropism cannot be related to gravitropism, since, in contrast to the case
of graviperception, no magnetically induced intracellular displacement of
statoliths was observed nor did roots and shoots show a different sign of the
*magneto*response.

Therefore one might speculate that magnetotropism is caused by a magnetic field dependence of the local growth rate. The negative magnetotropism mentioned above could originate from an accelerated cell elongation or an enhanced cell division rate in the region of stronger magnetic field strength; in addition, both elongation and division could depend on the direction of the applied magnetic field.

Hence, it seems very important for the understanding of Audus' experiment to search for magnetotropism in less complex, monocellular or non-dividing systems. On these lines *Schwarzbacher* and *Audus* have investigated [4.318] magnetotropism of the monocellular spore *phycomyces* in strong field gradients. Recently magnetotropism was observed even in homogeneous magnetic fields. As demonstrated by Fig. 4.25, *Sperber* et al. [4.49] found that the elongational growth of monocellular pollen tubes of lily is almost completely oriented parallel to a homogeneous strong magnetic field. The orientational distribution function is bimodal and symmetric, i.e. magnetic north and south are equivalent. In addition, in inhomogeneous strong fields, a preferential growth of the tubes into regions of lower magnetic fields was found (Fig. 4.26). These observations on non-dividing cells indicate that magnetotropism is not necessarily related to magnetic orientation of ordered macromolecular arrays as occurring during cell

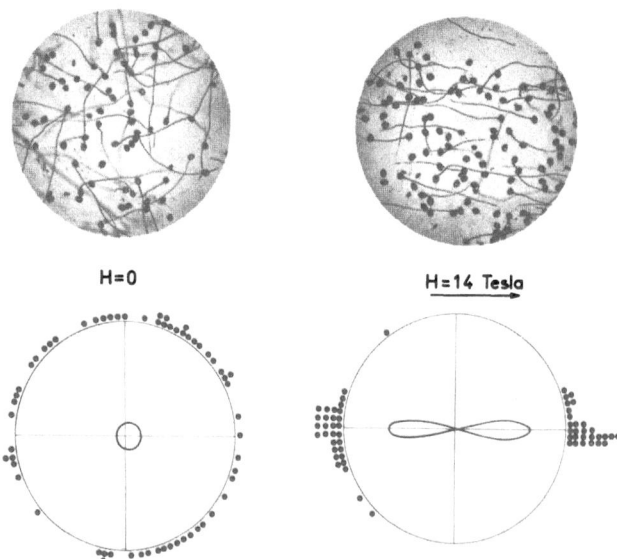

Fig. 4.25. Pollen tubes of lily show a strongly oriented elongational growth in *homogeneous* high magnetic fields (magnetotropism). The photographed tubes and grains (diameter typically some 100 μm) have grown for 3 h at 30 °C on the surface of agar gel inside a horizontal Bitter magnet without magnetic field (*left*) and in 14.0 T horizontal field (*right*). A polar representation of the orientational distribution function (best fit through the black data points) is given underneath for each sample. The radial distance from the origin is proportional to the number of tubes found with a tip direction within a 10° interval around the azimuthal angle θ [4.49]

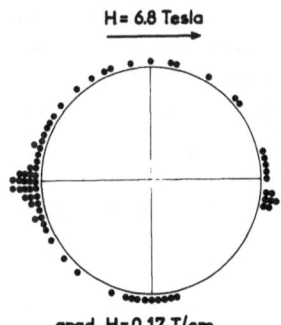

Fig. 4.26. Magnetotropism of pollen tubes of lily in *inhomogeneous* fields. Growth conditions were similar to those indicated in Fig. 4.25. The sample was placed on the axis of a horizontal solenoid but off centre, i.e. in the region of the axial field gradient ($H = 9.7$ T, grad $H = 140$ T/m). Superimposed on the bimodal orientation (of Fig. 4.25) is a unimodal component indicating the tendency of growth into regions of smaller magnetic fields [4.49]

division. Perhaps the mechanisms involved are much simpler than expected previously. A full field strength dependence of the magnetotropism of pollen tubes has not been evaluated yet, but this phenomenon is a typical high field effect: tubes grown for 9 h in only 3 T showed hardly any orientation.

We add some speculative remarks on magnetic field induced modification of growth which could be involved in magnetotropism.

First the *elongational growth* of single cells could possibly be enhanced because of accelerated *metabolism* in the magnetic field: some indirect evidence for modified ion transport across membranes has been discussed by *Aceto* et al. [4.278] and *Amer* and *Tobias* [4.319]. More recent findings that action potential [4.320] and conduction velocity [4.321] of frog sciatic nerve are enhanced in fields around 1 T might also be interpreted in terms of structural changes in membranes. This latter field has been critically reviewed by *Schwartz* [4.321, 322]. Since such effects should depend on the direction of the magnetic field with respect to the membrane, they possibly also could imply some anisotropic elongational growth. On the other hand, the membrane material built in the cell wall during elongational growth is supplied by intracellular vesicles which undergo fusion with the membrane in the regions of growth, which is the tip region in the case of pollen tubes. When deformed during the *fusion process* these large (~ several micron) vesicles presumably have a sufficiently high diamagnetic anisotropy to imply different probabilities for fusion into parts of the cell wall parallel and perpendicular to the magnetic field; as a result the elongational growth should become anisotropic.

There are some indications from biological experiments that *cell division* can be influenced by magnetic fields. *Mericle* et al. [Ref. 4.34, Vol. 1, p. 183] reported evidence for enhanced cell division rate in particular when a magnetic field was applied to the zone of cell multiplication (meristem) of barley roots; furthermore effects of steady magnetic fields have been observed on mitosis [4.323] and meiosis [4.324], embryonic development [4.325–327], growth of yeast (summarised in [4.328]) and on the fertility of small fish during several life cycles [4.329]. Keeping in mind the anisotropic diamagnetic properties of various biological macromolecules, and the existence of highly ordered liquid crystalline states *in vivo* particularly during cell division, one might speculate that magnetic fields, imposing orientations for instance on metaphase chromosomes, spindles, membrane lamellae and the cell wall, favour certain polarities of the dividing cell. Obviously, microscopic observations of cell multiplication in a strong magnetic field could help to explain some magnetobiological effects.

Finally we draw attention to the prediction made by several authors [4.330–333] that the stream of electrically conducting blood in large arteries can be reduced dramatically by a strong transverse magnetic field. It follows from classical magnetohydrodynamics that, for example, in a tube of 1 cm diameter and in 20 T the natural mean streaming speed should be slowed down by a factor of two. It should be easy to verify this effect directly by experiment, which promises various applications.

Recently *Gremmel* et al. [4.334] discovered that magnetic fields as commonly used in NMR tomography can cause sizeable temperature changes ($\Delta T \approx 5\,°C$) in those parts of the human body which are exposed to the magnetic field. Magnetic-field-induced temperature changes ($\Delta T \approx 2.5\,°C$) were also observed by *Sperber* et al. [4.335] in mice exposed to strong magnetic fields. Interestingly, in inhomogeneous fields the sign of the temperature change could be inverted if the gradient of the field relative to the animals' polarity was reversed.

Acknowledgements. The experimental work cited here was performed in close cooperation with many colleagues. They include A. Mayer, R. Oldenbourg, R. Ranvaud, M. v. Schickfus, E. Sénéchal, D. Sperber, M. Stamm and J. Torbet from our own laboratory. A. Domard, Y. Filippini, G. Fillion, J. M. Freyssinet, G. Hudry-Clergeon, J. J. Lawrence, B. Malraison, C. Meyer, H. Milas, R. Oberthür, Y. Poggi, F. Porte-Cuault, G. Porte, M. Rinaudo, K. Simpson and F. Volino are from various research institutions at Grenoble. Furthermore A. Blumstein, R. Blumstein (University of Lowell, Mass.), E. Fischer, M. Happ and H. Ringsdorf (University of Mainz), J. Kiepenheuer and K. Schmidt-Koenig (University of Tübingen), D. Marvin and his colleagues (European Molecular Biology Laboratory, Heidelberg), M. Showe (Biozentrum, Basel), G. Weill (CNRS, Strassbourg) and M. H. Weisenseel (University of Karlsruhe) were involved in parts of this work. Their contributions are gratefully acknowledged.

We should also like to thank our French colleagues from the Service National des Champs Intenses and M. Bichler, H. Dresler and A. Köhler for technical help.

Fruitful discussions with G. Eilenberger (KFA, Jülich), P. G. de Gennes (College de France, Paris), R. Klein (University of Konstanz), M. Papoular (CNRS, Grenoble), P. Pincus and P. Chaikin (University of California, LA) are acknowledged as well.

We are grateful to A. A. Bothner-By (Carnegie Mellon University Pittsburgh), E. Oldfield (University of Illinois, Urbana), H. Rüterjans (University of Münster), and K. Wüthrich (ETH, Zürich) for very instructive correspondence and communication of some work including photographs prior to publication. Thanks are due also to Mrs. Landsberg and Mrs. Rössler for carefully typing the final manuscript.

References

4.1   H.Kelker, R.Hatz, C.Schumann: *Handbook of Liquid Crystals*, (Verlag Chemie, Weinheim 1980)

4.2   G.Maret, M.v.Schickfus, A.Mayer, K.Dransfeld: Phys. Rev. Lett. **35**, 397 (1975)

4.3   G.Maret, J.Torbet, E.Sénéchal, A.Domard, M.Rinaudo, H.Milas: *Nonlinear Behaviour of Molecules, Atoms and Ions in Electric, Magnetic and Electromagnetic Fields*, ed. by L.Neel (Elsevier, Amsterdam 1979) p. 477

4.4   E.Sénéchal, G.Maret, K.Dransfeld: Int. J. Biol. Macromol. **2**, 256 (1980)

4.5   J.Torbet, G.Maret: J. Mol. Biol. **134**, 843 (1979)

4.6   G.Maret, K.Dransfeld: Physica **86–88B**, 1077 (1977)

4.7   N.Chalazonitis, R.Chagneux, A.Arvanitaki: C. R. Acad. Sci. **271**, 130 (1970)

4.8   E.Perplies, H.Ringsdorf, J.H.Wendorff: Polym. Lett. Ed. **13**, 243 (1975)

4.9   J.Torbet, J.M.Freyssinet, G.Hudry-Clergeon: Nature **289**, 91 (1981)

4.10   M.E.Michel-Beyerle, H.Scheer, H.Seidlitz, D.Tempus, R.Haberkorn: FEBS Lett. **100**, 9 (1979)

4.11   H.Rademaker, A.J.Hoff, L.N.M.Duysens: Biochim. Biophys. Acta **546**, 248 (1979)

4.12   A.J.Hoff, H.Rademaker, R. van Grondelle, L.N.M.Duysens: Biochim. Biophys. Acta **460**, 547 (1977)

4.13   M.C.Thurnauer, J.R.Norris: Biochim. Biophys. Res. Comm. **73**, 501 (1976)

4.14   J.Kalinowski, J.Godlewski: Chem. Phys. Lett. **36**, 345 (1975)

4.15   P.W.Atkins, G.T.Evans: Mol. Phys. **29**, 921 (1975)

4.16   P.W.Atkins: Chem. Phys. Lett. **18**, 355 (1973)

4.17   A.Gupta, G.S.Hammond: J. Chem. Phys. **57**, 1789 (1972)

4.18   L.R.Faulkner, A.J.Bard: J. Am. Chem. Soc. **91**, 6495 (1969)

4.19   H.Krath, H.Alms, J.M.D.Coey: J. Mater. Sci. **11**, 2283 (1976)

4.20   P.W.Selwood: Adv. Catal. **27**, 23 (1978)

4.21   N.Hata, Y.Yamanda: Chem. Lett. Jpn. 989 (Aug. 1980)

4.22   M.Sonneveld, L.N.M.Duysens, A.Moerdijk: Biochim. Biophys. Acta **636**, 39 (1981)

4.23   H.Hayashi, Y.Sakaguchi, S.Nagakura: Chem. Lett. 1149 (Sept. 1980)

4.24   K.Piotrowska, D.Edwards, A.Mitch, R.Dougherty: Naturwissenschaften **67**, 442 (1980)

4.25   P.W.Atkins: Recherche **10**, 118 (1979)

4.26   P.W.Atkins: *Physical Chemistry* (Oxford Univ. Press, Oxford 1978)

4.27   N.J.Turro, B.Kraeutler: Acc. Chem. Res. **13**, 369 (1980)

4.28   L.N.Mulay, I.L.Mulay: Anal. Chem. **52**, 199R (1980)

4.29   R.Stösser: Z. Chemie **17**, 201 (1977)

4.30   W.J.Simonsen, S.J.Gill: Rev. Sci. Instrum. **45**, 1425 (1974)

4.31   D.Melville, F.Paul, S.Roath: Nature **255**, 706 (1975)

4.32   C.S.Owen: Biophys. J. **22**, 171 (1978)

4.33   W.Haberditzl: Nature **213**, 72 (1967)

4.34   M.F.Barnothy (ed.): *Biological Effects of Magnetic Fields*, Vols. I, II (Plenum Press, New York, Vol. I 1964, and Vol. II 1969)

4.35   M.F.Barnothy, J.M.Barnothy: "Magnetobiology", in *Environmental Physiology* ed. by N.Balfour Slonim (Moshby, St. Louis 1974) p. 313

4.36   T.S.Tenforde (ed.): *Magnetic Field Effects on Biological Systems* (Plenum, New York 1979)

4.37   A.S.Presman: *Electromagnetic Fields and Life* (Plenum, New York 1970)

4.38   I.L.Gould: Am. Sci. **68**, 256 (1980)

4.39   R.B.Blakemore, R.B.Frankel: Sci. Am. **245**, 42 (Dec. 1981)

4.40   K.Schmidt-Koenig, W.T.Keeton (eds.): *Animal Migration, Navigation and Homing* (Springer, Berlin, Heidelberg, New York 1978)
       F.Papi, H.G.Wallraff (eds.): Avian Navigation (Springer, Berlin, Heidelberg, New York 1982)

4.41   A.J.Kalmijn: IEEE Trans. Mag. **17**, 1113 (1981)

4.42   W.F.Keeton: Sci. Am. **231**, 96 (Dec. 1974)

4.43   R.B.Frankel, R.P.Blakemore, R.S.Wolfe: Science **203**, 1355 (1979); see also: Phys. Today **22** (Nov. 1979)

4.44   J.L.Gould, J.L.Kirschvink, K.S.Deffeyes: Science **201**, 1026 (1978)

4.45   J.L.Kirschvink, H.A.Lowenstam: Earth Planet. Sci. Lett. **20**, 193 (1978)

4.46   C.Walcott, J.L.Gould, J.L.Kirschvink: Science **205**, 1027 (1979)

4.47   L.J.Audus, J.C.Wish: In [Ref. 4.34, Vol. I, p. 170]

4.48   D.Sperber: Unpublished
4.49   D.Sperber, G.Maret, M.H.Weisenseel, K.Dransfeld: Naturwissenschaften **68**, 40 (1981)
4.50   R.A.Dwek: *NMR in Biochemistry* (Oxford Univ. Press, Oxford 1973)
4.51   G.Levy (ed.): *Topics in Carbon-13 NMR Spectroscopy*, Vol. I (Wiley Interscience, New York 1974)
4.52   K.Wüthrich: *NMR in Biological Research: Peptides and Proteins* (North-Holland, Amsterdam 1976)
4.53   B.Pullmann (ed.): *Proceedings of the 33th Jerusalem Symposium on NMR*-Spectroscopy in Molecular Biology (Reidel, Dordrecht 1980)
4.54   K.Wüthrich, G.Wagner: Trends in Biochem. Sci. **3**, 227 (1978)
4.55   G.B.Lubkin: Phys. Today **32**, No. 11 (Nov. 1979) p. 17
4.56   A.A.Bothner-By, J.Dadok, submitted
4.57   E.Oldfield, T.M.Rothgeb: J. Am. Soc. **102**, 3635 (1980)
4.58   E.Oldfield: Private communication
4.59   F.Keilmann: J. Collect. Phenomena **3**, 169 (1981) and references therein
4.60   P.W.Selwood: *Magnetochemistry*, 2nd ed. (Interscience Publ. New York 1965)
4.61   W.Haberditzl: *Magnetochemie* (Akademie, Berlin 1968)
4.62   A.A.Bothner-By, J.A.Pople: Ann. Rev. Phys. Chem. **16**, 43 (1965)
4.63   M.R.Battagha, G.L.D.Ritchie: J. Chem. Soc. Faraday Trans. 2, **73**, 209 (1977)
4.64   K.Lonsdale: Proc. R. Soc. **171A**, 541 (1939)
4.65   K.Lonsdale, K.S.Krishnan: Proc. R. Soc. **156A**, 597 (1936)
4.66   R.Ditchfield: "Magnetic Suspectibility of Diamagnetic Molecules", in *MTP Int. Rev. Sci., Molecular Structure and Properties, Physical Chemistry*, Ser. 1, Vol. 2, ed. by A.D.Buckingham, Allen (Butterworth, London 1972)
4.67   J.W.Beams: Rev. Mod. Phys. **4**, 133 (1932)
4.68   A.D.Buckingham, J.A.Polple: Proc. Phys. Soc. **69**, 1133 (1956)
4.69   G.H.Meeten: J. Chim. Physique **7–8**, 1175 (1972)
4.70   L.Pauling: J. Chem. Phys. **4**, 673 (1936)
4.71   F.London: J. Chem. Phys. **5**, 837 (1937)
4.72   S.Yamaguchi; S.Okuda, N.Nakagawa: Chem. Pharm. Bull. **11**, 1465 (1963)
4.73   W.Zeil, H.Burchert: Z. Physik. Chem. Ser. 2, **38**, 47 (1963)
4.74   P.T.Marasimham, M.T.Rogers: J. Phys. Chem. **63**, 1388 (1959)
4.75   J.Guy, J.Tillieu: J. Chem. Phys. **24**, 1117 (1956)
4.76   T.Schaefer, Y.Yonemoto: Can. J. Chem. **42**, 2318 (1964)
4.77   A.Veillard, B.Pullman, G.Berthier: C. R. Acad. Sci. **252**, 2321 (1961)
4.78   G.Fillion, G.Maret: Unpublished
4.79   D.L.Worcester: Proc. Natl. Acad. Sci. (USA) **75**, 5475 (1978)
4.80   L.Pauling: Proc. Natl. Acad. Sci. (USA) **76**, 2293 (1979)
4.81   N.S.Murthy, I.R.Knox, E.T.Samulski: J. Chem. Phys. **65**, 4835 (1976)
4.82   K.Tohyama, N.Miyata: J. Phys. Soc. Jpn. **34**, 1699 (1973)
4.83   M.Kotani: Adv. Quant. Chem. **4**, 227 (1968)
4.84   N.Nakano, J.Otsuka, A.Tasaki: Biochem. Biophys. Acta **278**, 355 (1972)
4.85   P.G. de Gennes: *The Physics of Liquid Crystals* (Clarendon, Oxford 1974)
4.86   E.Cotton-Feytis, M.E.Faure-Frenniet: C. R. Acad. Sci. **214**, 996 (1942)
4.87   W.Arnold, R.Steele, H.Müller: Proc. Natl. Acad. Sci. **44**, 1 (1958)
4.88   N.E.Geacintov, F.V.Nostrand, M.Pope, J.B.Tinkel: Biochem. Biophys. Acta **226**, 486 (1971)
4.89   N.E.Geacintov, F.V.Nostrand, J.F.Becker, J.B.Tinkel: Biochem. Biophys. Acta **267**, 65 (1972)
4.90   F.T.Hong, D.Mauzerall, A.Mauro: Proc. Natl. Acad. Sci. **68**, 1283 (1971)
4.91   S.Sobajima: J. Phys. Soc. Jpn. **23**, 1070 (1967)
4.92   J.Torbet, G.Maret: Biopolymers **20**, 2657 (1981)
4.93   G.Weill: In *Molecular Electrooptics*, ed. by S.Krause (Plenum, New York 1981) p. 473
4.94   G.Maret, G.Weill: Biopolymers **22**, 2727 (1983)
4.95   M.P.Langevin: C. R. Acad. Sci. Paris **151**, 475 (1910)

4.96   A.Peterlin, H.A.Stuart: Z. Physik **112**, 1 and 129 (1939); and in *Optik* (Springer, Berlin, Heidelberg, New York 1933)

4.97   M.Born: Ann. Phys. **55**, 177 (1918)

4.98   W.Schütz: "Magnetooptik", in *Handbuch der Experimentalphysik*, Vol. 16, ed. by Wien-Harms (Akadem. Verlagsges., Leipzig 1936)

4.99   C.T.O'Konski, K.Yoshioka, W.H.Orttung: J. Phys. Chem. **63**, 1558 (1959)

4.100   M.J.Shah: J. Phys. Chem. **67**, 2215 (1963)

4.101   H.Geschka: Ph. D. Thesis, University of Ulm, FRG (1980)

4.102   K.Nagai, T.Ishikawa: J. Chem. Phys. **43**, 4508 (1965)

4.103   J.P.Flory: *Statistical Mechanics of Chain Molecules* (Wiley, New York 1969)

4.104   H.A.Stuart, A.Peterlin: J.Polym. Sci. **5**, 551 (1950)

4.105   L.D.Landau, E.M.Lifschitz: *Statistische Physik* (Akademie, Berlin 1970)

4.106   R.W.Wilson: Biopolymers **17**, 1811 (1978)

4.107   M.Stamm: Ph.D.Thesis, University of Mainz (1979)

4.108   E.W.Fischer, G.R.Strobel, M.Dettenmaier, M.Stamm, N.Staidle: Faraday Discuss. R. Soc. Chem. **68**, 26 (1979)

4.109   J.V.Champion, R.A.Desson, G.H.Meeton: Polymer **15**, 301 (1974)

4.110   M.Stamm, E.W.Fischer, G.Maret: Unpublished

4.111   A.E.Tonelli, Y.Abe, P.J.Flory: Macromolecules **3**, 294, 303 (1970)

4.112   A.Blumstein, G.Maret, S.Vilasagar: Macromolecules **14**, 1543 (1981)

4.113   G.Maret, A.Domard, M.Rinaudo: Biopolymers **18**, 101 (1979)

4.114   G.Weill, G.Maret: Polymer **23**, 1990 (1982)
        G.Weill, G.Maret, T.Odijk: Polym. Comm. **25**, 147 (1984)

4.115   R.B.Meyer: Appl. Phys. Lett. **12**, 281 (1968)

4.116   P.J.Batchelor, J.V.Champion, G.H.Meeton: J. Chem. Soc., Faraday Trans. 2, **76**, 1610 (1980)

4.117   P.C.Nägele, L.Beck: Macromolecules **10**, 213 (1977)

4.118   A.A.Jones: J. Polym. Sci. Polym. Phys. Ed. **15**, 863 (1977)

4.119   G.H.Meeton: Polym. Lett. Ed. **15**, 187 (1974)

4.120   G.Maret, M. v. Schickfus, J.H.Wendorff: Coll. Int. CNRS, *Physique Sous Champs Magnétiques Intenses* **242**, 71, CNRS, Paris (1975)

4.121   V.N.Tsvetkov, G.I.Kudryavtsev, E.I.Ryumtsev, V.Nicolaev, V.D.Kalmykova, A.V. Volokhina: Dokl. Acad. Nauk. SSSR **224**, 398 (1975)

4.122   T.Odijk: Polymer **19**, 989 (1978)

4.123   P.Pfeuty: J. Phys. (Paris) **39**, C2-149 (1978)

4.124   J.Skolnick, M.Fixman: Macromolecules **10**, 944 (1977)

4.125   R.E.Harrington: Biopolymers **17**, 919 (1978)

4.126   K.L.Cairney, R.E.Harrington: Biopolymers **21**, 923 (1982)

4.127   P.J.Hagerman: Biopolymers **20**, 1503 (1981)

4.128   V.Rizzo, J.Schellman: Biopolymers **20**, 2143 (1981)

4.129   G.Maret: Ph.D. Thesis, University of Konstanz, FRG (1976)

4.130   B.Vollmert: *Polymer Chemistry* (Springer, Berlin, Heidelberg, New York 1973) p. 478

4.131   S.Sokerov, G.Weill: Biophys. Chem. **10**, 161 (1979)

4.132   J.E.Godfrey, H.Eisenberg: Biophys. Chem. **5**, 301 (1976)

4.133   R.Williams, R.Crandall: Phys. Lett. **A48**, 225 (1974)

4.134   A.Klug, R.E.Franklin, S.P.F.Humphreys-Owen: Biochim. Biophys. Acta **32**, 203 (1959)

4.135   J.D.Bernal, J.Fankuchen: J. Gen. Physiol. **25**, 111 (1941)

4.136   G.J.Oster: J. Gen. Physiol. **33**, 445 (1950)

4.137   R.J.Goldacre: Nature **174**, 732 (1954)

4.138   J.H.M.Willison: J.Ultrastruct. Res. **54**, 176 (1976)

4.139   For example: C.Robinson, J.C.Ward, R.B.Beevers: Discuss. Faraday Soc. **25**, 29 (1958)

4.140   L.Onsager: Ann. N. Y. Acad. Sci. **57**, 627 (1949)

4.141   A.Yu.Grossberg, A.R.Khokhlov: Adv. Polym. Sci. **41**, 53 (1981)

4.142   P.J.Flory, G.Ronca: Mol. Cryst. Liq. Cryst. **54**, 289 and 311 (1979)

4.143   M.Nierlich, C.E.Williams, F.Boné, J.P.Cotton, M.Daoud, B.Farnoux, G.Jannink, C.Picot, M.Moan, C.Wolff, M.Rinaudo, P.G. de Gennes: J. Physique **40**, 701 (1979)

4.144 T.Odijk: Macromolecules **12**, 688 (1979)
4.145 J.Mahanty, B.W.Ninham: *Dispersion Forces in Colloid Science*, ed. by R.H.Ottewill, R.L.Rowell (Academic, New York 1976)
4.146 V.A.Parsegian: Ann. Rev. Biophys. Bioeng. **2**, 221 (1973)
4.147 F.Oosawa: *Polyelectrolytes* (Marcel Decker, New York 1971)
4.148 G.Maret, J.Torbet: Unpublished
4.149 R.Oldenbourg: Ph.D. Thesis, University of Konstanz (1981)
4.150 R.Oldenbourg, G.Maret, K.Dransfeld: Proceed. 27th Int. Symp. Macromolec. (IUPAC) Strasbourg (1981)
4.151 L.Lapointe, D.A.Marvin: Mol. Cryst. Liq. Cryst. **19**, 269 (1973)
4.152 E.Jizuka, J. Tsi Yang: In *Liquid Crystals and Ordered Fluids*, Vol. 3, ed. by J.F.Johnson, R.S.Porter, (Plenum New York 1978) p. 197
4.153 E.Jizuka: Polym. J. **10**, 235 (1978)
4.154 E.Jizuka: Polym. J. **10**, 293 (1978)
4.155 E.Jizuka, Y.Kondo: Mol. Cryst. Liq. Cryst. **51**, 285 (1979)
4.156 G.Maret, M.Milas, M.Rinaudo: Polym. Bull. **4**, 291 (1981)
4.157 P.Pincus: J. Appl. Phys. **41**, 947 (1970)
4.158 E.T.Samulski, A.V.Tobolsky: Macromolecules **6**, 555 (1968)
4.159 Y.Go, S.Ejiri, E.Fukada: Biochim. Biophys. Acta **175**, 454 (1969)
4.160 E.T.Samulski, H.J.C.Berendsen: J. Chem. Phys. **56**, 3920 (1971)
4.161 E.T.Samulski, A.V.Tobolsky: Biopolymers **10**, 1013 (1971)
4.162 E.Jizuka: J. Phys. Soc. Jpn. **31**, 1205 (1971)
4.163 E.Jizuka: Polym. J. **4**, 401 (1973)
4.164 N.Miyata, K.Tohyama, Y.Go: J. Phys. Soc. Jpn. **33**, 1180 (1972)
4.165 E.Jizuka: Polym. J. **5**, 62 (1973)
4.166 E.T.Samulski: J.Physique Suppl. **40**, C3-471 (1979)
4.167 E.Jizuka: Mol. Cryst. Liq. Cryst. **27**, 161 (1973)
4.168 C.Guha-Shridhar, W.A.Hines, E.T.Samulski: J. Physique Suppl. **36**, C1-269 (1975)
4.169 J.R.Fernandes, D.B.DuPré: Mol. Cryst. Liq. Cryst. Lett. **72**, 67 (1981), and references therein
4.170 E.G.Finer, A.Darke: J. Chem. Soc. Faraday Trans., Sect. 1 **71**, 984 (1975)
4.171 A.Blumstein (ed.): *Liquid Crystalline Order in Polymers* (Academic, New York 1978)
4.172 A.Blumstein (ed.): *Mesomorphic Order in Polymers and Polymerization in Liquid Crystalline Media* (Am. Chem. Soc. Symp. Ser. 74, Washington 1978)
4.173 A.Ciferri, W.R.Krigbaum, R.B.Meyer (eds.): *Polymer Liquid Crystals* (Academic, New York 1982)
4.174 M.Happ, H.Ringsdorf, G.Maret: Unpublished
4.175 W.Pechold, S.Blasenbrey: Kolloid Z. Z. Polym. **241**, 955 (1970)
4.176 V.Tsvetkov, E.Frisman: Acta Phys. Chim. URSS **19**, 7 (1944)
4.177 J.V.Champion, A.Dandridge, G.H.Meeton: Faraday Discuss. Chem. Soc. **66**, 266 (1978)
4.178 L.Liebert, L.Strzelecki, D.VanLuyen, A.M.Levelut: Eur. Polym. J. **17**, 71 (1981)
4.179 G.Maret, A.Blumstein, S.Vilasager: Polym. Prepr. Am. Chem. Soc. Dir. Polym. Chem. **22**, 246 (1981)
4.180 C.Noel, L.Monnevie, M.F.Achard, F.Hardouin, G.Sigaud, H.Gasparoux: Polym. **22**, 578 (1981)
4.181 F.Volino, A.F.Martins, R.B.Blumstein, A.Blumstein: C. R. Acad. Sci. Paris **292**, 829 (1981)
4.182 G.Maret, F.Volino, R.B.Blumstein, A.F.Martins, A.Blumstein: Proceed. 27th Int. Symp. Macromolec. (IUPAC), Strasbourg 1981
       G.Maret: Polym. Prepr. **24**(2), 249 (1983)
4.183 A.Blumstein, S.Vilasagar, S.Ponrathnam, S.B.Clough, R.B.Blumstein, G.Maret: J. Polym. Sci. Polym. Phys. Ed. **20**, 877 (1982)
4.184 F.Hardouin, M.F.Achard, H.Gasparoux, L.Liebert, L.Strzelecki: J. Polym. Sci. Polym. Phys. Ed. **20**, 975 (1982)
4.185 G.Maret, A.Blumstein: Molec. Cryst. Liq. Cryst.: **88**, 295 (1982)
4.186 S.Chandrasekhar: *Liquid Crystals* (Cambridge Univ. Press, Cambridge 1977)
4.187 V.Tsvetkov: Acta Phys. Chim. **19**, 86 (1944)

4.188 T.W.Stinson, J.D.Litster: Phys. Rev. Lett. **25**, 503 (1970)

4.189 J.C.Filippini, Y.Poggi, G.Maret: Colloq. Int. CNRS, *Physique Sous Champs Magnétiques Intenses*, **242**, 67 (1975)

4.190 J.C.Filippini, Y.Poggi: J. Physique Lett. **37**, L17, L97 (1976)

4.191 B.Malraison, Y.Poggi, J.C.Filippini: Solid. State Commun. **31**, 843 (1979)

4.192 P.H.Keyes, J.R.Shane: Phys. Rev. Lett. **42**, 722 (1979)

4.193 C.Rosenblatt: Phys. Rev. **A24**, 2236 (1981)

4.194 Y.Poggi, J.C.Filippini: Phys. Rev. Lett. **39**, 150 (1977)

4.195 B.Malraison, Y.Poggi, E.Guyon: Phys. Rev. **A21**, 1012 (1980)

4.196 B.Malraison: C. R. Acad. Sci. Paris **286B**, 307 (1978)

4.197 Y.Poggi, A.Aleonard, J.Robert: Phys. Lett. **54A**, 393 (1975)

4.198 I.Sakurai, Y.Kawamura, A.Ikegami, S.Iwayangi: Proc. Natl. Acad. Sci. **77**, 7232 (1980)

4.199 E.Boroske, W.Helfrich: Biophys. J. **24**, 863 (1976)

4.200 P.G. de Gennes, C.Taupin: Private communication

4.201 W.Helfrich: Phys. Lett. **43A**, 409 (1973)

4.202 W.Helfrich: Z. Naturforsch. **28C**, 693 (1973)

4.203 C.T.Meyer, Y.Poggi, G.Maret: J. Physique **43**, 827 (1982)

4.204 B.C.Gaffney, H.M.McConnell: Chem. Phys. Let. **24**, 310 (1974)

4.205 E.Meirovitch, J.H.Freed: J. Chem. Phys. **84**, 3295 (1980)

4.206 K.D.Lawson, T.J.Flautt: J. Am. Chem. Soc. **89**, 5490 (1967)

4.207 P.C.Isolani, L.W.Reeves, J.A.Vanin: Can. J. Chem. **57**, 1108 (1979)

4.208 J.Charvolin, A.M.Levelut, E.T.Samulski: J. Physique Lett. **40**, L587 (1979)

4.209 L.Q.Amaral, C.A.Pimentel, M.R.Tavares, J.A.Vanin: J. Chem. Phys. **71**, 2940 (1979)

4.210 J.Charvolin, Y.Hendrikx: J. Physique Lett. **41**, L597 (1980)

4.211 L.B.Johansson, O.Söderman, K.Fontell, G.Lindblom: J. Phys. Chem. **85**, 3694 (1981)

4.212 L.Liebert; A.Martinet: J. Physique Lett. **40**, L363 (1979)

4.213 L.Auvray: C. R. Acad. Sci. Paris **292**, 821 (1981)

4.214 G.Porte, Y.Poggi: In *Nonlinear Behaviour of Molecules, Atoms and Ions in Electric, Magnetic and Electromagnetic Fields*, ed. by L.Neel (Elsevier, Amsterdam 1979) p. 457

4.215 G.Porte, Y.Poggi: Phys. Rev. Lett. **41**, 1481 (1978)

4.216 G.Porte, J.Appell, Y.Poggi: J. Phys. Chem. **84**, 3105 (1980)

4.217 J.Appell, G.Porte, Y.Poggi: J. Colloid. Interface Sci. **87**, 492 (1982)

4.218 C.T.Meyer, Y.Poggi, G.Maret: J. Physique **43**, 827 (1982)

4.219 E.Meirovitch, J.H.Freed: J. Phys. Chem. **84**, 3295 (1980)

4.220 G.Maret, F.Cuault-Porte, A.Mayer, P.M.Vignais: Unpublished

4.221 R.Chagneux, N.Chalazonitis: C. R. Acad. Sci. **274D**, 317 (1972)

4.222 N.Chalazonitis, R.Chagneux: C. R. Acad. Sci. **275D**, 487 (1972)

4.223 R.Chagneux, H.Chagneux, N.Chalazonitis: Biophys. J. **18**, 125 (1977)

4.224 M.Chabre: Proc. Natl. Acad. Sci. USA **75**, 5471 (1978)

4.225 M.Michel-Villaz, H.R.Saibil, N.Chabre: Proc. Natl. Acad. Sci. USA **76**, 4405 (1979)

4.226 J.F.Becker, F.Trentacosti, N.E.Geacintov: Photochem. Photobiol. **27**, 51 (1978)

4.227 F.Hong: Biophys. J. **29**, 343 (1980)

4.228 H.Saibil, M.Chabre, D.Worcester: Nature **262**, 266 (1976)

4.229 M.Michel-Villaz, C.Roche, M.Chabre: Biophys. J. **37**, 603 (1982)

4.230 F.T.Hong: J. Coll. Interface Sci. **58**, 471 (1977)

4.231 R.S.Knox, M.A.Davidovich: Biophys. J. **24**, 689 (1978)

4.232 J.F.Becker, N.E.Geacintov, F. van Nostrand, R. van Metter: Biochem. Biophys. Res. Comm. **51**, 597 (1973)

4.233 J.Breton, M.Michel-Villaz, G.Paillotin: Biochim. Biophys. Acta **314**, 42 (1973)

4.234 N.E.Geacintov, F. van Nostrand, J.F.Becker: Biochim. Biophys. Acta **347**, 443 (1974)

4.235 J.Breton, P.Mathis: Biochem. Biophys. Res. Comm. **58**, 1071 (1974)

4.236 J.Breton, E.Roux, J.Whitmarsh: Biochem. Biophys. Res. Comm. **64**, 1274 (1975)

4.237 J.Garab, J.Breton: Biochem. Biophys. Res. Comm. **71**, 1095 (1976)

4.238 A.Vermeglio, J.Breton, P.Mathis: J. Supramol. Struct. **5**, 109 (1976)

4.239 J.F.Becker, J.Breton, N.E.Geacintov, F.Trentacosti: Biochim. Biophys. Acta **440**, 531 (1976)

4.240 J.Breton, G.Paillotin: Biochim. Biophys. Acta **459**, 58 (1977)
4.241 J.Breton: Biochim. Biophys. Acta **459**, 66 (1977)
4.242 J.D.Clement-Metral: FEBS Lett. **50**, 257 (1975)
4.243 D.M.Sadler: FEBS Lett. **67**, 289 (1976)
4.244 D.C.Neugebauer, A.E.Blaurock, D.L.Worcester: FEBS Lett. **78**, 31 (1977)
4.245 M.Tricot, C.Houssier: "Electrooptical Properties of Synthetic Polyelectrolytes", in *Polyelectrolytes*, ed. by K.C.Frisch, D.Klempner, A.V.Patsid (Technomic, Westport 1976)
4.246 a R.Oberthür, G.Maret, M.Showe: To be published
       b J.Torbet: EMBO J. **2**, 63 (1983)
4.247 A.Dobek, A.Patkowski, D.Labuda: J. Polym. Sci. Polym. Symp. **61**, 111 (1977)
4.248 G.Maret, R.Oldenbourg: Unpublished
4.249 J.M.Freyssinet, J.Torbet, G.Hudry Clergeon, G.Maret: Proc. Natl. Acad. Sci. USA **80**, 1616 (1983)
4.250 M.Murayama: Nature **206**, 420 (1965)
4.251 P.Costa Ribeiro, M.A.Davidovich, E.Wajnberg, G.Bemski, M.Kischinevsky: Biophys. J. **36**, 443 (1981)
4.252 A.Leitmannova, R.Stösser, R.Glaser: Stud. Biophys. **60**, 73 (1976)
4.253 G.Maret, A.Mayer: Unpublished
4.254 J.Torbet, M.Y.Norton: FEBS Lett. **147**, 201 (1982)
4.255 C.Nave, A.G.Fowler, S.Malsey, D.A.Marvin, H.Siegrist, E.J.Wachtel: Nature **281**, 232 (1979)
4.256 A.Avirom: Polymer Lett. Ed. **14**, 757 (1976)
4.257 L.Liebert, L.Strzelecki: C. R. Acad. Sci. Paris **276C**, 647 (1973)
4.258 S.B.Clough, A.Blumstein, E.C.Hsu: Macromolecules **9**, 123 (1976)
4.259 F.Cser: J. Physique Suppl. **40**, C3-459 (1979)
4.260 H.J.Lorkowski, F.Reuther: Plaste Kautsch. **23**, 81 (1976)
4.261 J.Wojtczak: Chem. Stosowana **4**, 387 (1958)
4.262 D.R.Kalkwarf, J.C.Langford: In *Biological Effects of External Low Frequency Electromagnetic Fields*, ed. by R.D.Phillips et al., DOE Symp. Ser. **50**, 408 (1979)
4.263 J.A.Oberteuffer: IEEE Trans. Magn. **9**, 303 (1973)
4.264 J.P.H.Watson: J. Appl. Phys. **44**, 4209 (1973)
4.265 S.J.Gill, C.P.Malone, M.Downing: Rev. Sci. Instrum. **31**, 1299 (1960)
4.266 S.J.Gill, C.P.Malone: Rev. Sci. Instrum. **34**, 788 (1963)
4.267 H.Kolm, J.Oberteuffer, D.Kelland: Sci. Am. **46** (Nov. 1975)
4.268 F.Paul, S.Roath, D.Melville: Br. J. Haematol. **38**, 273 (1978)
4.269 A.Porath-Furedi, P.Yanai: J. Histochem. Cyctochem. **27**, 371 (1979)
4.270 G.Bitton, R.Mitchell: Water Res. **8**, 549 (1974)
4.271 K.Mosbach, L.Andersson: Nature **270**, 259 (1978)
4.272 P.L.Kronick, G.L.Campbell, K.Joseph: Science **200**, 1074 (1978)
4.273 K.Mosbach, U.Schroeder: FEBS Lett. **102**, 112 (1979)
4.274 C.H.Evans, W.P.Tew: Science **213**, 653 (1981)
4.275 S.J.St.Lorant: "Biomagnetism, A Review", Proc. 6th Int. Conf. Magn. Techn. (MT6), (Alfa, Bratislava 1978) p. 337
4.276 D.D.Mahlum: Battelle Pacific Northwest Lab. Report, No. BNWL-1973/UC-20 (1976)
4.277 H.Martin, M.Lindauer: Fortschritte Zool. **21**, 210 (1973)
4.278 H.Aceto, C.A.Tobias, I.L.Silver: IEEE Trans. Magn. **6**, 368 (1970)
4.279 E.H.Frei: IEEE Trans. Magn. **8**, 407 (1972)
4.280 S.J.Williamson, L.Kaufmann, D.Brenner: *Superconductor Applications: Squids and Machines*, ed. by B.B.Schwartz, S.Foner (Plenum, New York 1977)
4.281 D.B.Geselowitz: IEEE Trans. Biomed. Eng. **26**, 497 (1979)
4.282 J.D.Palmer: Nature **198**, 1061 (1963)
4.283 R.Blakemore: Science **190**, 377 (1975)
       R.B.Frankel, R.P.Blakemore: J. Magn. Magn. Mat. **15–18**, 1562 (1980)
       R.B.Frankel, R.P.Blakemore, R.S.Wolfe: Science **203**, 1355 (1979)

4.284 R.B.Frankel, R.P.Blakemore: J. Magn. Magn. Mater. **15–18**, 1562 (1980)

4.285 J.L.Kirschvink: J. Exp. Biol. **86**, 345 (1980)

4.286 N.D.Gottlieb, W.E.Caldwell: J. Genet. Psychol. **111**, 85 (1967)

4.287 J.B.Phillips, K.Adler: In Ref. 4.130, p. 325

4.288 H.D.Picton: Nature **211**, 303 (1966)

4.289 M.C.Arendse: Nature **274**, 358 (1978)

4.290 M.Lindauer, H.Martin: In *Animal Orientation and Navigation*, ed. by S.R.Galler, K.Schmidt-Koenig, G.J.Jakobs, R.E.Belleville (Nasa SP-262, 1972) p. 559

4.291 M.Kisliuk, J.S.Ishay: Experientia **35**, 1041 (1979)

4.292 J.Zoeger, J.R.Dunn, M.Fuller: Science **213** 892 (1981)

4.293 D.P.O'Leary, J.Vilches-Troya, R.F.Dunn, A.Campoz-Munoz: Experientia **37**, 86 (1981)

4.294 A.J.Kalmijn: Oceanus **20**, 45 (1977)
       A.J.Kalmijn: IEEE Trans. Mag. **MAG-17**, 1113 (1981)

4.295 R.L.Jungermann, B.Rosenblum: J,. Theor. Biol. **87**, 25 (1980)

4.296 W.Wilschko, R.Wilschko: Science **176**, 62 (1971)

4.297 W.Wilschko, R.Wilschko: Z.Tierpsychologie **37**, 337 (1975)

4.298 S.T.Emlen, W.Wilschko, N.J.Demong, R.Wilschko, S.Bergmann: Science **193**, 505 (1976)

4.299 W.Viehmann: Behaviour **68**, 24 (1978)

4.300 D.Presti, J.D.Pettigrew: Nature **285**, 99 (1980)

4.301 a B.R.Moore: Nature **285**, 69 (1980)
       b J.L.Gould: Nature **296**, 205 (1982); see also Nature **300**, 293 (1982)

4.302 W.T.Keeton: Proc. Natl. Acad. Sci. **68**, 102 (1971)

4.303 C.Walcott, R.P.Green: Science **184**, 180 (1974)

4.304 J.Kiepenheuer: Naturwissenschaften **65**, 113 (1978)
       J.M.Barnothy: In [Ref. 4.34, Vol. II, P. 287]

4.305 C.Walcott: IEEE Transact. Magn. **16**, 1008 (1980)

4.306 R.R.Baker: New Scientist, p. 844 (Sept. 18, 1980)

4.307 R.R.Baker: Science **210**, 555 (1980)

4.308 J.L.Gould, K.B.Able: Science **212**, 1061 (1981)

4.309 J.L.Kirschvink: J. Exp. Biol. **92**, 333 (1981)

4.310 H.G.Wallraff: Oikos **30**, 188 (1978)

4.311 R.C.Gunter, S.Bamberger, G.Valvet, M.Crossin, G.Ruhenstroth-Bauer: Biophys. Struct. Mechanism **4**, 87 (1978)
       M.J.M.Leask: In [Ref. 4.40, p. 318]

4.312 E.D.Yorke: J. Theor. Biol. **77**, 101 (1979)

4.313 E.D.Yorke: J. Theor. Biol. **89**, 533 (1981)

4.314 J.L.Kirschvink, J.L.Gould: Biosystems **13**, 181 (1981)

4.315 J.L.Kirschvink: Biosystems **14**, 193 (1981)

4.316 R.Ranvaud, J.Kiepenheuer, G.Maret, K.Schmid-Koenig: Submitted to J. Comp. Phys.

4.317 L.J.Audus: Nature **185**, 132 (1960)

4.318 J.C.Schwarzbacher, L.J.Audus: J. Exp. Bot. **24**, 459 (1973)

4.319 N.M.Amer, C.A.Tobias: Radiat. Res. **31**, 644 (1967)

4.320 A.Edelman, J.Teulon, I.B.Puchnalska: Biochem. Biophys. Res. Commun. **91**, 118 (1979)

4.321 J.L.Schwartz: IEEE Trans. Biomed. Eng. **25**, 467 (1978)

4.322 J.L.Schwartz: IEEE Trans. Biomed. Eng. **26**, 238 (1979)

4.323 H.K.Goswami: Nucleus **16**, 24 (1973)

4.324 H.F.Linskens, P.S.G.M.Smeets: Experientia **34**, 754 (1978)

4.325 Levengood: In [Ref. 4.279]

4.326 P.W.Neurath: Nature **219**, 1358 (1968)

4.327 M.V.Joshi, M.Z.Khan, P.S.Damle: Differentiation **10**, 39 (1978)

4.328 B.Schaarschmidt: Naturw. Rundschau **30**, 365 (1977)

4.329 H.B.Brewer: Biophys. J. **28**, 305 (1979)

4.330 L.Y.Belousova: Biofizika **10**, 365 (1965)

4.331 E.M.Korchecskii, L.S.Marochnik: Biofizika **10**, 371 (1965)

4.332 V.A.Vardanyan: Biofizika **18**, 491 (1973)
4.333 V.Kumar: Stud. Biophys. **72**, 43 (1978)
4.334 H.Gremmel, H.Wendhausen, F.Wunsch: Wiss. Mitteilungen, Universität Kiel, Radiologische Klinik (1983)
4.335 D.Sperber, R.Oldenbourg, K.Dransfeld: Naturwiss. **71**, 100 (1984)

Additional Reference

Torbet, J., Dickens, M. J.: Orientation of skeletal muscle actin in strong magnetic fields. FEBS Lett. **173**, 403 (1984)
Torbet, J., Ronziere, M. C.: Magnetic alignment of collages during self-assembly. Biochem. J. **219**, 1057 (1984)

5. High Field Magnetic Confinement of Fusion Plasmas

D. Bruce Montgomery

With 27 Figures

The influence of the magnetic field strength on the criteria for the operation of nuclear fusion reactors is discussed. While it is generally desirable to use a very high field, there are other effects and technical problems that require trade-off considerations to be made in the planning and design of a fusion reactor. This is elaborated with the example of the Alcator high field tokamak experiments. The discussion is focused on practical aspects of high field magnet design with a view to mechanical stresses and to the cooling requirements. Special consideration is given to the ohmic heating transformer which is an essential element of the tokamak.

An alternative class of experiments are open confinement systems i.e. linear systems with open ends such as the theta-pinch or magnetic mirror machines. Different schemes are discussed and an outlook to future developments is given.

5.1 Introduction

The use of high magnetic fields to confine plasmas of thermonuclear interest has been pursued for more than 25 years. The magnetic field configurations necessary to confine plasmas have turned out to be more complex in many ways, and simpler in others, than was originally anticipated.

If plasmas had behaved as simple charged particles rather than exhibiting unstable collective effects, the problem would have been much easier. Much of the 1950s and 1960s was spent discovering new collective-effect instabilities which caused the plasma to leak rapidly from the test confinement device. Progress toward thermonuclear plasmas was very slow, but a base of understanding and a set of theoretical and experimental tools were gradually built. The earliest truly major devices were constructed in the late 1950s, such as the C stellarator at Princeton and a toroidal pinch (ZETA) at Culham. Both of these devices worked remarkably well from an engineering point of view, but much below expectation from a physics point of view.

In the late 1960s the tokamak began to gain recognition based on the successes of the Soviet program. It was in many ways simpler than the stellarator. The confinement transform, produced by a combination of toroidal and poloidal field, which is necessary to prevent rapid particle loss, was produced by current flow in the plasma, rather than by complex helical windings as in the stellarator. Many tokamaks have been built throughout the world, ranging in magnetic field from as little as 0.14 T to as high as 14 T, and from major toroidal radius of as little as 6.4 cm to 2.5 meters. The most recent tokamak construction, the TFTR at Princeton, represents a capital cost of some $ 300 million.

The recent achievement of significant plasma parameters, largely in tokamak devices, has given high confidence that reactor conditions can be achieved. Alcator C at MIT has achieved confinement times ($n\tau_E$), within a factor of about 5 of that required for ignition, but at temperatures 10 times too low. The PLT at Princeton has achieved the temperatures required for ignition, but at confinement $n\tau_E$, a factor of 20 too low. These two achievements will be combined in the new TFTR tokamak to achieve near ignition conditions, specifically "Scientific Breakeven", where the alpha heating from fusion reactions exceeds the heating necessary to produce the reactions. Devices beyond TFTR will be required to reach true ignition where sufficient fusion power is produced to overcome all losses and produce economic net power. The community confidently expects to produce such "demonstration reactors" by the turn of the century.

The Mirror Program is considered to be the principal backup to the Tokamak Program. While tokamaks have achieved the most notable success in advancing toward the necessary reactor conditions, the uncertainties still to be faced and the engineering complications of toroidal systems emphasize the wisdom of using multiple concepts. The largest of the current mirror constructions involves a large "Yin-Yang" superconducting coil, and that installation plus the many megawatts of neutral beam injection will cost some $ 100 million.

The Tokamak and the Mirror are both "main-line" programs. The remainder of the projects are generally referred to as "alternate concepts". One primary alternate is the Elmo Bumpy Torus, a series of linked mirror coils, and a second is the Reverse Field Pinch. Stellarators and torsatrons are also receiving renewed interest.

Fusion research has been a relatively "big science" since the early 1950s and now has a budget of 400 million dollars. It would be truly impossible to give a meaningful summary of the research and construction which has taken place over those years. A reasonable engineering survey can be made by examining the *Engineering Problems in Fusion Research* Conference proceedings [5.1–4] which are published every two years. They have now grown to a four-volume size and the conference is attended by approximately 1000 engineers working in magnetic confinement and laser fusion. Two relatively recent publications by the Office of Fusion Energy, Department of Energy [5.5,6], are useful in summarizing the status of fusion research today.

This volume concentrates on the use of high magnetic fields in various branches of science. The question arises as to what to consider a high field in plasma physics, in what is clearly a continuum. Most plasma devices must consider trade-offs between parameters such as size and field. Certain plasma characteristics favor scale, and others, fields. We have attempted to treat this trade-off aspect in some detail under our tokamak design considerations. Very often the stresses in the toroidal field coils are just as much determined by scale as by field. Most machines are designed to operate their coils near the stress limits no matter what the field level. We have therefore not made a sharp distinction, but we have tended to look at devices where the chief concern is with the field level and have looked mostly at devices where the central field is about 10 T. We should point out, however, that in tokamak toroidal devices the field is a function of radius, and the peak fields at the winding can be twice as high as on the centerline. Thus a reactor scale tokamak with a 5 T central field can have a peak field above 10 T.

Superconducting magnet technology is not treated in this chapter[1]. Most experimental devices are still made with copper coils cooled with either water or liquid nitrogen for the more compact high-field devices. In large projects like TFTR this can call for very large power supplies approaching 1000 MW peak power and 5000 MJ stored energy. However, no magnetic confinement reactor concept can expect to have an acceptably low circulating power without the use of superconducting coils. Such coils are now under development for use in future thermonuclear reactors.

The possible use of ultrastrong pulsed fields for dynamical plasma confinement is discussed in Section 6.6.

In this chapter we consider magnetic confinement devices in two broad categories: closed systems and open systems. Under closed systems we have considered the tokamak in some detail. Under open systems we have considered both truly linear devices and closed systems whose aspect ratio allows them to be "linearized".

5.2 Closed Confinement Systems

5.2.1 Background

The tokamak is the most widely utilized closed magnetic confinement device. A typical tokamak is shown in Fig. 5.1. The magnetic field system is composed of a toroidal field, an induction field which induces the plasma current and an equilibrium field which controls the position of the plasma column. The plasma current provides all the heating in so-called ohmically heated tokamaks, but is often supplemented by neutral-beam heating or radio-frequency heating to obtain higher temperatures. The parameters of a tokamak reactor large enough to produce net power [5.7] are given in Table 5.1. The peak magnetic field would be 13 T and the toroidal field coils some 6 by 9 meters. In a device of this scale, the great majority of the heating to reach ignition temperature is provided externally, generally as neutral-beam heating in current reference designs. Ignition temperature, taken as the point at which the alpha particles emitted in the fusion process provide enough heating for the process to be self-sustaining, depends on density, but is in the range of 10 keV. At these temperatures the

---

1 *Editor's note*: it was originally planned to include a chapter on superconducting high field magnets in this book, but this was not ready in time to be included.

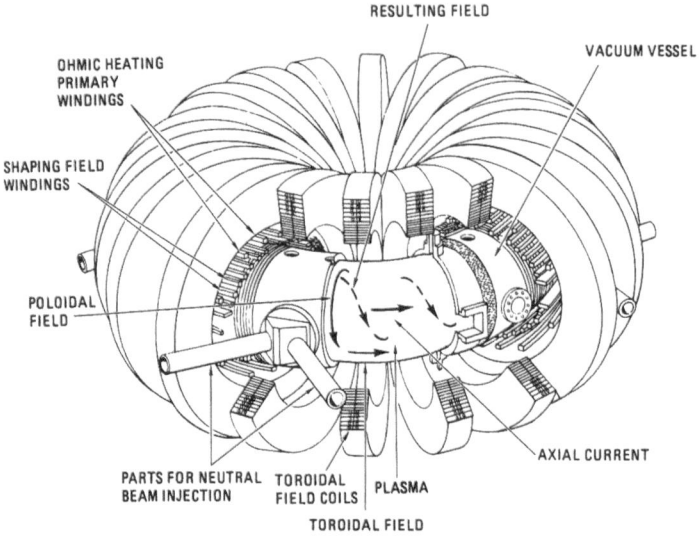

Fig. 5.1. Basic components of the tokamak

plasma has a conductivity nearly as good as copper at liquid helium temperature. This phenomenal conductivity of course explains why it is so difficult to heat plasmas to high temperatures with ohmic heating alone.

A number of research tokamaks produce temperatures of 1 keV where plasma conductivity equals that of copper at room temperature. At this temperature, ohmic heating can provide all the necessary heating if the device parameters are chosen to favor high plasma current density. This generally calls for the highest magnetic fields in compact devices. In the next section we examine the trade-off considerations which affect such high current density devices.

Table 5.1. Parameters of a High Field Tokamak Fusion Reactor

Major radius	6 . m
Plasma halfwidth	1.2 m
Field at plasma	7.4 T
Maximum field at winding	13.1 T
Plasma current	6.7 MA
Total fusion power	2440 MW
14 MeV neutron loading	3.4 MW/m²
Gross electric power	870 MW
Net electrical power	775 MW

5.2.2 Tokamak Plasma Trade-Off Considerations

In a tokamak, the confinement properties are a function of the poloidal magnetic field produced by the plasma current. There is a well-established stability requirement relating the poloidal field B_p, the toroidal field B_T, and the plasma minor and major radii, a and R:

$$B_p \leq B_T \frac{a}{R} \frac{1}{q}. \tag{5.1}$$

The q factor, called the safety factor, represents how closely one can approach the limiting value $B_p = B_T a/R$. The q value achievable in machines is in the range of 2.0 to 3. Most analysis assumes a value of 2.5.

A somewhat simplified expression for the maximum energy density which can be contained by the poloidal field can be written:

$$nkT \leq \frac{B_p^2}{2\mu_0} \leq \frac{1}{2\mu_0} \cdot \frac{1}{q^2} B_T^2 \left(\frac{a}{R}\right)^2. \tag{5.2}$$

In a toroidal device, it is difficult to increase (a/R) beyond a certain value, and all tokamaks have an a/R ratio between 0.3 and 0.15. The energy density achievable then depends crucially on the toroidal field value.

All tokamaks depend on ohmic heating for their initial start-up and to produce target plasmas for any subsequent auxiliary heating. The ohmic heating energy density is determined by the current density and the plasma resistivity, $\omega_v = J^2 \varrho$. The current which can be carried in the plasma is determined by the limitation on poloidal field expressed by (5.1), and consequently the maximum achievable current density for a given major and minor radius is

$$J < \frac{2}{\mu_0 q} \frac{B_T}{R}. \tag{5.3}$$

If achievement of high ohmic heating rates were the exclusive goal, one would raise the toroidal field as high as possible and build a machine of the minimum major radius.

Achievement of high ohmic heating temperature involves not only maximizing the heating rate, but also minimizing the energy loss rate by maximizing the energy confinement time. Tokamak experiments have revealed that confinement time scales with plasma density and the minor radius squared

$$\tau \sim na^2. \tag{5.4}$$

The constant of proportionality is 5×10^{-21} in many tokamaks [5.8].

The achievable density in ohmically heated tokamaks is proportional to the toroidal field and thus

$$\tau \sim B_T a^2. \tag{5.5}$$

In the Alcator tokamaks, the achievable temperature with ohmic heating is closely related to the poloidal field, and obeys [5.9]

$$T = 700 \left(\frac{a}{R} \frac{B_T}{q_0} \right)^{0.8}$$

where q_0 is the q on the axis, and generally is taken as 1.0.

The product of confinement time and particle density is a parameter of particular interest to the goal of ignition. The product must achieve a certain critical value (approximately 10^{20} s/m^3) known as the Lawson criterion, to produce more power from α heating than is consumed in producing it. The product of $n\tau$ can be written as

$$n\tau \sim n^2 a^2 \sim B_T^2 a^2. \tag{5.6}$$

One last figure of merit often used to describe tokamaks is the total plasma current multiplied by the inverse aspect ratio. This product controls the α particle confinement, a value of 8×10^6 A being required for adequate confinement:

$$I_p \frac{R}{a} \leq B_T \frac{a(2\pi)}{\mu_0 q}. \tag{5.7}$$

All the relationships we have developed show the benefit of a high toroidal field. Given a fixed investment, however, the criteria cannot all be satisfied because they are mutually incompatible. Large devices improve confinement time, while small devices improve the ohmic heating. To state that devices should be no larger than necessary and have as high a field as economically feasible is true, but not very helpful. We have plotted some of these figures of merit in Fig. 5.2 for a fixed aspect ratio $A = R/a$ of 3, which is very typical of tokamaks.

As an example of the use of Fig. 5.2, we take the $B_T \cdot a$ criterion. We calculate the necessary $B_T \cdot a$ value from (5.7), as the requirement on α confinement to sustain ignition is the most restrictive. We find that $B_T \cdot a$ must lie in the 4 to 5 range. Recent studies [5.10] indicate that if B_T/R is near 25 T/m and B_p near 2.5 T, an ohmically heated tokamak might achieve ignition. For a $q = 2.5$, Fig. 5.2 shows a central field near 20 T. Devices of this class have been labeled "Ignitors" [5.11]. By way of comparison, the most ambitious ohmically heated tokamak currently in operation is the Alcator C machine [5.12], where $B_T = 14$ T, $R = 0.64$ m and $a_m = 0.215$ m ($a_p = 0.17$ m).

If we relax the requirement of ohmic heating to ignition, and rely instead on some form of auxiliary heating, the requirements on B_T/R and B_p can be relaxed.

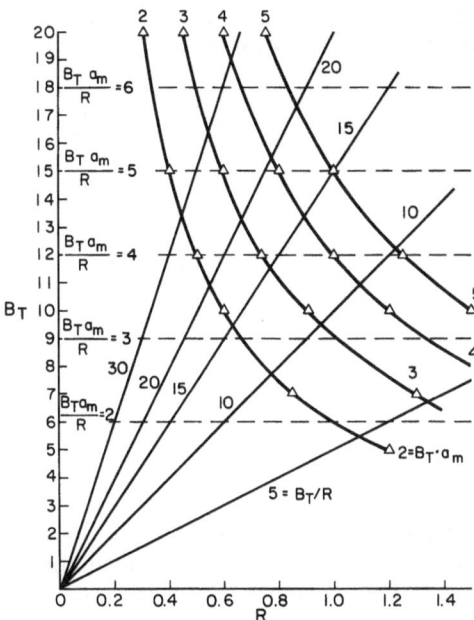

Fig. 5.2. Plasma figures of merit plotted for a fixed aspect ratio $A = 3$

Such a device might be the Ignition Test Reactor (ITR) [5.13, 14] being studied by IPP Garching and the Plasma Fusion Center at MIT in which the major radius is 1.3 m and the field 10 T. The $B_T \cdot a$ product exceeds 4 but the ohmic heating is down by an order of magnitude. The difference is made up with neutral-beam heating and major radius compression.

5.2.3 Tokamak Toroidal Field Trade-Off Considerations

Tokamak design projects invariably operate under some constraints, whether it be the size of an existing power supply, or the strength of materials available. It is therefore generally necessary to optimize plasma parameters under constrained conditions. Clearly one cannot move around the parameter space of Fig. 5.2 without regard for the consequences to magnet heating and stress, or power supply capacity. Since all aspects of the tokamak are interrelated, true optimization is often complex. To gain some insight into the process we have chosen to construct a unified figure for magnet-related parameters (Fig. 5.6 below) which parallels Fig. 5.2, but which is based on various simplifying assumptions. The next several sections derive these simplifying relations.

a) Toroidal Field Coil Heating

The critical zone in the TF coils is clearly in the inner throat of the coil. The cross section available for copper and structure is least in this region. To estimate the

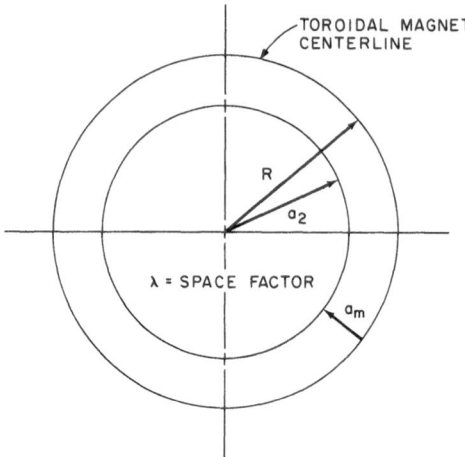

TOROIDAL MAGNET
CENTERLINE

R

a_2

λ = SPACE FACTOR

a_m

Fig. 5.3. Definition of variables in the tokamak core region

heating which will take place during a pulse, we first assume that all the cross section between the central axis and the inner bore of the TF coil is available on some fractional basis as shown in Fig. 5.3. This means that we shall temporarily ignore the space which is ultimately utilized by the central transformer. As we shall see later, this is an oversimplification and Sect. 5.2.4d derives suitable correction factors.

The toroidal field produced by the total TF coil current is

$$B = \frac{\mu_0 I}{2\pi R} = 2 \times 10^{-7} \frac{I}{R}$$

$$I = \frac{2\pi}{\mu_0} B_0 R_0 .$$

(5.8)

The cross section available for that current is then

$$A_c = \pi a_2^2 \lambda = \pi (R_0 - a_m)^2 \lambda,$$

(5.9)

where λ is the fractional space allotted to the conductor. The current density j and the length of the pulse τ determine the heating. It is convenient to write the heating parameter as $j^2\tau$, and combining (5.8.9) we can write

$$j^2\tau = \left(\frac{I}{A_c}\right)^2 \tau = \tau \left[\frac{2}{\mu_0} \frac{1}{\lambda} \frac{B_0}{R_0} \left(\frac{A}{A-1}\right)^2\right]^2,$$

(5.10)

where $A = R_0/a_m$. It is interesting to note that current density in the magnet is related to current density in the plasma by the same B/R relationship.

The temperature rise in the coil can be found directly from the $j^2\tau$ values [5.15]. Most high field compact tokamak experiments used liquid nitrogen for

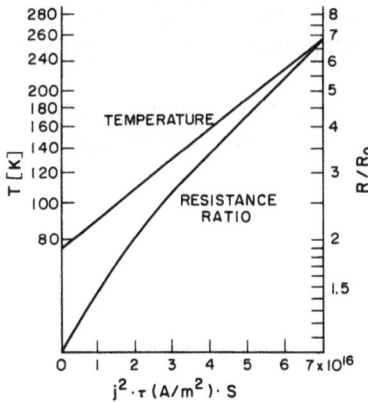

Fig. 5.4. Temperature rise and resistivity ratio as a function of the square of current density and pulse time, for copper at an initial temperature of 77 K

precooling. The resistivity of copper is lower by a factor of 7.5 at liquid nitrogen temperature, while the heat capacity is lower by a factor of only 2. Thus the heating rate is almost 4 times less near nitrogen temperature than near room temperature. A graph of ΔT and ϱ/ϱ_0 versus $j^2\tau$ for copper starting at 77 K is given in Fig. 5.4. We note that the initial resistivity doubles for $j^2\tau = 2 \times 10^{16}$ and essentially reaches its room temperature value at 6×10^{16}. If a typical magnet with $A = 3$ and $\lambda = 0.5$ must have an equivalent square-wave pulse of 1 s, and if we limit the resistance rise to a factor of 2, the B/R ratio must be less than 25 T/m. To achieve 10 s under the same conditions, it must be less than 8 T/m. These heating constraints are plotted in Fig. 5.6 below.

b) Toroidal Field Simplified Stress Analysis

The principal stress in the TF coil is the tension in the throat. If we ignore all bending stresses, and radial stresses due to the inward force on the coil (or assume a bending free shape) we can write a very simple expression for the tension in the throat [5.10]. Sophisticated finite-element analysis has shown that this simple tension value will in general be off by only less than 10% from the true tension value. Using these simplified assumptions, the upward force acting on the throat can be written as

$$F = \frac{\pi}{\mu_0} B_0^2 R_0^2 \left(\frac{1}{A-1} \right). \tag{5.11}$$

If we again assume that all the cross section between the central axis and the TF coil inner bore were available to carry this load, we can write an expression for the tension stress

$$\sigma_z = \frac{B_0^2}{2\mu_0} \frac{2}{A} \left(\frac{A}{A-1} \right)^3. \tag{5.12}$$

For a magnet of $A = 3$ and a field of 10 T, this stress would be 9.2 kg/mm². When combined with radial stress and bending, the combined stress level approaches the safe limit for nonreinforced copper coils.

An equally simple approximation can be made for the radial stress [5.10] due to the inward force on the coils by assuming that the inward magnetic pressure is supported on the inner circumference:

$$\sigma_r = \frac{B_0^2}{2\mu_0} \frac{A}{A-1}.$$ (5.13)

For the above example of $A = 3$, $B_0 = 10$ T, the radial stress would be 6.3 kg/mm², or about 2/3 of the tension stress.

It is customary to combine these orthogonal stresses into a von Mises stress. Ignoring shear stresses we can write

$$\sigma_{VM} = \sqrt{\sigma_z^2 + \sigma_r^2 - (\sigma_r \cdot \sigma_z)}.$$

For the above example, the combined stress would be 12.5 kg/mm².

Two additional simplifying formulas are useful in examining the toroidal field coil stress, namely the total inward and total upward forces on the coil based on magnetic pressure considerations [5.10]. To make this approximation, all the current in the coil is assumed to flow at a single radius a_s. Choosing this radius as one third of the way through the copper gives the most reasonable fit to true analysis. The total vertical and radial force can then be written as

$$F_z = \frac{\pi}{\mu_0} B_0^2 R_0^2 \ln \frac{R_0 + a_s}{R_0 - a_s}$$

$$F_r = \frac{2\pi}{\mu_0} B_0^2 R_0^2 \left[\frac{R_0}{(R_0^2 - a_s^2)^{1/2}} - 1 \right].$$ (5.14)

c) Toroidal Field Stored Energy and Resistive Power

The energy stored in the toroidal field can be estimated by assuming all the energy is stored in the bore of the coil:

$$U = \frac{B_0^2}{2\mu_0} \cdot V$$

$$V = 2\pi^2 a_m^2 R$$ (5.15)

$$U = \frac{\pi^2}{\mu_0} \frac{R^3}{A^2} B_0^2.$$

The correction is nontrivial however; for example, if the outer radius of the coil is 1.5 times the inner radius, the additional energy stored outside the bore is approximately 30 %.

As an example of energy stored, if a tokamak had a central field of 10 T and major radius of 1 m and an aspect ratio of 3, the field would store 85.6 MJ. Lines of constant stored energy are given in Fig. 5.6 below.

It is also possible to obtain a reasonable estimate of the resistive power in a tokamak by assuming that the great majority of the power is dissipated in the critical throat region. Outside this constricted region the copper cross section is generally increased until the resistance is negligible compared to that of the throat. To obtain an estimate, we assume that power is dissipated in a cylinder whose outer diameter is $(R_0 - a_m)$ and whose height is $2a_m$. The resistance for an equivalent single turn can then be written as

$$R = \varrho/\lambda \, \frac{2a_m}{\pi(R_0 - a_m)^2}$$

$$R = \frac{\varrho}{\lambda} \cdot \frac{2A}{\pi R_0 (A-1)^2} \,. \tag{5.16}$$

The resistive power can then be determined using (5.8) for the field and current, and (5.16):

$$W = I^2 R = \frac{\varrho}{\lambda} \, \frac{8\pi}{\mu_0^2} \, B_0^2 R_0 \, \frac{A}{(A-1)^2} \,. \tag{5.17}$$

The peak resistive power in the tokamak will depend on the amount of heating. If we take a case where $j^2\tau$ is 2×10^{16}, the resistance rise will be a factor of 2, Sect. 5.2.3a. If we look at a liquid nitrogen cooled copper tokamak of central field 10 T, major radius 1 m, and aspect ratio of 3, the peak dissipated power would be approximately 12 MW using (5.16).

The ratio of stored energy to energy dissipated per second can be written from (5.15, 17)

$$\frac{U}{W^1} = \frac{\pi\mu_0}{8} \, \frac{\varrho}{\lambda} \cdot R_0^2 \, \frac{(A-1)^2}{A^3} \,. \tag{5.18}$$

It is interesting to note that this ratio depends on the scale of the magnet but not on the magnetic field. The larger the tokamak, the larger the ratio of stored energy to dissipated energy per unit time, assuming similar aspect ratios.

If we make some assumptions about temperature rise, the time variable resistivity in (5.18) can be replaced by a peak value. Figure 5.5 plots the ratio of stored energy to dissipated energy/second for $j^2\tau = 2 \times 10^{16}$ and 6×10^{16}. The figure can be used to assess the resistive power requirements once the stored

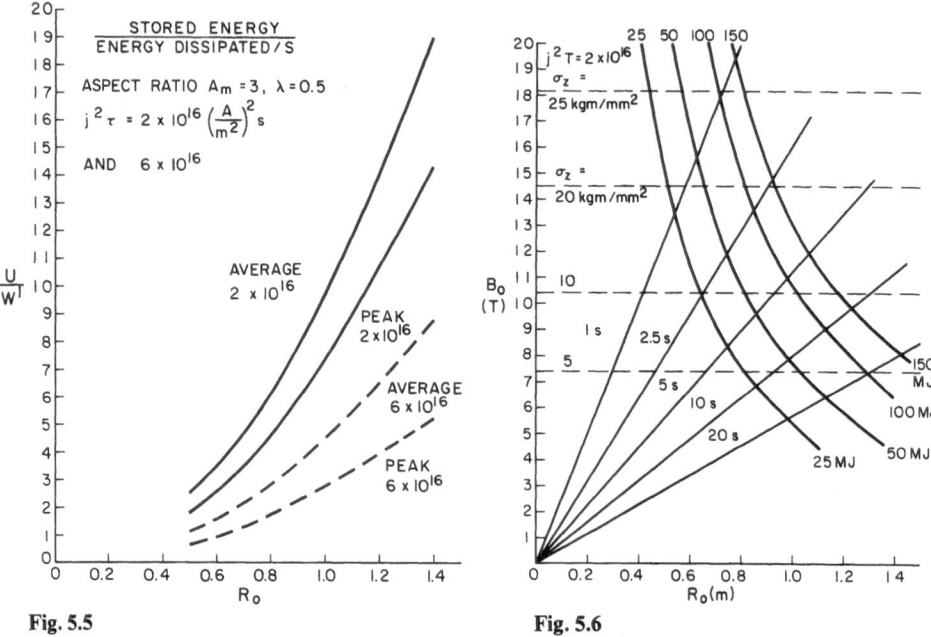

Fig. 5.5

Fig. 5.6

Fig. 5.5. Ratio of stored energy to dissipated power per unit length for a fixed aspect ratio $A = 3$, for two values of $j^2\tau$

Fig. 5.6. Magnet figures of merit plotted for a fixed aspect ratio of $A = 3$

energy is known. For example, if the tokamak has a 1 m radius and the $j^2\tau$ is limited to 2×10^{16}, the peak power/second would be 1/7.3 of the stored energy.

The average resistivity over the pulse can be used to estimate the total resistive energy extracted from the energy supply. For the same case as above, the average dissipated power/second is 1/9.7 of the stored energy.

d) Design Curves for Toroidal Field Trade-Off Considerations

The simplifying assumptions in the previous sections can be combined together into a single parametric presentation as has been done in Fig. 5.6 for a single value of aspect ratio, and a single temperature rise assumption.

It is significant to note how similar in form they are to the set of plasma parametrics chosen in Fig. 5.2. The lines of constant plasma current density B_T/R reflect the lines of constant pulse time limits set by magnet heating. The lines of constant poloidal field $B_T a/R$, which represent plasma pressure, reflect the lines of constant magnet stress. Lastly the lines of constant $B_T \cdot a$, which are related to the overall confinement quality, reflect the total stored energy in the system.

The most striking feature of Fig. 5.6 is the clear indication that larger systems will support longer pulse times and have lower stresses at fixed aspect ratio. As

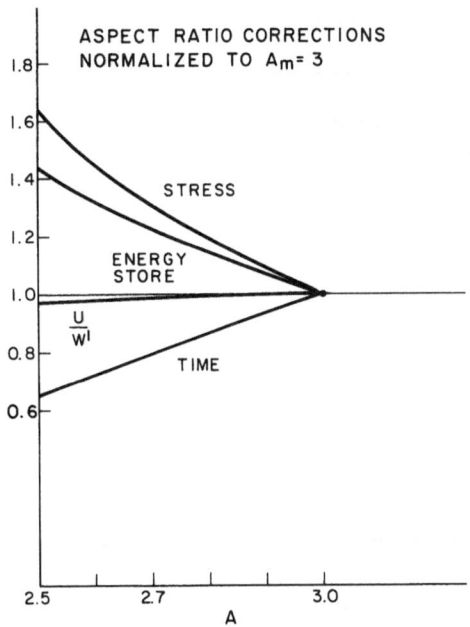

Fig. 5.7. Correction factors for the magnet figures of merit in Fig. 5.6 as a function of aspect ratio

pointed out in Sect. 5.2.1 however, these larger systems must depend on auxiliary heating.

Figures 5.5, 6 are constructed for a fixed aspect ratio of 3. The relationships used to construct the figures all contain the aspect ratio, and the corrections required as the aspect ratio changes from 3 have been plotted in Fig. 5.7. As the aspect ratio decreases at a fixed field, stresses and stored energy go up and the allowed pulse time goes down. The stored energy to dissipated energy/second ratio stays essentially constant.

With Figs. 5.2, 6, 7 one can explore optimizations under given constraints. Table 5.2 gives an example of what happens to a toroidal system as the aspect ratio is changed from 3 to 2.5 under the constraint of (a) constant stress and constant major radius, (b) constant stress and constant plasma current density, B_T/R and (c) constant stored energy and constant radius.

In two of the cases some parameters improve with the reduction in aspect ratio and others decline. In case (a) the allowable pulse time increases slightly, but all the plasma parameters for an ohmically heated tokamak diminish in comparison to the $A = 3$ case. For (b) all plasma parameters diminish, and the pulse time is shorter, but less stored energy is required.

In case (c) only one of the plasma parameters declines, but the stress has increased. For (a) and (b), we therefore conclude that an aspect ratio of 3 is more nearly optimum than an aspect ratio of 2.5, and for (c), 3 would also be advantageous if the stress level is a limiting factor.

Table 5.2. Scaling from base case $A = 3$, with a change to $A = 2.5$ under three different constraints

	(a) Constant σ_z Constant R	(b) Constant σ_z Constant B/R	(c) Constant U Constant R
B_T	0.78	0.78	0.83
R	$\boxed{1.0}$	0.78	$\boxed{1.0}$
τ_{pulse}	1.06	0.65	0.95
σ_2	$\boxed{1.0}$	$\boxed{1.0}$	1.14
U	0.87	0.415	$\boxed{1.0}$
$B_T \cdot a$	0.93	0.74	1.0
$\dfrac{B_T \cdot a}{R}$	0.93	0.93	1.0
$\dfrac{B_T}{R}$	0.78	$\boxed{1.0}$	0.83

This type of exercise can be pursued for other constraints and for aspect ratios greater than 3.0. An optimum choice can generally be found for any one parameter, but it seldom coincides with the optimum for another parameter of interest.

In constructing the ratios in Table 5.2 the aspect ratio for the magnet and plasma are the same, implying no radial build for the plasma limiter and chamber. This could be corrected if a more precise optimization search were being made.

5.2.4 Ohmic Heating Transformer Considerations

a) Background

The tokamak depends on plasma current for confinement and must induce that current through flux linkage to a transformer coil. There are three options for such a transformer; an iron core can be passed through the central portion of the machine, an air core winding can replace the iron core, or a winding wound on the vacuum chamber can parallel the plasma current. The last case has the best coupling but greatly complicates the tokamak construction; the iron core requires little excitation energy but is flux limited by saturation and produces nonlinear equilibrium complications. The middle case is generally chosen, even though the energy requirements are the greatest.

Transformer excitation is often accomplished by precharging the coil, then injecting resistance and rapidly collapsing the field. This generally provides sufficient flux to establish the plasma current rapidly, and the coil is then driven with an opposite direction current at a relatively slow rate of rise to maintain the

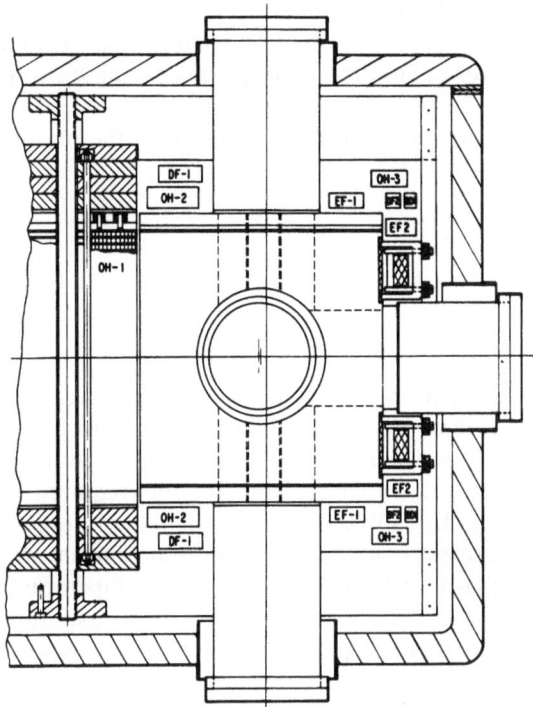

Fig. 5.8. Overview of the Alcator C ohmic and equilibrium coils. The ohmic coils are OH-1,2,3; the plasma coupled equilibrium coils are EF1,2 and the plasma decoupled equilibrium coils are DF1,2; Coil BD-1 provides radial fields for vertical stability

current. If a flux change greater than the precharge is required to establish the current, a parallel capacitor bank of energy storage comparable to the coil can be provided to promote a "double swing" [5.16]. As the transformers generally involve many megajoules of energy, such a bank represents a major investment. Most tokamaks are therefore designed to have a transformer with sufficient flux in the precharge to establish the current fully.

To initiate a plasma discharge, the ionized-particle orbits must close upon themselves. Therefore, the transformer is generally designed to have negligible fringe field in the plasma region. As soon as plasma current starts to flow, however, there must be a proportional vertical field (Sect. 5.2.5), but this is provided by an independent coil set. A typical air-core ohmic heating transformer is shown in Fig. 5.8. More than 80 % of the flux is produced by the central coil. The other coils are often referred to as "compensation" coils, as their main purpose is to compensate the fringe field in the plasma region.

b) Flux Requirements

The first step in transformer design is to establish the approximate flux requirements. The flux required is the sum of three components: (1), that required to establish the plasma, determined by the inductance of the plasma column and the maximum current; (2), that required to maintain the steady plasma current, which depends on the resistive voltage drop in the plasma and the pulse length; and (3), the resistive voltage drop in the plasma during start-up of the discharge.

The flux required to establish the plasma is determined from the plasma inductance and the peak current

$$\Phi = L \cdot I_p$$

$$L \cong \mu_0 R_0 \left(\ln \frac{8R}{a} - 1.25 \right).$$

(5.19)

assuming a parabolic current distribution across the plasma cross section. For a plasma aspect ratio of 3 and a major radius of 1 m, the inductance would be 2.4×10^{-6} H (henry). If the tokamak were to have a 10 T field and if we assume a $q = 2.5$, (5.7) gives the peak current as 2.2×10^6 A. Thus the flux required to establish the current would be 5.3 webers (Vs).

To determine the voltage necessary to maintain a steady-state current, we must know the resistivity of the plasma which is dependent on temperature, impurity content and other factors of increasing importance as the temperature approaches reactor conditions. For estimating purposes however, it is possible to approximate the resistivity as being equal to that of copper when the plasma is relatively free of impurities and the temperature is near 1 keV. Resistivity at higher or lower temperatures can be found from scaling with temperature to the $(-3/2)$ power. Thus a plasma at 15 keV would be 58 times more conductive and at 100 eV, 32 times less conductive. The effect of impurities is taken as a simple multiplier of the resistivity, an effective plasma "Z" of 2 implying twice the resistivity.

To estimate the voltage drop for our example of a 2.2×10^6 A current in a 1 m radius, aspect ratio 3 machine, we assume that the hottest part of the plasma occupies the region to half the minor radius or 0.165 m. If the plasma is assumed to be 1 keV, the resistivity is taken as 2×10^{-8} Ωm, equivalent to copper, and the resistance would be approximately 1.5×10^{-6} Ω. To maintain a current of 2.2×10^6 A would therefore require 3.3 V. If the pulse is assumed to be one second long, 3.3 Vs would be required for maintenance.

It is difficult to estimate the third component of the required flux, namely that required for resistive voltage during start-up. The impurity content and the temperature of the discharge at early stages is difficult to predict. A reasonable upper estimate can probably be made by assuming a temperature near 100 eV for some early period. For example, if we assume 100 eV for 10 ms, the volt-seconds required for our example would be comparable to the volt-seconds required to maintain the plasma. This illustrates why it is advantageous to establish the current as rapidly as possible to avoid costly early volt-second losses.

If in the above example we wish to build up the current in 100 ms we must supply an inductive loop voltage of

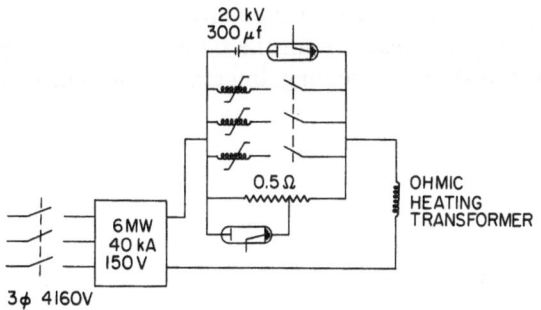

Fig. 5.9. Typical ohmic heating transformer excitation circuit

$$E = L\frac{dI}{dt} = 2.4 \times 10^{-6} \times \frac{2.2 \times 10^6}{0.1} = 50 \text{ V}.$$

To build up the current in 100 ms of course implies that the transformer current will be collapsed in a comparable time. If we assume that the transformer might store 6.5 MJ (Sect. 5.2.4d) and operate at a current level of 50×10^3 A, the transformer inductance would be 52×10^{-3} H, and a 100 ms discharge would require a resistance which would drop 26 kV. Thus the switch which must open and commutate the current into a resistance, as shown in Fig. 5.9, must carry 50 kA and open across 26 kV. These switches have now become major features of tokamak excitation systems [5.17, 18].

The transformer need not supply all the flux needed to build up the current. The equilibrium field needed to establish the current must be raised from zero to full value as the current is established. Integrating this field over the area from the machine centerline to the plasma centerline adds appreciably to the flux available for start-up. As derived in Sect. 5.2.5, the equilibrium field for our 2.2×10^6 A, 1 m radius plasma would be 0.67 T, and if we assume that the value is uniform out to the 1 m radius, it would produce 2.1 Vs of additional flux.

c) Simplified Transformer Design

If we assume that the central transformer is long compared with its diameter, some simple approximations can be derived for flux, stress and peak field. The central field, for example, is essentially

$$B_t = \mu_0 \frac{NI}{l}$$

$$= \mu_0 j \lambda (a_2 - a_1) \tag{5.20}$$

$$= \mu_0 j \lambda a_2 (1 - \gamma),$$

where $\gamma = a_1/a_2$, the inside over the outside transformer radius, j the current density in the conductor and λ the volume fraction occupied by the conductor.

The flux produced by swinging such a transformer coil from a field of B_t to zero can be estimated if we assume the field within the bore is B_t, and then linearly decreases to zero at the outside of the winding. Integrating the field area we obtain

$$\Phi = \frac{\mu_0 \pi}{3} j \lambda a_2^3 (1 - \gamma^3).$$ (5.21)

For typical cases where a_1 is of the order of 1/3 to 1/2 of a_2, γ^3 can be neglected.

If we want to estimate the pulse time capability of the transformer, we can rewrite (5.21) in the $j^2\tau$ form used with the toroidal field coils

$$j^2\tau = \left(\frac{\Phi}{a_2^3}\right)^2 \left(\frac{3}{\pi \mu_0 \lambda}\right)^2 \tau.$$ (5.22)

Lines of constant pulse time are plotted versus required Φ and available a_2 in Fig. 5.10. The pulse time plotted is the equivalent square wave at full current made up from the initial charging current ramp, the field collapse, and the negative swing ramp.

It is also relatively simple to estimate the peak stress in the transformer. If each turn in the transformer is assumed to be locally self-supported by a hoop tension σ_{hoop}, the approximate stress can be written as

$$\sigma_{hoop} = j \times B \cdot r = \mu_0 \lambda j^2 (a_2 r - r^2).$$ (5.23)

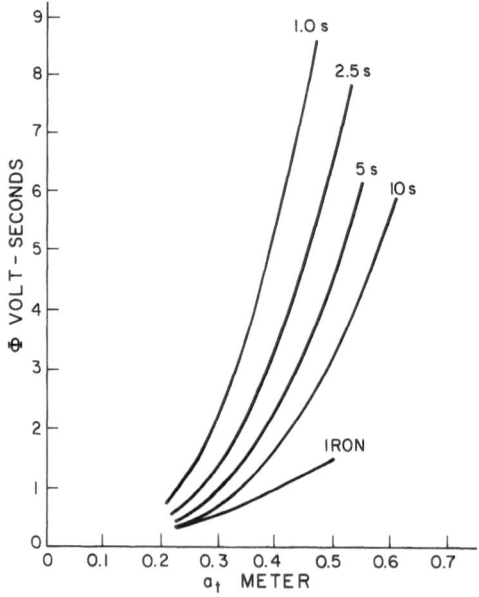

Fig. 5.10. Allowable equivalent pulse time and available single swing flux as a function of transformer outer diameter

The maximum of this hoop tension occurs at the radius at which the derivative with respect to r is zero, that is $a_2/2$, provided that the coil extends radially inward at least this far. Evaluating (5.23) for $a_2/2$ leads to an expression for peak stress:

$$\sigma_{max} = \frac{\mu_0 \lambda}{4} j^2 a_2^2 . \tag{5.24}$$

We can substitute (5.24) into (5.21) to obtain stress in terms of the required flux and the available outer radius; we have neglected γ^3:

$$\sigma_{max} = \frac{1}{\lambda} \left(\frac{\Phi}{a_2} \right)^2 \frac{1}{(\frac{2}{3}\pi)^2 \mu_0} . \tag{5.25}$$

Lines of constant stress are plotted in Fig. 5.11.

The stored energy in the transformer can be evaluated from the flux produced and the ampere-turns necessary to produce the flux

$$\Phi = L \cdot I$$

$$U = \tfrac{1}{2} L I^2 = \tfrac{1}{2} \Phi \cdot I.$$

To evaluate the net I, we must make an assumption about the transformer length, which we take as a typical value of $1.5\, R_0$. If we further assume that the

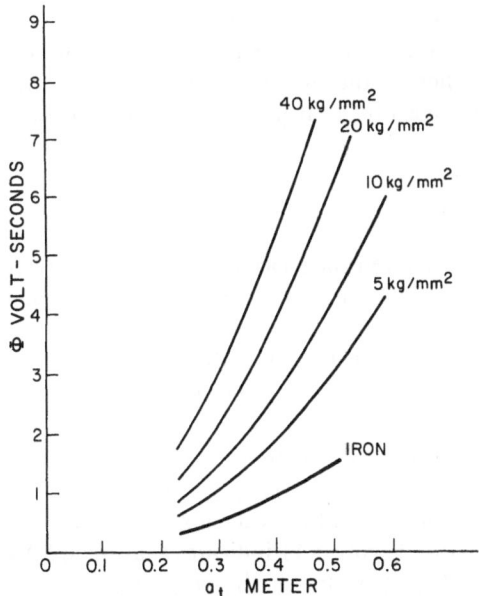

Fig. 5.11. Lines of constant stress as a function of single swing flux and available outer transformer radius

inner radius of the transformer is $a_2/2$ then

$$I \simeq \frac{a_2}{2} \, 1.5 \, R_0 \lambda j. \tag{5.26}$$

If we now combine (5.21, 25, 26) and again neglect γ^3, then

$$U = \frac{4.5 \, R_0}{4 \pi \mu_0} \left(\frac{\Phi}{a_2} \right)^2. \tag{5.27}$$

This relation yields results within 20 % of more sophisticated calculations. As an example, the Alcator C transformer with a 22 cm effective outer radius produces 1.8 Vs of flux with a stored energy of 13 MJ. Relation (5.27), using the Alcator R_0 of 64 cm, yields a result of 13.1 MJ.

In most high field tokamaks, the transformer will be found to have a peak field approximately that in the tokamak. The peak field is sensitive to the actual bore in the transformer (unlike the peak stress which occurs at $a_2/2$). The peak field can be derived from (5.20, 21) to be

$$B_{max} = \frac{3\Phi(1-\gamma)}{\pi a_2^2 (1-\gamma^3)}. \tag{5.28}$$

For an Alcator-type case, where $\gamma = 0.4$, $a_2 = 22$ cm, and $\Phi = 1.8$ Vs, the peak field is 20.5 T. The peak field in the Alcator tokamak coil is 14 T at the centerline and 21 T at the winding edge, almost exactly that in the transformer.

The large peak fields in such air-core transformers clarify why they produce more flux than iron core transformers which are limited by saturation to 1.8 T. The increase in flux is not 10 to 1, however, as the iron core can have a uniform field over its entire cross section, whereas the air core field drops linearly throughout the winding. The flux produced by an iron core at 1.8 T is given in Figs. 5.10, 11.

d) Transformer Interface with TF Coil

In the tokamak simplified analysis, we ignored the space at the machine center that is to be occupied by the transformer. We can now examine that simplifying assumption.

Let us use as an example the familiar 10 T tokamak with its 1 m radius. As calculated in the last section for a 2.2×10^6 A plasma, we need 5.3 Vs to charge, 3.3 Vs to maintain the discharge for 1 s, and perhaps as much as 3.3 Vs for early initiation losses. If we wish to supply all the "fast" volt-seconds for charge and initiation from the transformer collapse, we need to supply 8.6 Vs during the current rise. As calculated, the rising equilibrium field will supply 2.1 Vs so we require a balance of 6.5 Vs from the core. Figure 5.11 indicates that the 6.5 Vs can be supplied by a transformer of 0.44 m radius if the peak stress can be allowed to reach 20 kg/mm². At 10 kg/mm², the radius would have to be 0.51 m.

It is interesting to note that having selected a transformer capable of 6.5 Vs per half-swing, the negative swing could sustain the plasma for nearly two seconds at 1 keV, and for nearly 6 s if the temperature were 2 keV.

To choose the outer radius of the transformer in our example above, we need to examine the consequence for the TF coil. The heating and stress relationships in Sects. 5.2.3a, b can be recalculated for the reduced area available when the transformer must be accommodated. The required design multipler has been plotted in Fig. 5.12 for two values of aspect ratio. For our example of $A = 3$, we note that if the transformer outer radius a_t is taken as 0.45 m for the $R_0 = 1$ m tokamak, the stresses calculated ignoring the transformer must be multiplied by 2.25, and the pulse time must be reduced by 5. The combined von Mises stress, previously calculated as 13.5 kg/mm², becomes 30 kg/mm², and the 10 s pulse time allowed before the resistance increased by a factor of 2 must be reduced to 2 s.

Thus we have a situation where even choosing a high transformer stress level of 40 kg/mm² the transformer must occupy a significant fraction of the available space. It also further illustrates the fact that high field, high performance tokamaks place roughly equal burdens on the TF coil and the ohmic heating coil.

The energy stored in this transformer, using (5.27), would be approximately 65 MJ.

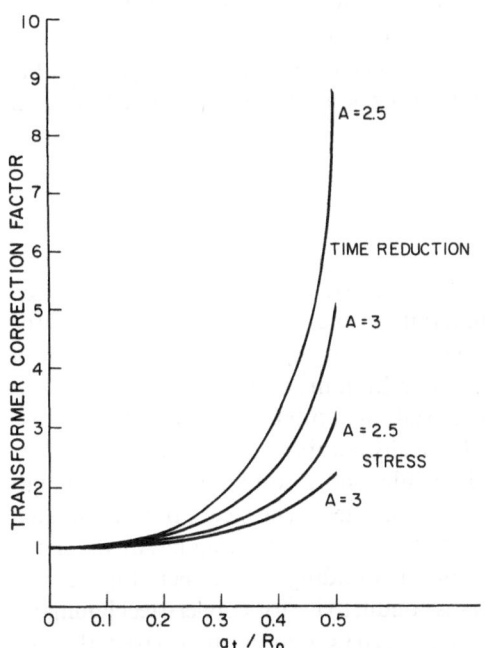

Fig. 5.12. Correction factors to the TF coil heating and stress to account for the space required for the ohmic transformer

5.2.5 Equilibrium Field Considerations

To hold the plasma in the center of the vacuum chamber requires a vertical, or equilibrium, field. The required value of the field to hold the plasma at a radius R_0 should be

$$B_v = \frac{\mu_0 I_p}{4\pi R_0}\left[\beta_p + \ln \frac{8R}{a} - \frac{5}{4}\right]. \tag{5.29}$$

The poloidal beta β_p appears in the expression and varies from nearly zero to approximately one depending on the plasma temperature and density. A machine with a plasma current of 2.2×10^6 A, a major radius of 1 m and a poloidal beta of 1.0 would require an equilibrium field of 0.67 T.

If the gradient of equilibrium field with major radius falls within a certain range, the plasma will have a natural up-down stability. The field should decrease as the major radius increases, and should have a field index N in the range of 0 to -1.5:

$$0 < N = \frac{R}{B}\frac{\Delta B}{\Delta R} < -1.5. \tag{5.30}$$

It is customary to design tokamaks not to depend entirely on natural up-down stability because small error fields due to nonsymmetries can bias the plasma up or down. A low field pair of radial field coils is generally included in the coil sets.

The equilibrium field coils, typified by those in Fig. 5.8, generally have stored energies in the megajoule range. The Alcator C coils, for example, store 1 MJ. To provide dynamic feedback to improve position control generally requires rather substantial power supplies. Alcator C utilizes a 15 MW controlled rectifier supply.

5.2.6 Materials Considerations

The magnet designer must always be concerned with the trade-off between strength and conductivity of the materials available for construction of the toroidal field. Figure 5.13 collates materials ranging from 100 % to 25 % of copper conductivity and strengths from 35 kg/mm^2 to 150 kg/mm^2.

It is often more useful to utilize materials in combination rather than to use the alloys shown in Fig. 5.13. By mechanically combining steel with copper, for example, a composite structure can be made whose conductivity and strength depend on the relative amounts of copper and steel. The steel can be laminated with copper by roll-bonding or by some form of interlocking keyways, or the steel can be in the form of a casement surrounding the copper. Figure 5.14 illustrates the trade-off of strength versus conductivity for copper-steel laminates for 35 kg/mm^2 ultimate strength copper and 210 kg/mm^2 ultimate strength steel.

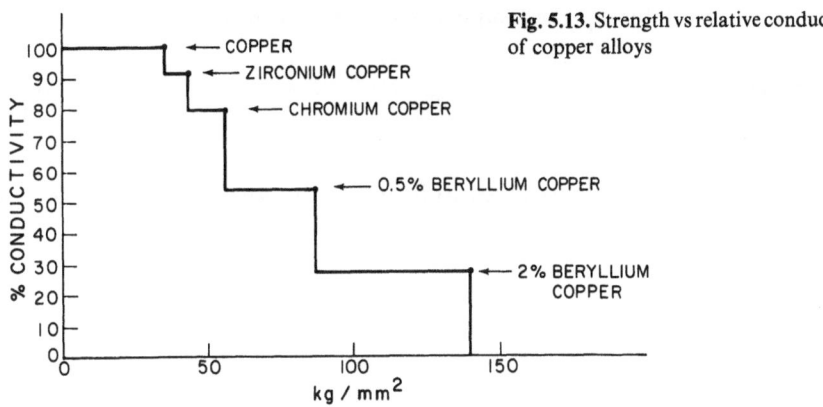

Fig. 5.13. Strength vs relative conductivity of copper alloys

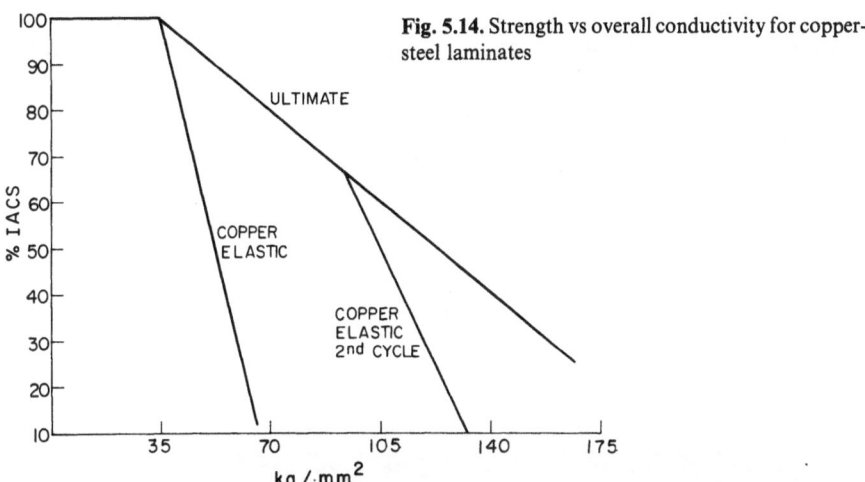

Fig. 5.14. Strength vs overall conductivity for copper-steel laminates

The figure illustrates that if sufficient steel is included, the copper will always remain within its elastic range. If less steel is used, the copper will yield during the first pulse, and go into compression following the pulse. Subsequent cycles will alternately cycle the copper from compression to tension but the copper will remain within the elastic range.

Magnets such as Alcator A and Alcator C have used copper-steel laminates within the elastic range of the copper. No magnets have yet been built which take advantage of driving the copper beyond its elastic limit on the first cycle, but such a technique could be used if one can ensure that the steel and copper will not slip with respect to each other on each successive pulse. Roll-bonded laminates would be particularly useful in such an application [5.19].

When a two-material laminate is used, the two materials will share the load according to the law of mixtures. If the materials have a cross-sectional area A_i

and an elastic modulus E_i, the load carried by each will be

$$P_1 = A_1 E_1 \varepsilon$$
$$P_2 = A_2 E_2 \varepsilon, \tag{5.31}$$

and since the strains are constrained to be equal, the stress in each will be

$$\sigma_1 = E_1 \varepsilon$$
$$\sigma_2 = E_2 \varepsilon \tag{5.32}$$
$$\frac{\sigma_1}{\sigma_2} = \frac{E_1}{E_2}.$$

Thus in the case of a copper-steel laminate where the modulus of steel is approximately twice as high as copper, the stress level in the steel will always be twice that in the copper as long as the copper remains in its elastic range. If the

Table 5.3. Resistivities normalized to copper at 20 °C

Material	20 °C	−196 °C (77 K)
Copper	1.0	0.125
Be-Copper (0.5 %)	1.8	0.79
Be-Copper (2 %)	4.8	3.6
Cr-Copper	1.17 − 1.35	0.48 − 0.56
Zr-Copper	1.1	0.33

Fig. 5.15. One of 240 tapered Bitter turn assemblies for Alcator C. The copper plate carries steel keys which match keyways cut in the stainless steel reinforcing sheet. The steel sheet is insulated on its upper surface by a bonded 0.4 mm glass-epoxy laminate

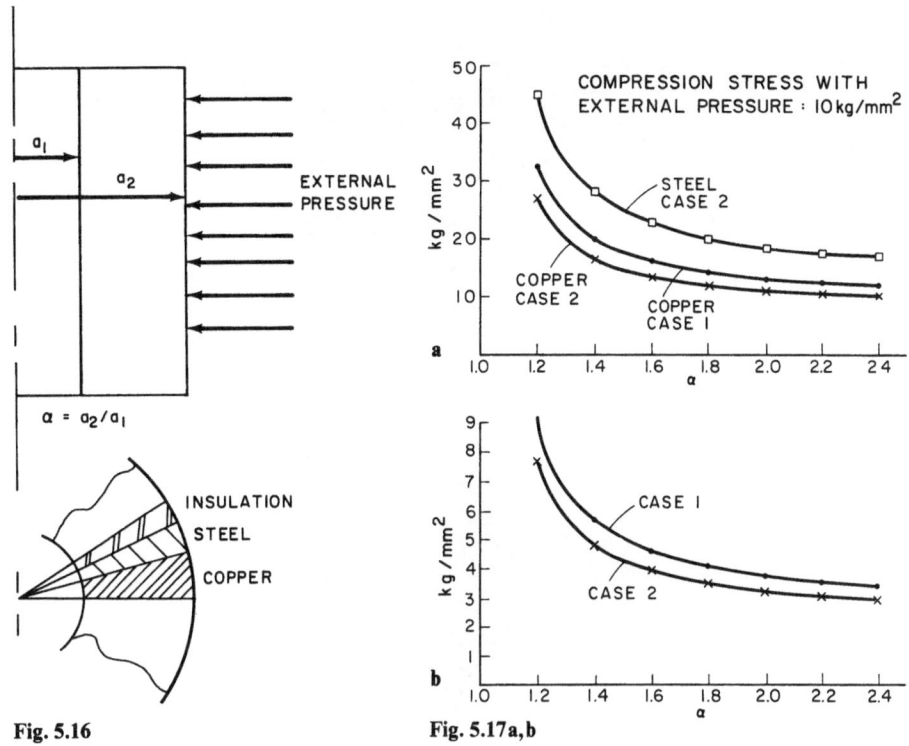

Fig. 5.16   **Fig. 5.17a,b**

Fig. 5.16. Definition of simplified variables for the tokamak core subjected to an inward magnetic pressure

Fig. 5.17a,b. Compressional stresses in the tokamak core for (a) 95 % copper, 5 % insulation, and (b) 50 % copper, 45 % steel, and 5 % insulation

cross section of the steel is 1/3 that of the copper, the loads carried by each would be the same.

If the magnet is to be liquid nitrogen cooled, the use of pure copper as opposed to copper alloys is particularly important. Only copper exhibits a substantial decrease in resistivity at the temperature of liquid nitrogen, as illustrated in Table 5.3.

The magnet structures must also contain insulation. For the Bitter plate type toroidal field coils, Fig. 5.15, the insulation is in the form of sheets sandwiched between each of the copper-steel laminate plates. The inward forces on the plates exert compression forces on those insulators.

The face pressure on the insulation in a toroidal field coil depends on the magnetic field strength and the area in the magnet to support the inward force. To be conservative, one might assume that "wedging" takes place only over the inner throat region of the magnet. In the case of the Alcator magnet, for example, the manufacturing tolerances were in fact chosen so that the wedging would take

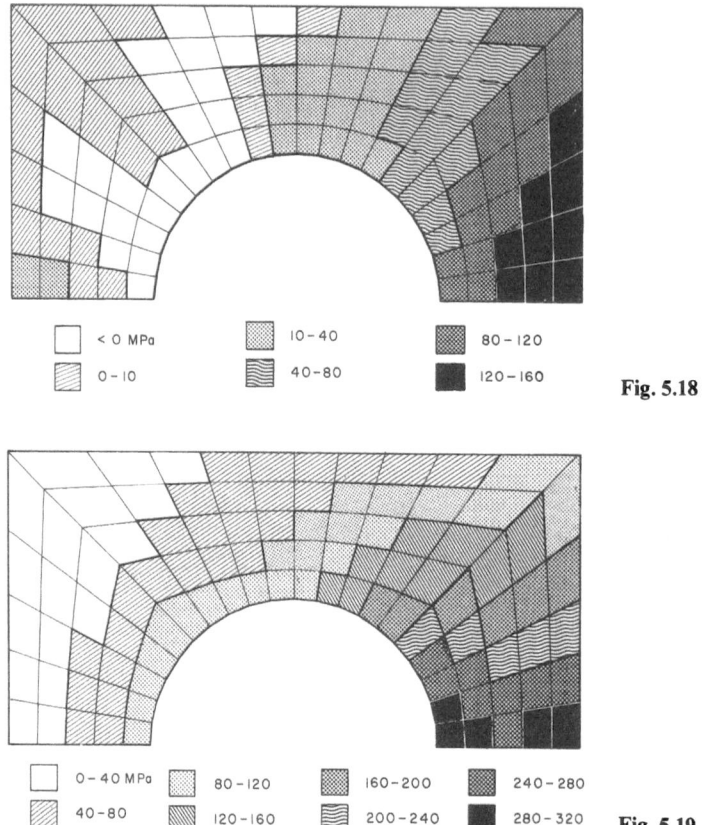

☐	< 0 MPa	▨	10–40	▨	80–120
▨	0–10	▨	40–80	■	120–160

Fig. 5.18

☐	0–40 MPa	▨	80–120	▨	160–200	▨	240–280
▨	40–80	▨	120–160	▨	200–240	■	280–320

Fig. 5.19

Fig. 5.18. Resultant face pressure on the TF insulator in Alcator C at 14 T field. The throat region is to the right

Fig. 5.19. Combined von Mises stress in the copper-steel laminate Alcator C plate. The peak calculated strain is a bulge of 3.8 mm at the upper and lower magnet surfaces. The throat region is to the right

place first in that more critical area. If we assume wedging only at the inside, we can find the compressional stresses as a function of the parameters defined in Fig. 5.16. The compressional stresses for these assumptions are plotted in Fig. 5.17 for 95 % copper, 5 % insulation, and for 50 % copper, 45 % steel and 5 % insulation.

Fig. 5.18 shows the face pressures [5.12] on the insulation in the Alcator C magnet, where the peak compressional forces reach 160 MPa (16 kg/mm^2). This is still within the capability of fiberglass reinforced epoxy sheets whose safe limits are in range of 30 kg/mm^2. Fig. 5.19 illustrates the combined von Mises stresses in the copper-steel laminate of the Alcator C plates arising from the face compression, tension and bending [5.12].

5.2.7 Liquid Nitrogen Cooling Considerations

As discussed in Sect. 5.3.1, it is customary to consider liquid nitrogen precooling for compact high field pulse magnet systems. Both Alcator machines [5.12,15], the Frascati FT tokamak [5.20], the ORMAK tokamak [5.21], the DITE tokamak [5.22], and the proposed ZEPHYR compact ignition experiment at Garching [5.13,14] are all based on liquid nitrogen precooling. The ORMAK coils were of hollow conductor construction with internal LN_2 circulation, the FT and DITE tokamaks are made from individual coils with external cases containing the LN_2, and the Alcator machines are cooled from their upper surface only by flooding the upper coil surface with a shallow puddle of LN_2. We describe several typical aspects of the Alcator liquid nitrogen operations.

a) Cooldown

The Alcator A machine has been running with liquid nitrogen cooling since 1973 and has been cooled from room temperature to liquid nitrogen temperature approximately 250 times during that period. The toroidal field magnet, auxiliary coils and structure weigh 8000 kg. A typical cooldown takes six hours and uses 2500 liters of LN_2 at a linear rate over the period. This total consumption is nearly exactly consistent with the assumption that only the heat of vaporization of the liquid contributes to the heat removal [5.23].

The TF magnet is cooled from its upper surface by allowing a puddle of LN_2 about 2 cm deep to lie on the surface of the coil. The barrier which determined the liquid height is shown in the photograph in Fig. 5.20. Excess liquid spills over into the bottom of the dewar shown in Fig. 5.21 and is pumped back over the top of the coil.

Fig. 5.20. Alcator A before the LN_2 dewars were in place. Cooling is provided by flooding the top of the magnet with LN_2

Fig. 5.21. Alcator A with its polyfoam cryostat in place

Table 5.4. q/A versus ΔT for pool-boiling LN_2

ΔT [K]	q/A [W/cm^2]
Film Boiling	
300	5.5
250	3.5
200	3.0
150	2.0
100	1.5
60	0.9
Nucleate Boiling	
13	19
10	13
6	3.5
4	1.4
2	0.25

The temperature of the TF coil drops linearly over the six-hour period, suggesting an approximately constant heat-transfer rate. The average heat-transfer rate over the period is very nearly 1.0 W/cm^2. Typical measured q/A values as a function of surface to liquid temperature difference are given in Table 5.4 and suggest that somewhat faster cooling rates might have been anticipated [5.23]. The discrepancy occurs principally because liquid is not supplied sufficiently rapidly to keep the cooling surface completely covered during the higher temperature portion of the cooldown. A sample mass cooled with sufficient liquid coverage exhibited an average q/A of 3 W/cm^2 over the 300–77 K interval.

The maximum cooldown rate is determined by the heat-transfer rate and the liquid supply rate. The cooldown rate chosen in many systems, however, often depends on the thermal gradients within the magnet and the resultant thermal stresses. The Alcator devices are cooled over 100 % of their upper surfaces and therefore are cooled symmetrically in a circumferential sense, but asymmetrically top to bottom. The top-to-bottom temperature gradient at a surface heat flux of 1 W/cm^2 is 0.25 °C per centimeter of height. For the 53 cm A machine plate, this results in a maximum temperature difference of 13 °C, much less than the temperature differences present at the end of a field pulse.

b) Recool

The current distribution in the Alcator plates is not uniform, the plates being tapered to form the toroidal magnet. During the short pulse, the magnet behaves locally adiabatically and little heat leaves the area in which it is generated. A calculated temperature distribution [5.12] at the end of a pulse in Alcator C is given in Fig. 5.22.

Immediately following the pulse, heat will diffuse out from the hot spot, and the coil will reach an equilibrium temperature. Typical thermal diffusion times are on the order of 60 s. The equilibrium temperature rise resulting from the distribution in Fig. 5.22 is only 6 K, in marked contrast to the local hot spot temperature rise of 160 K. At this 6 K temperature head above the bath, the heat fluxes are below the film boiling transition, and typical q/A can be 4–10 W/cm^2. The 36×10^6 J of deposited heat can thus be removed through the 3.6×10^4 cm^2 of magnet surface in approximately 5 min. The 36 MJ input boils off approximately 250 l of liquid during the 5-min period.

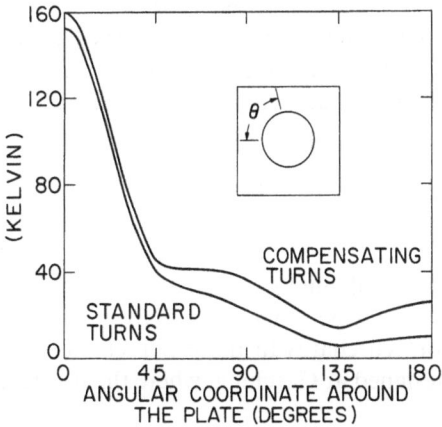

Fig. 5.22. Calculated temperature distribution immediately following a 14 T pulse in Alcator C

c) Nitrogen Cooling System

The nitrogen cooling aspects of the Alcator machine are relatively simple. The main TF coil is cooled by flooding the upper surface. Excess liquid runs off into a sump dewar and is pumped back over the magnet. The liquid level in the sump dewar is controlled by measuring the pressure head of liquid (nominally 90 cm) and adding more liquid as required from the external 30,000 l tank. The pump is 6 HP which typically delivers 50 m of head at 500 l/min.

The other coil systems are also liquid nitrogen cooled. The central OH transformer is made up of 36 pancake windings whose surfaces are flooded with LN_2. The central coil is completely immersed in its own fiberglass containment vessel. In Alcator A, the equilibrium coils were hollow conductor coils cooled by circulating nitrogen. This has proved satisfactory, but the multiplicity of cold, insulated, manifold connections represents a complication. The Alcator C uses strip wound coils cooled from the surface and appears to be a simpler construction technique.

The dewars which surround the machines are made of polyurethane foam covered inside and outside with fiberglass. Gas-tight seals are made by rubber boots, and liquid seals are made by utilizing a stand-pipe principle around all the penetrations in the bottom of the dewar. A slight overpressure is automatically maintained in the dewar by vaporizing LN_2 through a valve servo-controlled from the overpressure monitor. The dewar is made in four sections, so the top and bottom are easily removable for maintenance, and the two central sections can be put in place around the diagnostic and support penetrations. The diagnostic penetration covers are maintained at room temperature by controlled silicone-rubber pad heaters.

The loss rate of the dewar is approximately 1.5 kW, and thus consumes approximately 40 liters of LN_2 per hour. The machine support arms shown in Fig. 5.23 must carry 40 tonnes and, being short, would be potential loss paths if not made of low conductivity material. The three G-10 arms actually represent less than 0.1 kW loss and are hence negligible. The 240 cm² of copper bus penetrations into the dewar, 64 cm² for the 200 kA TF, 24 cm² for the 50 kA OH, and 152 cm² for the 20 pairs of 5 kA conductors for the equilibrium systems, are held to a loss of 0.6 kW by making a long transition length. The bus length is about 4 meters between room temperature and LN_2. By limiting the loss, the warm end also stays at room temperature, eliminating condensation problems.

5.2.8 Examples of High Field Tokamaks

Several representative examples of high field tokamaks are listed in Table 5.5; in order of their major radii, they cover a very wide range of scale from the very small capacitor-driven unit designed at the Kurchatov Institute of Technology in 1976 to the very large unit now being designed at Garching, where the toroidal field coils alone would weigh 300 tonnes.

Fig. 5.23. Alcator C showing tripod support and short horizontal G-10 fiberglass support arms. The tube ends which will carry the long bus transitions are visible at the upper left

Table 5.5. High field tokamak examples

Device	R_0	A_p	B_0 [T]	τ [ms]	Refs.
Kurchatov 20 T	6.4	1.6	20	0.6	[5.24]
Triam-2	50	12	10	300	[5.25]
Alcator A	54	9.5	10	300	[5.15]
Alcator C	64	17	14	500	[5.12]
Ignitor (CNEN)	80	21	21	1300	[5.10]
Frascati FT	83	29	10	1000	[5.20]
Garching ITR	130	50	9.5	8000	[5.13,14]

A number of design examples and figures based on the Alcator machines have been used already in this chapter. The Frascati tokamak is roughly comparable to the Alcator C in scale, but is constructed rather differently. Figure 5.24 shows a cut-away model of the Frascati machine. Instead of using copper-steel laminates as in the Alcator machines, the steel reinforcing is in the form of steel cases surrounding the 24 individual Bitter coils.

The small Kurchatov designed coil [5.24] is the smallest tokamak that has come to our attention and has potential applicability to relatively modest experimental programs due to its small scale and small energy requirements. A 320 kJ capacitor bank is used for the toroidal field and 20 kJ for the air-core transformer system. To accommodate the very high fields (the peak field at the coil boundary reaches 34 T) the "single-turn" construction shown in Fig. 5.25

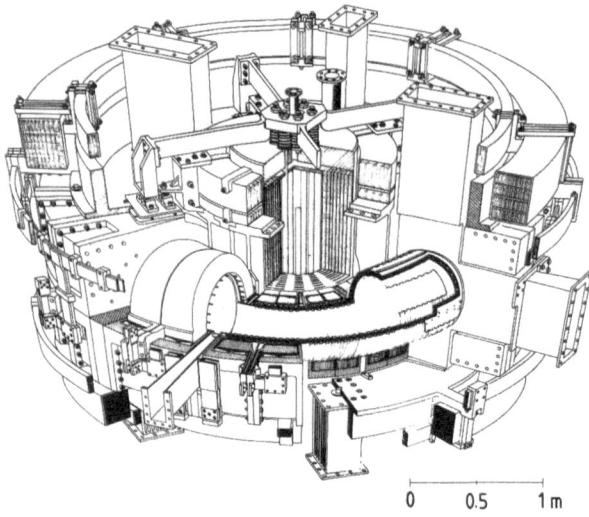

Fig. 5.24. Cut-away view of the Frascati TF toka-mak

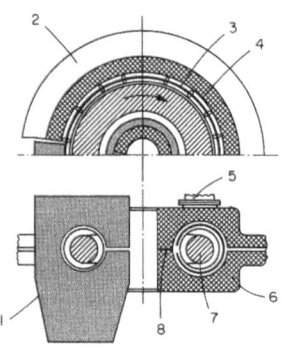

Fig. 5.25. Single-turn TF coil utilized in the small Kurchatov tokamak. Construction of the installation: (*1*) collector of the inductor, (*2*) collector of the longitudinal field coil, (*3*) inductor, (*4*) liner, (*5*) evacuation of the chamber, (*6*) coil of the longitudinal field, (*7*) plasma column, (*8*) contact of the coil blocks

was chosen and several copper-alloy and steel alloys evaluated. For very short pulses conductivity is not of major concern.

Because of the single-turn nature of the coil, a very large current of 6.4 MA must be fed to the TF coil. A toroidal coupling transformer is used to match the 30 kV capacitor bank to the single-turn coil. The air-core coupling transformer had a major radius of 45 cm and a minor radius of 30 cm, and is thus considerably larger than the tokamak. A coupling coefficient of 0.99 is reported with a primary current of 230 kA and a secondary current of 6.4 MA.

The Kurchatov transformer used to induce the 200 kA plasma current is located around the plasma column as shown in Fig. 5.25. This location for a transformer is more efficient than the conventional centerline location, but is chosen in this case principally to avoid field penetration problems arising from the massive nature of the single-turn TF coil.

The Kurchatov 20 T project was intended as a model for a much more ambitious project leading to an ignition grade plasma.

5.2.9 Other Closed Systems

Tokamaks are not the only high field toroidal (closed) systems receiving attention today. There is considerable interest in so-called toroidal pinch experiments, in particular reverse-field pinches (RFPs). Pinch experiments often have very large aspect ratios (R/a) and consequently they appear essentially linear over any local region. We shall therefore treat them as special cases of open systems in the next section.

Closed systems like stellarators and torsatrons differ from tokamaks in that all or part of the confinement is provided by external helical windings rather than from the plasma current [5.26]. Interest in these machines preceded tokamaks but was superceded by successes in the "simpler" tokamaks. With the greater understanding of plasma physics available today, interest is returning to these more sophisticated devices. If the transform is supplied by external coils, the plasma can be "currentless", so avoiding the pulse nature of the tokamak.

At present there are no proposals for very "high field" torsatrons or stellarators, the proposed field being nearer to 5 T.

5.3 Open Confinement Systems

5.3.1 Background

Many plasma physics experiments as well as reactor concepts involve magnetic systems which are open-ended rather than closed as in toroidal confinement. Traditionally these devices have called for very large magnetic fields. Losses from the ends of the device are compensated by either making the systems very long or by "plugging" the ends by creating end plasma cell conditions which reduce end losses.

There is a class of closed devices which have very large aspect ratios, that is, their major diameter is many times their minor diameter. Because of the lack of curvature effects in these devices, much of their physics can be investigated in linear systems. When the experiments reach the point where confinement is limited by end losses, they can be made longer or made into a large aspect ratio torus. This has been the general history of the theta pinch experiments, for example, where 1, 3, 5 and 8 m linear experiments were run before going to the 4 m radius, 25 m circumference torus. The plasma radius in these devices was of the order of 1 cm, or an aspect ratio of 400 for the toroidal experiment.

Mirrors are a major example of open systems. The current designs involve a central solenoid section with mirror coils at the end to create a plasma plug. The reactor concepts profit by having magnetic fields in the end region as high as possible.

We shall briefly discuss several of the open and "linearized" closed systems. Some of these concepts, like the reverse field pinch, are inherently pulse devices as they require dynamic interaction with the plasma, such as compression. Others, such as the tandem mirror, are potentially steady state. Steady-state devices depend on superconducting magnets, whereas pulse devices utilize resistive magnets, although often in "hybrid" use with steady-state superconducting coils.

5.3.2 Simplified Stress and Stored Energy Considerations

Linear field systems have been widely treated in the literature, so we give here the simplest considerations to aid in scaling. Certain coil heating and stress considerations developed for the TF coil and the OH transformer in the previous sections can equally well be used here.

The coil stress in a linear magnetic system can also be estimated from the simple magnetic pressure [5.27]

$$P_m = \frac{B^2}{2\mu_0}.$$
(5.33)

At a field of 10 T, the equivalent pressure is 4 kg/mm^2 (40 MPa). If we assume that all the magnetic pressure were concentrated on the inner surface, and if we represent the coil by a thick cylinder whose outer radius to inner radius is α, the tangential, radial and shear stresses at the inner edge would be

$$\sigma_t = P_m \left(\frac{\alpha^2 + 1}{\alpha^2 - 1} \right)$$

$$\sigma_r = P_m \left(\frac{1 - \alpha^2}{\alpha^2 - 1} \right)$$
(5.34)

$$\sigma_s = P_m \left(\frac{\alpha^2}{\alpha^2 - 1} \right).$$

These simple formulas tend to overestimate the true stress by a factor ranging from 20% at $\alpha = 1.1$ to 50% at $\alpha = 3$. The expressions are nonetheless useful for estimating practical field limits. For example, a coil with $\alpha = 1.5$ and a field of 20 T would have a shear stress near 30 kg/mm^2, essentially the limiting stress for a copper coil.

The energy stored per unit length in the bore of a long coil of inner radius a_1 can be simply estimated from

$$U_b' = \frac{B_0^2 V}{2\mu_0} = \frac{B_0^2 a_1^2}{8 \times 10^{-7}}.$$
(5.35)

Energy is also stored within the field windings. The total energy per unit length in a uniform current density coil can be found by integrating:

$$U'_{\text{Total}} = U'_b \left(1 + \frac{(\alpha - 1)\,(\alpha + 3)}{6} \right). \tag{5.36}$$

For a coil of $\alpha = 1.5$ the total energy is 1.38 times that stored in the bore and for a coil of $\alpha = 3$, 3.0 times. For a Bitter type coil, where the current does not flow throughout the winding, an estimate of the energy outside the bore can be found by defining the equivalent radius as 1/3 of the way to the equivalent skin depth.

Stored energies can be quite large in some of the magnetic systems described. In a particular reference theta-pinch reactor (RTPR) [5.28], the compression coils have an inner radius of 0.95 m and a field of 11 T. The stored energy would be 178 MJ/m. The operating scenario calls for the compression field to rise in 20 ms.

5.3.3 Laser and E-Beam Heated Solenoid Devices

Several proposals [5.6] have been made to heat a plasma contained in a long solenoid by means of CO_2 laser heating or E-beam heating. In both cases the reference reactors have solenoids which are very long in order to cut down end losses.

In one study of a laser-heated solenoid reactor [5.28, 29], the reactor, which was chosen to have a length of 1000 m, had a magnetic system consisting of a 20 T superconducting solenoid of 1 m inner radius which would surround ten 2.8 cm bore pulse solenoids each of which would produce an additional 24 T. The superconducting magnet would be steady state and be shielded from neutron heating. The pulse coils would be sufficiently thin walled (5.6 cm) to allow the bulk of the neutrons to pass through and enter the surrounding blanket area.

This reactor suffers from a high circulating power requirement for the pulse coils as well as from considerable technological difficulties. The pulse magnets, which are designed in zirconium copper to a 42 kg/mm² stress level, require 1.7 MJ/m of energy storage, would rise in 100 μs and decay (after crowbarring) with a time constant of 20 ms. The exposure of the coils to a full hard neutron flux clearly will severely limit the life of the coils and their insulation.

The superconducting background field coil is used to reduce the power required for the resistive pulse coils. This "hybrid" concept has been used in steady-state applications to boost the field above the limitations set by the available power. In the above reactor case, the stored energy, (and the dissipated energy) are reduced by the ratio

$$\frac{U}{U_0} = \left(\frac{B_{\text{resistive}} - B_{\text{superconducting}}}{B_{\text{total}}} \right)^2. \tag{5.37}$$

To produce 20 T in a 1 m radius coil from superconductors would be a major challenge. The Laser-Heated Solenoid design team has suggested a novel support structure utilizing a "pressure bag" approach [5.30] to limit the strain on the superconducting windings. Small superconducting coils have reached fields of 17.5 T and undoubtedly could produce 20 T given sufficient investment.

Potential interest in the laser-heated solenoid approach has led to several short-term experiments. For example, a 12.5 T solenoid 1 m long with a 5 cm bore was constructed at Princeton to utilize a 3 kV, 350 KJ capacitor bank [5.31]. A second 25 T coil 23 cm long operated for approximately 1000 pulses before failure [5.32].

A 2 cm, 30 T coil in two 50 cm long modules has been constructed at Mathematical Sciences Northwest [5.33]. The magnet will ultimately be extended to a 3 m length. A larger proof-of-principle experiment has also been proposed with a 30 m length coil with fields in the 30 to 45 T range [5.34]. If we assume a coil inner radius for such an experiment of 2.5 cm, the stored energy in the 30 m length bore at 45 T would be 50 MJ.

5.3.4 Pinch Experiments

a) Theta Pinch

Pinch experiments have been underway for years and they have traditionally been performed at "high" magnetic field levels.

The Scyllac series at Los Alamos is one of the most famous of the theta pinch experiments [5.35]; major experiments in this area were also carried out at the Naval Research Laboratory [5.36], Culham Laboratory [5.37] and IPP, Garching [5.38]. Several of these experiments involved capacitor banks in the megajoule range and coils many meters in length.

The Scyllac 5 m linear theta pinch experiment serves as a good example of this class of experiments [5.39, 40]. The 5 m section is composed of 1 m "single-turn" modules of 5.5 cm inner radius. They are each driven by 700 kJ, 60 kV capacitor banks to produce fields of 10 T. The field's rise is about 5 ms and decay after crowbarring is in 250 ms. The 5 m section was "end plugged" by 5.5 cm radius, 16 cm long mirror coils capable of producing 20 T when each is driven by a 175 kJ, 60 kV capacitor bank [5.41].

Several of the Scyllac experiments utilized coils whose inner bore was machined in a long pitch helical form in an attempt to stabilize certain destructive plasma modes [5.42]. The 25 m torus, for example, was shaped to produce $l=1$ and $l=0$ "bumpy" equilibrium fields of 50 cm pitch and had an average inner radius of 7.2 cm. The torus was driven from a 10.5 MJ, 60 kV bank and could achieve fields of 6 T.

A number of reference theta pinch reactor designs were evolved over the years and all involved the requirement for very large stored energies. The RTPR example [5.28] in Sect. 5.3.2 required 178 MJ per meter of length and would have had a circumference of 350 m. Considerable design and experimental work was invested in superconducting energy storage and rotating transfer "capacitors" to store and supply this energy at high efficiency.

b) Reverse Field Pinch

One of the most active current areas of pinch research is in the reverse field pinch area [5.43]. It had been noted repeatedly in the pinch experiments that there was a tendency for currents to form in the plasma so as to produce a self-made reverse field at the outside of the plasma, and that this was a stable mode. Machines have subsequently been designed to program the fields to produce a reverse field at the outside. The suggested parameters [5.44] for the next generation RFP device, expected to be built on an international cooperative scale, are given in Table 5.6. The magnetic fields in this device are relatively low.

Table 5.6. Parameters of a proposed RFP experimental device

Major Radius	1.8 m
Minor Radius	0.6 m
Peak Plasma Current	2 MA
Rise Time	1 – 12 ms
Decay Time	15 ms
Forward Field	1 T
Reverse Field	1.5 T
Bank Energy	30 MJ

c) Fast Liner Experiments

The use of collapsing liners to compress plasmas provided a driving interest for much of the early megagauss magnetic field experimental work. Considerable work is carried out today, with the LINUS Program at the Naval Research Laboratory [5.45] and the Fast Liner work at Los Alamos [5.46] being notable examples. In a LINUS reference reactor [5.47] utilizing 2 m diameter gas-piston driven liquid lithium liners, peak fields inside the collapsed liners are predicted to exceed 60 T. Seed fields in the order of 7.5 T would be required. This is discussed further in Sect. 6.6.

5.3.5 Mirror Confinement

The simple magnetic mirror is one of the earliest confinement geometries investigated. The large end losses, however, have pushed the devices toward increasingly complex arrangements. The latest configuration is shown in Fig. 5.26, where the central solenoid section is "end stopped" by mirror coils and by barrier coils [5.48]. To plug the central section effectively, the end fields must be higher by a factor of 2 to 3 than in the central region. Since the economic

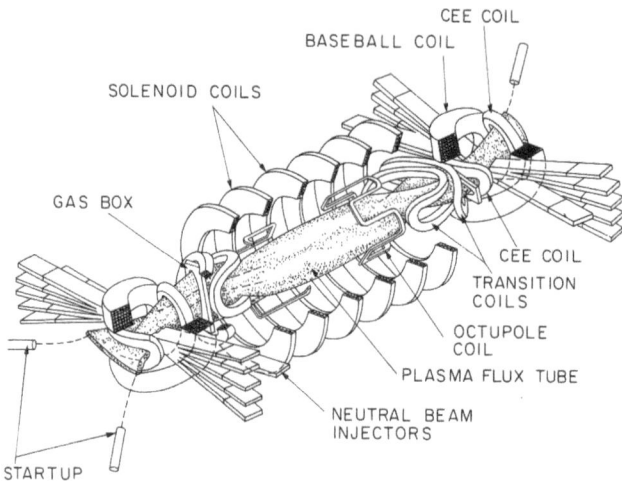

CEE COIL
BASEBALL COIL
SOLENOID COILS
GAS BOX
CEE COIL
TRANSITION
COILS
OCTUPOLE
COIL
PLASMA FLUX TUBE
NEUTRAL BEAM
INJECTORS
STARTUP

Fig. 5.26. Tandem mirror concept with end plugs thermal barriers

viability of a reactor would depend on the energy density in the central region, there is a major premium on producing as large an end cell field as possible. The current designs call for barrier fields of the order of 15 T and fields at the mirror coil winding of 12 T. Proposals have been made to make the barrier coils in a hybrid form, where the inner elements would be resistive and the outer superconducting. This can extend the field range as resistive coils have no critical field limit, but clearly increases the requirements for circulating power.

Mirrors can also be linked together, either linearly as in the "multiple mirror" proposals, or toroidally to form a so-called bumpy torus. One device of current interest is the Elmo Bumpy Torus (EBT), where electron rings are produced by RF between the mirror coils and stabilize against the toroidal

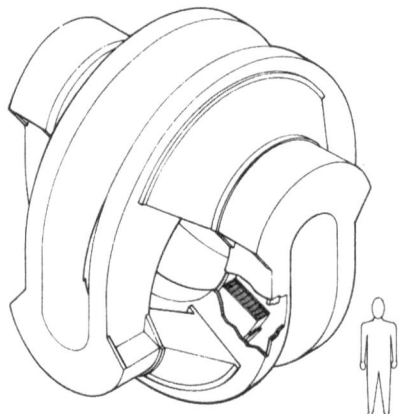

Fig. 5.27. Mirror Fusion Test Facility (MFTF) magnet with supporting structure

effects [5.49]. The fields of interest are limited only by the availability of resonant RF equipment. The current upper limit of 110 GHz is matched to a 4 T field.

The mirrors are inherently steady-state devices and consequently even current experimental programs profit particularly from superconducting technology. The Mirror Program is presently building a very large superconducting magnet [5.50] whose complex geometry is shown in Fig. 5.27. The peak field at the winding will be 8 T, and the central field, 3 T.

5.4 Future Prospects

Fusion research will continue to grow in the future and certainly in two modes. In the main line, the devices will grow ever larger and will utilize superconducting magnets to the maximum extent to effect a favorable energy balance. Peak fields near 12 T appear to be a reasonable limit of requirement and probably of realistic achievement. These main line projects will involve very large teams of engineers and will continue to challenge the best magnetic design talent. High field requirements will show up in the main confinement coils, and in the case of the tokamak, in the ohmic system and quite possibly in localized "divertor coils" used to exhaust impurities [5.51].

A fusion research comes closer to reactor development we can expect the alternate paths to see increased activity also. The growing interest in fusion, and the resultant budgets, will support a healthy climate for new ideas. Many of these ideas will be first explored on a modest scale, and will very likely often utilize the highest practical magnetic fields.

Acknowledgement. The author is indebted to Carl Weggel for several contributions to this chapter, and for his fundamental contributions to the high field Alcator tokamak programs.

References

5.1   Proc. 5th Symp. on Engineering Problems of Fusion Research, Princeton, NJ (1973) IEEE Pub. No. 73CH0843-3-NPS
5.2   Proc. 6th Symp. on Engineering Problems of Fusion Research, San Diego, CA (1975) IEEE Pub. No. 75CH1097-5-NPS
5.3   M.S.Lubell, C.Whitmire, Jr. (eds): Proc. 7th Symp. on Engineering Problems of Fusion Research, Knoxville, TN (1977) IEEE Pub. No. 77CH1267-4-NPS
5.4   C.K.McGregor, T.H.Batzer (eds): Proc. 8th Symp. on Engineering Problems of Fusion Research, 4 Vol., San Francisco, CA (1979) IEEE Pub. No. 79CH1441-5-NPS
5.5   U.S. Department of Energy, Office of Fusion Energy, Office of Energy Research, Status of Tokamak Research, DOE/ER-0034, National Technical Information Service, Springfield, VA (October 1979)
5.6   U.S. Department of Energy, Office of Fusion Energy, An Evaluation of Alternate Magnetic Fusion Concepts 1977. DOE/ER-0047, National Technical Information Service, Springfield, VA (May 1978)

5.7   D.R.Cohn, J.H.Schultz: High Field Compact Tokamak Reactor (HFCTR), Conceptual Design, MIT, Plasma Fusion Center Research Report RR79-2, Cambridge, MA (January 1979)

5.8   W.M.Stacey: Fusion Reactor Development: A Review, *Adv. Nuclear Sci. Techn. 15*, 129 (1981)

5.9   S.Fairfax: Alcator Group, Energy and Particle Confinement in the Alcator Tokamaks, Proc. 8th International Conference on Plasma Physics and Controlled Nuclear Fusion Research, International Atomic Energy Agency, IAEA-CN-38/N-6, Brussels (July 1980)

5.10   L.Anzidei, L.Lovisetto, G.Malavasi: Structural Analysis of a Toroidal Magnet for a High Field Poloidal Tokamak, [translated from Italian by Kyra Hall, PFC November 1979], EURATOM-CNEN Association on Fusion, Frascati, Italy (April 1979)

5.11   L.De Menna, G.Rubinacci, F.Esposito: Structural Analysis of a Compact Toroidal Ignition Experiment, Proc. 8th Symp. on Engineering Problems of Fusion Research, 4 Vols., San Francisco, CA (1979) IEEE Pub. No. 79CH1441-5-NPS, p. 1479

5.12   C.Weggel, W.Hamburger, B.Montgomery, N.Pierce: The Alcator C Magnetic Coil System, Proc. 7th Symp. on Engineering Problems of Fusion Research, Knoxville, TN (1977) IEEE Pub. No. 77CH1267-4-NPS, p. 54

5.13   U.Brossman, J.E.Gruber, W.Hamburger, O.Jandl, M.Soll, B.Streibl: Tape-Wound 'Toroidal (TF) Magnet for ZEPHYR, IPP-Report No. IPP 1/176, EURATOM Association, Garching, Germany (November 1979)

5.14   H.Becker, L.Bromberg, D.Cohn, J.Lettvin, J.E.C.Williams: Engineering Design Considerations for Compact Ignition Test Reactors, Proc. 8th Symp. on Engineering Problems of Fusion Research, 4 Vols., San Francisco, CA (1979) IEEE Pub. No. 79CH1441-5-NPS, p. 1097

5.15   D.B.Montgomery: The Alcator Project, Colloques Intern. C.N.R.S., No. 242 – Physique Sous Champs Magnétiques Intenses, Grenoble (1974)

5.16   R.W.Callis, H.J.Varga, J.C.Wesley: Toroidal and Ohmic Heating Power Supplies for Doublet III, Proc. 6th Symp. on Engineering Problems of Fusion Research, San Diego, CA (1975) IEEE Pub. No. 75CH1097-5-NPS, p. 458

5.17   C.E.Swannack, R.A.Haarman, J.D.Lindsay, D.M.Weldom: HVDC Interrupter Experiments for Large Magnetic Energy Transfer and Storage (METS) Systems, Proc. 6th Symp. on Engineering Problems of Fusion Research, San Diego, CA (1975) IEEE Pub. No. 75CH1097-5-NPS, p. 662

5.18   W.M.Parsons, R.W.Warren, E.M.Hoenig, J.D.Lindsay, P.Bellamo, R.L.Cassel: Interrupter and Hybrid-Switch Testing for Fusion Devices, Proc. 8th Symp. on Engineering Problems of Fusion Research, 4 Vols., San Francisco, CA (1979) IEEE Pub. No. 79CH1441-5-NPS, p. 689

5.19   C.F.Weggel: A New Ultra-High-Strength Laminated Electrical Conductor, *Adv. Cryogenic Engineering 25*, Madison, WI (1980)

5.20   L.Anzidei, L.Bettinali, G.Celentano, G.B.Malavasi, G.B.Righetti, B.Rumi, E.Salpietro, R.Toschi: Technological Aspects of the Cryogenic Magnet for the Frascati Tokamak, Proc. 5th Symp. on Engineering Problems of Fusion Research, Princeton, NJ (1973) IEEE Pub. No. 73CH0843-3-NPS, p. 337

5.21   H.M.Long, J.N.Luton: Cryogenic Engineering for *Ormak*, Proc. 5th Symp. on Engineering Problems of Fusion Research, Princeton NJ (1973) IEEE Pub. No. 73CH0843-3-NPS, p. 380

5.22   K.M.Plummer, D.V.Bayes, D.Bell, J.Burt, F.Galloway, B.C.Sanders, D.E.Skelton, G.L.Varley: The Engineering of the "Divertor Injection Tokamak Experiment" (DITE), Proc. 6th Symp. on Engineering Problems of Fusion Research, San Diego, CA (1975) IEEE Pub. No. 75CH1097-5-NPS, p. 361

5.23   D.B.Montgomery, N.T.Pierce: The Alcator Liquid Nitrogen-Cooled Tokamaks, in *Applications of Cryogenic Technology*, 7, ed. by J.R.Missig and R.W.Vance, Scholium International, Inc., 1978

5.24   Yu.A.Alekseev, A.M.Andrianov, V.L.Baryshev, V.I.Vasilev, V.F.Demichev, M.N.Kazeev, V.V.Kisula: Toroidal Magnetic Trap With Longitudinal Field of 200 kOe, Kurchatov Institute of Atomic Energy, Moscow (1976) [translation from Russian, B.Bialocki, 1977]

5.25   Triam-Project, Plasma Physics Group, Research Institute for Applied Mechanics, Kyushu University, Japan (1978)

5.26 S.Glasstone, R.H.Lovberg: Controlled Thermonuclear Reactions, Robert E. Krieger Publishing Company, Huntington, NY (1975)

5.27 D.B.Montgomery, R.J.Weggel: Solenoid Magnet Design, Robert E. Krieger Publishing Company, Huntington, NY (1980)

5.28 An Engineering Design Study of A Reference Theta-Pinch Reactor (RTPR), Los Alamos Scientific Laboratory Report 5336, Argonne National Laboratory Report 8019

5.29 Mathematical Sciences Northwest, Inc., A Feasibility Study of a Linear Laser-Heated Solenoid Fusion Reactor, EPRI ER-171 (1976)
      L.C.Steinhauer: Conceptual Fusion Reactor Designs Based on the Laser Heated Solenoid, Proc. 7th Symp. on Engineering Problems of Fusion Research, San Diego, CA (1975) IEEE Pub. No. 75CH1097-5-NPS, p. 979

5.30 P.G.Marston, J.J.Nolan, R.J.Averill: The Magnet System for a Laser Heated Solenoid Fusion Reactor, Proc. 7th Symp. on Engineering Problems of Fusion Research, San Diego, CA (1975) IEEE Pub. No. 75CH1097-5-NPS, p. 1123

5.31 A.H.Bohr, F.W.Kloiber, K.B.Silverman: Design of a 125 Kilogauss One Meter Magnet Coil, Proc. 7th Symp. on Engineering Problems of Fusion Research, Knoxville, TN (1977) IEEE Pub. No. 77CH1267-4-NPS, p. 837

5.32 P.Bonanos: Mechanical Design of a 250 Kilogauss Solenoidal Magnet, Proc. 6th Symp. on Engineering Problems of Fusion Research, San Diego, CA (1975) IEEE Pub. No. 75CH1097-5-NPS, p. 1127

5.33 T.E.DeHart, J.F.Zumdieck, A.L.Hoffman, D.D.Lowenthal, E.A.Crawford, B.Perry: The Laser Heated Solenoid Proof-of-Concept Experiment (PCX) Facility, Proc. 7th Symp. on Engineering Problems of Fusion Research, Knoxville, TN (1977) IEEE Pub. No. 77CH1267-4-NPS, p. 1619

5.34 P.H.Rose, L.C.Steinhauer, R.T.Taussig: Fusion Reactor Development Scenarios for the Laser Solenoid Concept, Proc. 7th Symp. on Engineering Problems of Fusion Research, Knoxville, TN (1977) IEEE Pub. No. 77CH1267-4-NPS, p. 609

5.35 W.E.Quinn: Linear Theta-Pinch Experiments, Proc. of the High Beta Workshop, ERDA-76/108, Los Alamos, NM (1975)

5.36 A.C.Kolb, H.R.Griem, W.H.Lupton, D.T.Phillips, S.A.Ramsden, E.A.McLean, W.R.Faust, M.Swartz: Nuclear Fusion — Suppl. Part 2 (1962) p. 553

5.37 A.D.Beach, A.B.Bodin, C.A.Bunting, D.J.Dancy, G.C.H.Heywood, M.R.Kenward, J. McCartan, A.A.Newton, I.K.Pasco, R.Peacock, J.L.Watson: Nuclear Fusion 9, 215–222 (1969)

5.38 E.Fünfer: Proc. 6th European Conference on Contr. Fusion and Plasma Phys., Moscow (1973) p. 109

5.39 A.G.Bailey, M.D.Machalek: Scylla IV-P Linear Theta Pinch, A Design and Construction Overlook, Proc. 6th Symp. on Engineering Problems of Fusion Research, San Diego, CA (1975) IEEE Pub. No. 75CH1097-5-NPS, p. 387

5.40 L.D.Hansborough, C.F.Hammer, K.W.Hanks, T.E.McDonald, W.C.Nunnally: Engineering Prototypes for Theta-Pinch Devices, Proc. 6th Symp. on Engineering Problems of Fusion Research, San Diego, CA (1975) IEEE Pub. No. 75CH1097-5-NPS, p. 391

5.41 K.W.Hanks, G.P.Boicourt, A.G.Bailey: 250-kG Mirror Coil Development for the Linear Scyllac Experiment, Proc. 5th Symp. on Engineering Problems of Fusion Research, Princeton, NJ (1973) IEEE Pub. No. 73CH0843-3-NPS, p. 314

5.42 W.H.Borkenhagen, W.R.Ellis, H.W.Harris, E.L.Zimmerman: Auxiliary Coil System for Scyllac, LA 4815-MS, LASL, Los Alamos, NM (1971)

5.43 C.F.Hammer, R.S.Dike, W.C.Nunnally: Engineering Description of the LASL ZT-40 Toroidal Z-Pinch, Proc. 7th Symp. on Engineering Problems of Fusion Research, Knoxville, TN (1977) IEEE Pub. No. 77CH1267-4-NPS, p. 655

5.44 T.E.James, H.A.B.Bodin, S.Skellett: Some Design Aspects of a Large Reverse Field Pinch Experiment, Proc. 6th Symp. on Engineering Problems of Fusion Research, San Diego, CA (1975) IEEE Pub. No. 75CH1097-5-NPS, p. 383

5.45 P.J.Turchi, A.E.Robson: Electromagnetic Implosion of Large Diameter Liners, *Pulsed High Beta Plasmas, Plasma Physics*, ed. by D.E.Evans (Pergamon Press 1975)

5.46 D.L.Book: NRL Group, SAI Group, LASL Group, Experimental and Theoretical Liner Fusion Studies at NRL, SAI and LASL, NRL Memorandum Report 3826, Washington, D.C. (1978)

5.47 P.J.Turchi, A.E.Robson: Conceptual Design of Imploding Liner Fusion Reactors, Proc. 6th Symp. on Engineering Problems of Fusion Research, San Diego, CA (1975) IEEE Pub. No. 75CH1097-5-NPS, p. 983

5.48 T. H.Batzer: The Technology of Mirror Machines — LLL Facilities for Magnetic Mirror Fusion Experiments, Proc. 7th Symp. on Engineering Problems of Fusion Research, Knoxville, TN (1977) IEEE Pub. No. 77CH1267-4-NPS, p. 2

5.49 A.L.Boch: Oak Ridge National Laboratory Report ORNL/TM-7191 (1980)

5.50 C.D.Henning, A.J.Hodges, J.H.Van Sant, R.E.Hinkle, J.A.Horvath, R.E.Heintz, E.Dalder, R.Baldi, R.Tatro: Mirror Fusion Test Facility Magnet, Proc. 8th Symp. on Engineering Problems of Fusion Research, 4 Vols., San Francisco, CA (1979) IEEE Pub. No. 79CH1441-5-NPS, p. 739

5.51 T.F.Yang, A.V.Lee, G.W.Ruck, T.Prevenslik, G.S.Meltzer: Design of an Advanced Bundle Divertor for the Demonstration Tokamak Hybrid Reactors, Proc. 8th Symp. on Engineering Problems, of Fusion Research 4 Vols., San Francisco, CA (1979) IEEE Pub. No. 79CH1441-5-NPS, p. 615

6. Pulsed and Ultrastrong Magnetic Fields

Noboru Miura and Fritz Herlach

With 69 Figures

Pulsed magnetic fields are divided into two broad categories: "nondestructive" fields below 100 T with a pulse duration of the order milliseconds to seconds, and "destructive" fields above 100 T with a pulse duration of the order of microseconds or less. Different methods to generate these fields are discussed in practical detail and with a view to further development. The destruction of the conductor material by Joule heating and mechanical deformation (shock waves) determines the highest field that can be obtained with a given device. In most systems the second effect is dominant and for megagauss fields this can be expressed by a simple semi-empirical relation between a typical speed in the system (implosion speed, speed of energy transfer) and the peak field.

So far, experiments have been done mainly in solid state physics and mostly with semiconductors: magnetotransport phenomena including the magneto-phonon effect and photoconductivity, cyclotron and electron paramagnetic resonances, magnetoreflection and magnetotransmission, magnetostriction, Faraday- and Zeeman-effects, exciton spectra, spin-flip transitions, high field superconductors and the effects of magnetically induced high pressure. In megagauss fields, many of these effects become nonlinear and thus provide a more stringent test for theory.

Research on megagauss fields was initiated for the purpose of short term confinement of high density plasma. This was not immediately successful but a recent revival of this scheme has led to the well-founded design proposal of a reversible-implosion fusion reactor with unique design features.

6.1 An Invitation to Use Pulsed Magnetic Fields

The basic apparatus for generating pulsed magnetic fields is fairly compact and inexpensive. This has induced quite a few researchers to build a small pulsed field facility for their experiments, but many have abandoned it again in favor of the more convenient experimentation with superconducting magnets and an occasional guest experiment at a national magnet laboratory. Single-shot experiments can indeed be frustrating: while the equipment cannot be adjusted and tested under actual working conditions, a small mistake often results in complete failure. Then, painstaking detective work is needed to find out what happened, in particular with explosive-driven experiments. For that matter, "nondestructive" coils also have a tendency to blow up occasionally. The short pulse time and low repetition rate restrict the use of signal-enhancing integration techniques; the field volume is usually rather small; conducting and paramagnetic samples are heated, and sensitive electrical experiments are disturbed by induced voltages and other electrical transients.

However, the ultimate rewards and possibilities inherent in pulsed field experiments are well worth the effort needed to overcome initial difficulties. Once the transient data recording technique has been mastered, large amounts of data can be taken in reasonable time, and parameter studies such as a temperature dependence can be rapidly completed. Problems of long-term · stability are greatly reduced; it is sufficient to know the parameters at the instant of the shot. In the large field range, even broad resonance lines become discernible. This allows an extension of measurements from cryogenic to higher temperatures and to obtain results with highly disordered or impure samples. The restriction on integration techniques is at least partially offset by the increased magnitude of the effects to be measured. In some cases it may turn out that integration had only been used for convenience and that there are other ways to boost the signal. Experiments with pulsed fields are bound to stimulate innovation. To cope with long relaxation times and eddy current heating, pulsed field techniques are now extended to pulses of sufficient duration to accommodate most experiments of interest. As a basis for further development, know-how and technical skill are presently accumulated at dedicated facilities.

Experiments with pulsed magnetic fields were pioneered and brought to a lonely first peak by *Kapitza* [6.1,2]. His bold beautiful experiments in the late twenties are still exemplary. Kapitza used heavy equipment, i.e., a 100 MW flywheel alternator (with 50 MW output into a matched load), from which he switched out a single half-cycle of 10 ms duration. The top of the pulse was flattened by the special design of the excitation winding. Using ingenious and innovative experimental methods, he studied magnetoresistance, magnetization, magnetostriction, Zeeman effect and particle tracks.

Preceded by an experiment by *de Haas* and *Westerdijk* (20 T with a storage battery and hydrogen cooling) [6.3] and an early 45 T capacitor discharge experiment by *Wall* [6.4], the revival of the technique in the fifties began at the lower end of the energy scale. *Raoult* [6.5], *Champion* [6.6], *Myers* [6.7], *Shoenberg* [6.8] who studied the de Haas-van Alphen effect, and in particular *Olsen* and his group [6.9] used capacitor discharges of the order of 1 kJ into small solenoids cooled by liquid nitrogen or helium to obtain peak fields of the order of 20 T. The magnetoresistance of metals was comprehensively studied [6.10–13], as well as size effects [6.14,15] and critical fields of superconductors [6.16]. There was a series of magnetization measurements, mainly to study magnetic transitions, by *Jacobs* [6.17], *De Blois* [6.18], *Flippen* [6.19] and *Rohrer* and colleagues [6.20], who also investigated the magnetoresistance of dilute magnetic (i.e., Kondo) alloys [6.21]. With the foundations laid by *Furth* and *Waniek* [6.22], *Foner* and *Kolm* [6.23] initiated the move towards higher energies and fields. Their field generator with a pulse duration (half-period) of 120 μs was used for the first pulsed field cyclotron resonance experiments [6.24]. In the sixties, capacitor banks with energies in the range 10–200 kJ came into use [6.25] with occasional record peak fields up to 50 T [6.26] and fields for experimental use mostly below 40 T. Facilities dedicated to experiments with these non-destructive fields are now established in several countries [6.27–35], with pulse times in the range 0.1 ms–1 s.

The magnetic stress related to a field of 50 T is 10 kb, close to the yield point of the strongest alloys. It is possible to make massive coils with distributed stress that will work in the range 50–100 T but the danger of destruction is sharply increased. Fields above 100 T (= 1 megagauss) may be called "ultrastrong" with good reason. They are in a different category in that destruction is inevitable and proceeds at a rate related to the velocity of sound in the conductor material, restricting the pulse time to a few microseconds or less. The big step across this limit was taken in the late fifties along two separate tracks: *Furth* et al. [6.36] used a fast capacitor discharge into a massive single-turn coil; *Fowler* et al. [6.37] introduced the method of explosive-driven flux compression in the open literature. This inspired physicists in several countries to undertake similar experiments. Soon it was realized that the reproducible generation of megagauss fields under conditions suitable for experimentation was much more difficult than had been anticipated. This was the subject of lively discussions at the first megagauss conference, held at Frascati in 1965 [6.38]. It took a number more years until the first experimental results of general interest were obtained. In the seventies, three different techniques had matured to this point: explosive-driven flux compression, mainly in plane geometry with "bellows" or "strip" devices [6.39], fast capacitor discharges into light-weight single-turn coils [6.40], and electromagnetic flux compression [6.41,42]. The course of the development is illustrated in Fig. 6.1 with the peak fields reported by various authors in the open literature. Part of the work with explosives had been classified and thus may have been completed earlier. Possibly for the same reason, the information given in some of the papers is fragmentary and not sufficient to analyze or reproduce the results at other laboratories.

A stationary magnetic field is a convenience that can be used without worrying much about the equipment by which it is generated. To the contrary, a pulsed field – and in particular a megagauss field – is rather an integral part of the experiment. Many experiments are dominated by the conditions imposed by the pulsed field generator, which may in fact require more attention from the experimentalist than the experiment itself. Therefore we begin this chapter with a discussion of basic working principles for the efficient design and operation of pulsed field apparatus. General magnet design has been elaborately treated in [6.72–74], the latter dealing with pulsed fields only.

Nondestructive fields with long pulse duration are well suited for most types of experiments, in fact much better than is generally believed. It is mainly a matter of the available resources to adapt the pulsed field generator to a given experiment. With megagauss fields, much effort is needed for setting up a reliable field generator, and experiments are selected with a view to the stringent conditions imposed by the megagauss field environment.

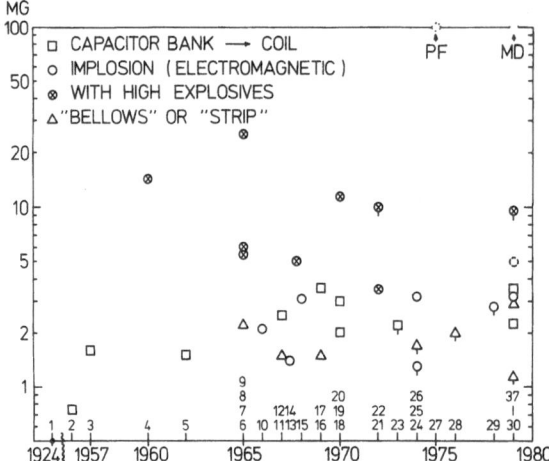

Fig. 6.1. The highest peak fields in chronological order of publication. A vertical dash indicates experimental applications, which are usually at lower fields than reported maxima. Authors: *(1) Kapitza* [6.1], *(2) Foner* and *Kolm* [6.23], *(3) Furth* et al. [6.36], *(4) Fowler* et al. [6.37], *(5) Shneerson* [6.43], *(6–7) Herlach* and *Knoepfel* [6.44, 45], *(8) Fowler* et al. [6.46], *(9) Sakharov* et al. [6.47], *(10) Cnare* [6.41], *(11) Bichenkov* et al. [6.48], *(12) Forster* and *Martin* [6.49], *(13–14) Alikhanov* et al. [6.50], *(15) Drew* et al. at A.W.R.E. [private communication], *(16) Babarina* et al. [6.51], *(17) Shearer* [6.52], *(18) Shneerson* [6.53], *(19) Andrianov* et al. [6.54], *(20) Besançon* et al. [6.55], *(21) Herlach* and *Kennedy* [6.56], *(22) Hawke* et al. [6.57], *(23) Herlach* and *Mc Broom* [6.40], *(24) Miura* et al. [6.42, 58], *(25) Pavlovskii* et al. [6.59], *(26) Mikhkel'soo* et al. [6.60], *(27) Nardi* et al. (Plasma Focus) [6.61], *(28) Fowler* et al. [6.62], *(29) Miura* et al. [6.42], *(30) Caird* et al. [6.63], *(31) Gennadiev* et al. [6.64], *(32) Bichenkov* et al. [6.65], *(33) Andrianov* et al. [6.66], *(34) Botcharov* et al. [6.67], *(35) Mc Daniel* et al. [6.68] (private communication), *(36) Pavlovskii* et al. [6.69], *(37) Alikhanov* et al. ("magneto-pressed discharge") [6.70, 71]

Most of the trouble is related to the sample and its electrical connections. Therefore the best experiments might be those which do not require a sample at all: interaction of megagauss fields with elementary particles [6.75], with electromagnetic radiation or, ultimately, with the vacuum. Some of these experiments are related to astrophysical problems in that ultrastrong magnetic fields are present in some types of stars. Plasma physics is in the same category, although in an implosion a plasma can be "poisoned" by particles emitted from the imploding walls. New ideas presently under investigation may even lead to practical fusion reactors [6.76].

Solid-state physics is a vast area for experiments with megagauss fields. Between 100–1000 T, the magnetic energy becomes comparable to fundamental interactions in the solid, the cyclotron orbit shrinks to molecular dimensions, the spacing of the Landau levels becomes larger than kT at room temperature and above, and extreme quantum limits are exceeded, e.g., where the lowest Landau level approaches the Fermi energy. The most suitable samples are insulators and semiconductors, and optical detection methods are the first choice. It is a lucky coincidence that with megagauss fields, the frequencies of resonance phenomena (e.g., cyclotron resonance) are shifted into the optical region and that the development of optical techniques is presently receiving a strong boost from laser technology. Picosecond laser spectroscopy [6.77] will ideally fit the short pulse duration of the fields. In addition, there are the common nonresonant magnetooptical effects which can be driven into nonlinearity. Megagauss physics is the domain of nonlinear behavior, where solid-state theories can be critically tested. Effects of similar magnitude are obtained by hydrostatic compression of the crystal lattice, resulting in substantial deformation of the electron orbits. Megagauss fields are well suited as a pressure-transmitting medium to obtain quasi-hydrostatic pressures much in excess of 1 megabar ($=100$ GPa) [6.57]. This entire field of research is a challenge for the future.

6.2 Submegagauss Fields with Pulse Times from Milliseconds to Seconds

6.2.1 Design and Operation of Nondestructive Coils

The design principles for nondestructive pulsed field coils are closely related to those for stationary fields [6.73]. Pulsed field coils are heated adiabatically during the current pulse. For fields with long duration (~ 1 s) it is conceivable to add a pulsed cooling system, but this has not yet been tried and may indeed not be worth the effort. The dominating factor is the mechanical strength for containing the magnetic forces. This becomes quite critical between 30 and 40 T; at present 50 T is an upper limit for the repeated operation of wire-wound, high inductance coils with a typical pulse duration of 10 ms. Higher fields up to 75 T have been

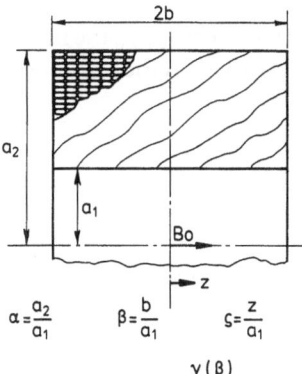

$$\alpha = \frac{a_2}{a_1} \qquad \beta = \frac{b}{a_1} \qquad \varsigma = \frac{z}{a_1}$$

$$B_0 = \mu_0 \, \frac{NI}{a_1} \, \overbrace{\frac{1}{2(\alpha-1)} \, \ln \frac{\alpha + \sqrt{\alpha^2 + \beta^2}}{1 + \sqrt{1 + \beta^2}}}^{\gamma(\beta)}$$

Fig. 6.2. Definition of the coil parameters and magnetic field of a coil with constant current density. N is the number of turns

obtained with solid helix coils [6.22] at the expense of pulse duration ($\simeq 0.1$ ms). If a very large energy supply is available, of the order of a megajoule, the magnetic forces can be spread over a larger volume. *Date* [6.31] is using coaxial free-standing helices made of maraging steel with the aim of reaching the megagauss limit. A critical problem is the possible destructive release of this large energy in case something goes wrong, which is always possible with coils that are so close to the edge of destruction.

Within the limitations imposed by the mechanical strength, the design of the coil and choice of power supply depend on the desired peak field, useful volume, pulse duration and field homogeneity. It is practical to discuss a coil in terms of the inner radius and the two dimensionless parameters α and β defining the shape of the coil. With the definitions given in Fig. 6.2, the field/current ratio at a point z on the axis is given by

$$B(\zeta) = \mu_0 \frac{NI}{a_1} \frac{1}{2\beta} [(\beta + \zeta)\gamma_{(\beta + \zeta)} + (\beta - \zeta)\gamma_{(\beta - \zeta)}] \tag{6.1}$$

for a coil with constant current density.

Wire-wound coils are usually precooled by liquid nitrogen and allowed to heat up to no more than 100 °C during the pulse, thus absorbing about 100 kJ per kg copper. The maximum pulse duration Δt at field B is related to the total energy W and thus to the mass of the coil. This can be derived from the integral

$$\int_0^{\Delta t} (I/A)^2 dt = D \int_{T_0}^T (c_p/\varrho) dT = A^{-2} \int_0^W dW/R \tag{6.2}$$

and is given by

$$\int_0^{\Delta t} B^2 dt = \frac{2\mu_0^2}{\pi} \frac{1}{a_1} \gamma^2 \beta \frac{\alpha - 1}{\alpha + 1} f\theta(T_0, T) W, \tag{6.3}$$

where I is the current, A the cross section of the wire, D the density, T_0 the initial and T the final temperature, c_p the specific heat, ϱ the resistivity, R the resistance, f the filling factor and

$$\theta(T_0, T) = \int_{T_0}^T (c_p/\varrho) dT / \int_{T_0}^T c_p dT \tag{6.4}$$

is a function typical of the conductor material. Examples are given in Table 6.1. The peak magnetic field is determined by the fraction W_m of the total energy transferred to the coil inductance:

$$W_m = \frac{1}{2} LI^2 = \frac{\pi}{16\mu_0} a_1^3 g \frac{(\alpha + 1)}{\beta\gamma^2} B^2. \tag{6.5}$$

Table 6.1. Electrical and thermal properties of copper (assuming a resistance ratio $\varrho_{273}/\varrho_4 = 200$)

Temperature initial [K]	final [K]	Enthalpy change [J/g]	θ [$10^8\,\Omega^{-1}\,m^{-1}$]	$\int \frac{c_p}{\varrho}\,dT$ [$10^{13}\,Jkg^{-1}\,\Omega^{-1}\,m^{-1}$]	Resistivity [$10^{-8}\,\Omega m$]
4	100	11	9	1	0.38
	150	25	5	1.3	0.83
	200	42	3.4	1.4	1.2
	300	80	2.1	1.7	1.7
	400	119	1.6	1.9	2.3
27	100	11	8.4	0.9	
	150	25	4.5	1.1	
	200	42	3.1	1.3	specific heat
	300	80	2.0	1.6	[$Jg^{-1}\,deg^{-1}$]
	400	119	1.5	1.8	
77	100	5	3.5	0.18	0.25
	150	20	2.2	0.4	0.32
	200	37	1.6	0.6	0.36
	300	74	1.2	0.9	0.39
	400	114	0.9	1.1	0.4

The factor that accounts for the edge effects in the inductance is g,

$$L = \frac{\mu_0 \pi}{8}\, a_1\, \frac{(\alpha+1)^2}{\beta}\, N^2 g. \tag{6.6}$$

An approximate relation for g has been given by *Welsby* [6.78]:

$$g = 1 + 0.225\,\frac{\alpha+1}{\beta} + 0.64\,\frac{\alpha-1}{\alpha+1} + 0.42\,\frac{\alpha-1}{\beta}. \tag{6.7}$$

More precise values can be taken from the tables compiled by *Grover* [6.79]. A two-dimensional least-squares fit to these tables is incorporated in the BASIC routine given in Fig. 6.3.

The resistance of the coil in terms of the geometrical parameters is

$$R = \varrho \pi a_1 (\alpha+1)\,\frac{N}{A} = \varrho\,\frac{\pi}{2}\,\frac{\alpha+1}{\alpha-1}\,\frac{1}{\beta}\,\frac{N^2}{a_1 f}, \tag{6.8}$$

where A is the cross section of the wire and the corresponding time constant is

$$\tau = \frac{L}{R} = \frac{\mu_0}{4}\,\frac{1}{\varrho}\, a_1^2 (\alpha^2 - 1) f g. \tag{6.9}$$

```
100    REM       INDUCTANCE     A=ALPHA,    B=BETA,   R: INNER RADIUS
110    Y=LGT(2*B/(A+1))
120    IF Y<1 AND Y>-.7 THEN 150
130    G=0
140    GOTO 340
150    G=6884+5018*Y-1945*Y*Y-1783*Y`3+1410*Y 4+615*Y 5-612*Y 6
160    Y=(A-1)/2/B
170    X=(A-1)/(A+1)
180    IF X>1 OR Y>1 THEN 130
190    FOR I=1 TO 5
200    Z=0
210    FOR J=1 TO 4
220    IF X<>0 THEN 240
230    X=1.E-10
240    IF Y<>0 OR J<>1 THEN 260
250    Y=1.E-10
260    READ E
270    Z=Z+E*Y`(J-1)
280    NEXT J
290    G=G-Z*X`(I-1)
300    NEXT I
310    RESTORE
320    IF R=0 THEN 340
330    RETURN G*5.E-12*PI 2*R*(A+1)`2/B
340    RETURN (G/10000)
350    DATA -1,-14,41,-26,6711,-3739,1688,-216,-3445,11999,-14383,5747
360    DATA 92,-12426,20737,-9720,-23,4831,-9154,4585
```

Fig. 6.3. A BASIC routine for calculating the inductance of wire-wound coils, based on a 2d polynomial fit to Tables 22 and 37 in *Grover* [6.79]. For R=0, the factor g is returned

The peak magnetic energy W_m is related to the total energy W by the duration and shape of the pulse and by the resistivity function, which can be illustrated by the equations for a capacitor discharge with constant resistance. The discharge waveform (in square brackets for an aperiodic discharge, $\omega_0 < \delta$) is given by

$$I = I_0 \frac{\omega_0}{\omega} e^{-\delta t} \sin \omega t; \qquad \left[= I_0 \frac{\omega_0}{\omega_a} e^{-\delta t} \sinh \omega_a t \right]$$

$$\omega_0 = \sqrt{\frac{1}{LC}}$$

$$\delta = \frac{R}{2L} = \frac{1}{2\tau} \tag{6.10}$$

$$\omega = \sqrt{\omega_0^2 - \delta^2}; \qquad \left[\omega_a = \sqrt{\delta^2 - \omega_0^2} \right]$$

$$I_0 = U_0 \sqrt{\frac{C}{L}} = \sqrt{\frac{2W}{L}}.$$

To simplify calculations further one can introduce a dimensionless "damping parameter" in several ways. We prefer

$$q = \frac{\delta}{\omega} = \frac{1}{2\omega\tau} = \frac{t_0}{2\pi\tau}; \qquad \left[= \frac{\delta}{\omega_a} \right]. \tag{6.11}$$

Among others, this is most practical for developing equations that determine circuit parameters from a measured waveform. The half-period is $t_0 = \pi/\omega$ and for the first peak we get

$$t_1 = \frac{1}{\omega} \arctan \frac{1}{q} \qquad \left[= \frac{1}{2\omega_a} \ln \frac{q+1}{q-1} \right]$$

$$I_1 = I_0 \exp(-\delta t_1) = I_0 \exp\left(-q \arctan \frac{1}{q}\right) \qquad \left[= I_0 \left(\frac{q-1}{q+1}\right)^{q/2} \right] \qquad (6.12)$$

$$W_m = \frac{1}{2} L I_1^2 = W \exp\left(-2q \arctan \frac{1}{q}\right) \qquad \left[= W \left(\frac{q-1}{q+1}\right)^{q} \right].$$

It is instructive to consider the energy ratio as a function of t_0/τ (in this age of pocket computers, a figure is deemed unnecessary). For short pulses with $t_0/\tau < 0.1$, almost all of the energy is available to build up the magnetic field and is resistively dissipated only afterwards. In this regime, mechanically strong conductor materials with high resistivity can be used without precooling. If a long pulse duration is required, an allowance must be made for the resistive loss.

Figure 6.4 illustrates the design of a coil with respect to the geometrical parameters for a typical value of $W_m/W = 0.7$ at $t_0/\tau \gtrsim 1$. The parameters α and β are related by the condition of constant winding volume,

$$V = 2\pi a_1^3 (\alpha^2 - 1)\beta = \pi a_1 (\alpha + 1) NA/f. \qquad (6.13)$$

As the maximum in the field is rather flat, the dimensioning is not critical; in most cases the shape of the coil will be determined by the desired homogeneity. The remaining free parameter is the cross section of the wire. Apart from a possible small influence on the filling factor, this does not have any direct effect on the efficiency of the coil because it does not affect the time constant. It determines only the voltage and current needed to energize the coil; for a given capacitor bank, it determines the duration and shape of the pulse as illustrated in Fig. 6.5.

For the resistivity and specific heat, the following functions have been used in Fig. 6.5:

$$\varrho_0 = -3.41 \times 10^{-9} + 7.2 \times 10^{-11} T \qquad \text{at} \quad B=0$$

$$\varrho_{(B)} = \varrho_0 \left(1 + 10^{-6.821 + 2.537 \log x - 0.1853 (\log x)^2}\right)$$

$$x = |B \cdot 1.668 \times 10^{-7}/\varrho|$$

$$c_p = -1.6608 - 0.05457 y + 1.742 y^2 - 0.8621 y^3 + 0.1221 y^4$$

$$y = \log_{10} T.$$

Examples illustrating the practical range of coils are given in Table 6.2.

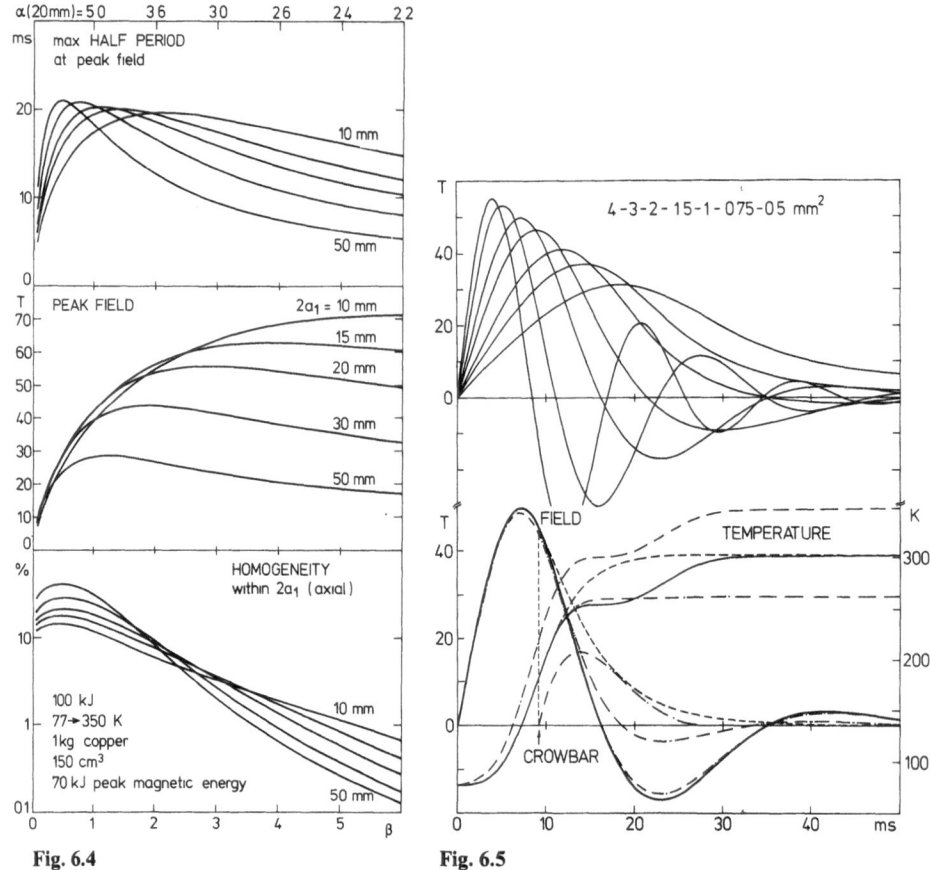

Fig. 6.4   **Fig. 6.5**

Fig. 6.4. Limiting design parameters for a 100 kJ copper coil, precooled by liquid nitrogen and allowed to heat up to an average temperature of 350 K

Fig. 6.5. Computer simulation of a capacitor discharge into a wire-wound coil cooled by liquid N_2. (*Upper part*) influence of the wire cross section on the waveform, coil parameters from Example 1 in Table 6.2. (*Lower part*) for the coil with 2 mm² wire [(——) same as in upper part] the effects of 0 Ω crowbar (– – –) and 0.68 Ω crowbar for maximum external energy dissipation (–··–··–) are shown; crowbar current in units of equivalent magnetic field. (–·–·–) shows the effects of magnetoresistance and eddy current heating (Example 2), neglected in all other curves. Here, *T* at the inner radius of the coil is given; the average temperature is similar to Example 1 (——)

After determining the approximate parameters for the desired performance, one may integrate the waveform numerically, with resistance varying as a function of temperature and magnetic field. The result of such integrations is given in Fig. 6.5, which illustrates the effect of a crowbar circuit and the magnetoresistance and eddy current heating. The two latter effects result in an inhomogeneous temperature distribution (Table 6.2). The additional heating of

Table 6.2. Calculated performance of wire-wound coils with constant current density, precooled by liquid N_2

Example		1	2[a]	3	4	5	6
Energy	MJ	0.1	0.1	1	10	10	100
Peak field	T	49.8	48.7	50	40	40	50
Pulse duration[b]	ms	16.3	16.3	145	330–110	110–400	480–1200
Final temperature	K	303	348[a]	300	160	305	304
Homogeneity[c]	%	0.95	0.95	0.98	0.3	0.3	0.09
Coil i.d.	mm	20	20	30	40	40	50
o.d.	mm	55	55	134	250	250	530
length	mm	90	90	150	320	320	700
Copper weight	kg	1.33	1.33	13.5	110	110	1025
Wire cross section	mm²	2	2	2.6	40	40	150
Filling factor		0.8	0.8	0.75	0.8	0.8	0.75
Inductance	mH	3.74	3.74	117	15.4	15.4	43.5
dc resistance at 77 K	mΩ	80	80	480	17	17	11
Voltage	kV	5	5	10	1	2.5	3.5
Peak current	kA	6.2	6.0	3.0	16.8	16.8	36
Peak power	MW	(20)	(20)	(22)	20	40	110

[a] Only this example includes estimates of magnetoresistance and skin effect. A wire thickness of 1 mm has been assumed. The temperature is given at the inner surface.
[b] Half-period of the sine wave. For controlled power (10 and 100 MJ), the rise time and the plateau are given.
[c] Axial homogeneity over a length equal to the inner diameter.

the inner layers must be considered when determining the maximum energy that the coil can absorb without heat damage. The eddy current heating, averaged over the thickness d of a flat conducting sheet, is calculated in the low-frequency approximation. This assumes an eddy current distribution linear with the distance from the surface, i.e., the magnetic field caused by the eddy currents is neglected. The Joule heating power is given by

$$\frac{dW_j}{dt} = \frac{d^2}{12\varrho} \left(\frac{dB}{dt}\right)^2 \cdot Vf. \tag{6.14}$$

The ac resistance of a coil with M layers is found by averaging the power dissipation over a cycle:

$$R_{ac} = R_{dc} \left\{ 1 + \left(\frac{d}{a}\right)^4 \left[\frac{4}{45} + \frac{1}{9}(M^2 - 1)\right] \right\}, \quad \text{where} \tag{6.15}$$

$$a = \sqrt{\frac{2\varrho}{\mu_0 \omega}} \tag{6.16}$$

is the skin depth and ω the angular frequency of the alternating current. The variation of the ac resistance during the transient pulse is approximated by

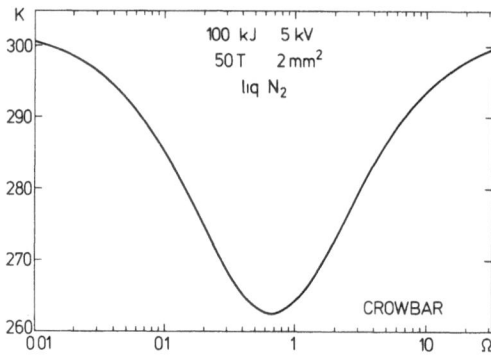

Fig. 6.6. Final temperature at the inner surface of the coil, as a function of the crowbar resistor, based on Example 1, Table 6.2

multiplying the second term in [6.15] with a normalized factor proportional to [6.14]. This is a fairly crude approximation, but it can be justified on the grounds that in most practical cases, the contribution to the heating is modest and the effect on the waveform is almost negligible. The crowbar circuit can be dimensioned for either maximum external energy dissipation (crowbar resistance $R_c = \sqrt{L/C}$) or for maximum extension of the pulse duration ($R_c \simeq 0$). Both modes are illustrated in Fig. 6.5, and the lowering of the final temperature as a function of the crowbar resistor is shown in Fig. 6.6. In these calculations it has been assumed that the crowbar diodes are connected across the capacitor to prevent diode failure caused by overvoltage peaks [6.30]. For all but the very small ratios of t_0/τ, the effects of the crowbar are modest, its main advantage being the protection of the capacitors from voltage reversal, which would shorten their lifetime. Apart from using a crowbar, the pulse duration can be lengthened by increasing the circuit resistance at the expense of peak field.

With a controlled power supply, the waveform depends on the available power and the relation of the peak voltage to the cross section of the wire. In particular, the ratio of the rise time to the plateau is determined by a combination of these factors; the ratio will usually be in the range 0.1–0.2. In the regime where the field rise is power limited, the waveform will rise approximately as $\sqrt{t - t_0}$.

High inductance coils are best wound from wire with rectangular cross section to obtain a good filling factor and to avoid the wires slipping under the influence of magnetic stress. The commercially available formvar-coated wire is not mechanically strong, thus the coils must be reinforced on the outside; for example, with a heavy steel cylinder or a fiberglass-epoxy composite. Compactness and additional insulation are achieved by potting the coils in epoxy (e.g., Stycast 2850 FT which is heavily filled to obtain a good heat conductivity and a thermal dilatation similar to that of copper). A safe but messy procedure consists in applying the epoxy while winding the coil. Vacuum impregnation is not always reliable. This can be improved by the application of pressure after impregnation. Nevertheless, at fields above 30 T there is some deformation due to compression of the winding; this can be of the order of one millimeter at the inner radius in the midplane. After a few "training" shots, the coil becomes stable, presumably by work hardening. It can then be used for a reasonable number of experiments (about 100); however, at field levels of the order of 40 T spontaneous failure will eventually occur. It has been found that copper gradually extrudes through small cracks in the insulation; failure may occur when the extruded copper makes contact with the next layer. Other effects are the initiation of a surface discharge, or the overheating of the inner layer due to the decreased cross section of the wire caused by the expansion.

To prevent the damage caused by a coil implosion, it is a good practice to monitor the inductance and resistance of the coils and check every magnetic field record (i.e., the sensitive dB/dt signal) for the onset of irregularities which may indicate impending failure. An original alternative that appears to work quite well consists in using a microphone to monitor the noise made by the coil during the pulse, and to compare this by computer to a prerecorded "standard" noise pulse [P.S. Flower, private communication]. There is no doubt that much stronger coils can be made with special wires such as copper wire with a steel core [6.80], (e.g., Copperweld) which is commerically available, but not yet with specifications well suited for high field coils. As an alternative, bimetallic wire as it is manufactured to make superconducting coils has been used in the USSR [6.81]. The superconductor material in the core of this wire is mechanically very strong, the coils are cooled as usual by liquid nitrogen. *Ozhogin* et al. [6.82] have made such coils that reliably generate 50 T with 15 ms pulse duration. At Leuven, coils have been wound with a stainless steel band on top of rectangular copper wire. As with the superconducting wire, there is an insulation problem because the steel wire is not furnished with insulation. Great care must be taken to obtain a good insulation by means of glassfiber cloth and epoxy. The steel causes small peaks to appear in the dB/dt signal in the close vicinity of the current reversal, in most experiments this will not be disturbing. In recent years, a number of new copper alloys (Table 6.3) have become commercially available that combine extra strength with good conductivity. Most of these require work hardening; this must be considered when comparing yield strength values given by the manufacturers.

The mechanical deformation of multilayer coils has been theoretically analyzed [6.83–86] and found to agree with experience within the relatively large margins of applicable failure criteria. For a long coil with constant current density and without radial transmission of stress between the windings, the relation between the center field and the maximum stress σ (which occurs at $r/a_1 = \alpha/2$) is given by

$$B = \sqrt{2\mu_0\sigma}\ \sqrt{2}\ (1 - 1/\alpha). \tag{6.17}$$

If there is radial transmission of stress, σ_r becomes negative below $r/a_1 \sim 1.8$. Unless there is a strong adhesion between the layers, they will be pulled apart and this part of the coil will behave like a coil with free windings. In a coil with strong adhesion and particularly in a massive coil (solid helix, Bitter coil) the inner layers are pulled outwards, which results in a substantial increase of the stress at

Table 6.3. Special copper alloys for electrical conductors (cf. Figs. 5.13, 14, Table 5.3)

Resistivity at room temp. [$\mu\Omega$ cm]	Yield strength [kb]	Alloy [%]	Trade name
2.02	2.8–5.6	Ag 0.04; P 0.05; Mg 0.11	SSC-155 Super Silver-Copper
1.9	4.2 (4.9 at 400 °C)	Zr 0.1–0.15	
2.8	4–5	Fe 2.4; P 0.03	Kronakupfer
2.2–3.4	1.6–5.4	Fe 2.1–7.6; P 0.0015–0.15; Zn 0.05–0.2; Pb < 0.03	HSM-Cu
1.87	3.2–4.7	Al_2O_3 0.2	Glidcop AL-10
	5.4–7	Al_2O_3 0.2 wire 0.5 mm	Glidcop AL-10
1.94	4.7–5.1	Al_2O_3 0.4	Glidcop AL-20
2.03	4.9–5.5	Al_2O_3 0.7	Glidcop AL-35
2.21	5.2–5.7	Al_2O_3 1.1	Glidcop AL-60
1.77	1–2.5	pure copper, hard drawn	

the inner radius, possibly by as much as a factor of the order of two [6.27, 83, 84]. Figure 6.7 shows a crack that has developed in such a coil after a large number of shots.

A complete stress calculation requires the computation of the axial and radial magnetic field components throughout the winding volume [6.88]. Such calculations are useful to predict general trends in coil behavior but are of limited value in the design of specific coils. As a particular result, it was found that an internal reinforcing cylinder is most effective when placed at a radius between $0.35\ a_2$ and $0.39\ a_2$ [6.27]. A wire-wound coil is an inhomogeneous composite structure and the properties of the materials, modified by the initial plastic

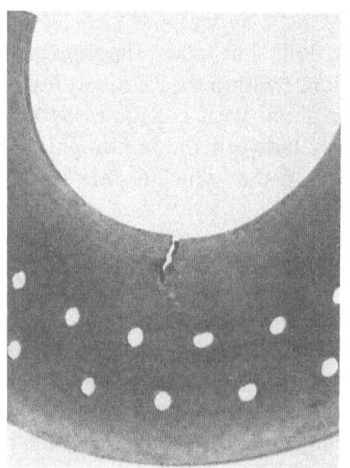

Fig. 6.7. A radial crack in a pulsed Bitter coil with an inner diameter of 50 mm [6.87]. These cracks develop gradually and coils with small cracks have been operated for large numbers of shots until flashover finally occurred. *Furth* et al. [6.36] called this the "saw effect" because it seems that the magnetic field "saws" its way into the coil. Photograph by courtesy of C.P. Flower, now at CERN

deformation and by the application of simultaneous mechanical and thermal shock, are not well known. In practice, $\sqrt{2\mu_0\sigma}$ in (6.17) gives a reasonable estimate for the peak field to be obtained with a well-made coil of average proportions ($\alpha \sim 3$–4). This is 22 T at 0.2 GPa (soft copper), 33 T at 0.45 GPa (work-hardened copper), > 35 T at > 5 GPa (ordinary stainless steel, special copper alloys), 50 T at 1 GPa (beryllium-copper) and 70 T at 2 GPa (very hard maraging steel).

The actual manufacture of successful coils is much more an art than a science, and the most valuable research tool may be the saw used for cutting up the coils that have failed. It is certain that better coils could be made using sophisticated modern techniques and materials. A hindrance to their application is the modest resources of most pulsed field laboratories, where researchers would rather pursue their experiments than spend time on further improvements to the coils.

The wire-wound coil with a capacitor bank of the order of 100 kJ is an elegant, compact laboratory instrument. Its main restriction is in the pulse duration (half-period) of the order of 10–20 ms. It is possible to obtain a pulse with a flattened top by arranging the capacitor bank as a transmission line [6.89]. At the ISSP (Tokyo), such a combination of capacitors and inductors has been set up with the following characteristics: 112 kJ at 4 kV in four sections, generating 40 T in 15 mm diameter with a flat top of 1 ms duration. For pulses with essentially longer duration, large facilities with huge coils are required, where only a small fraction of the stored energy ends up in the experimental volume (Table 6.2, Fig. 6.8). An advantage of the large coils is the greater overall mechanical strength. As power sources, a megajoule capacitor bank [6.30], direct power from the mains by special arrangement with the utility [6.27, 28] and energy storage in a flywheel [6.29] have been used.

The technological development in this area is presently receiving a boost from nuclear fusion experiments where there is a growing trend for switching from capacitor banks to flywheels, e.g., at the Max Planck Institute in Munich and the Joint European Torus in Culham. On the whole, pulsed power is now becoming of interest for quite a few applications [6.90]. It has been proposed to

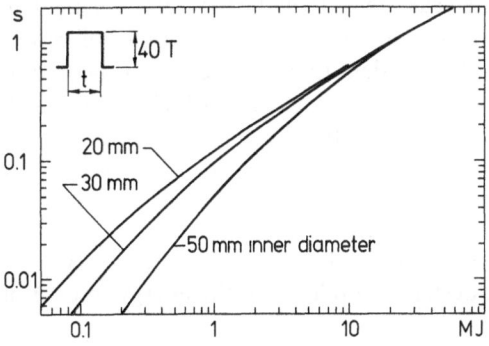

Fig. 6.8. The pulse duration at a constant field level of 40 T, as a function of the total energy delivered by the power supply. In practice, an allowance must be made for the finite rise and fall times of the field pulse (cf. Table 6.2)

Fig. 6.9. A machined copper-beryllium helix as used at the ISSP [6.95]; i.d. 15 mm, o.d. 50 mm, length 42 mm, peak field 45 T with a half period of 0.4 ms, using a 32 kJ, 3.3 kV capacitor bank

use the 200 MJ, 200 MW machine of the ALCATOR experiment at MIT to generate fields up to 75 T for 0.5 s [6.91]. Thus we are finally back to the system introduced by Kapitza!

Fields in excess of 50 T can be obtained nondestructively with coils machined from hard materials, at the expense of pulse duration. This can be either a machined helix (Fig. 6.9) as introduced by *Myers* [6.7] and further developed by *Foner* and *Kolm* (75 T with 120 μs half-period) [6.23], a solid single-turn coil driven by a pulse transformer (70 T with 1.25 μs quarter period) [6.92], or a flux concentrator [6.93, 94]. This is a solid block of conducting material with slits and cavities that guide the magnetic flux from a multiturn coil to the experimental volume by means of induced currents in the skin layer. It may be regarded as a pulse transformer incorporated into the single-turn coil. In a flux concentrator, the force on the primary winding can be partially compensated by the currents in the concentrator which is the secondary winding.

Now and then there have been proposals for force-free coil configurations where at every point the current is parallel to the local field [6.96]. However, these ideas have never been applied in practice. A straightforward example of a force free coil is a helix with a low pitch calculated to make the azimuthal field on the outside equal to the axial field inside. The force at the ends of the coil can be eliminated by winding it as a torus. It is evident that much energy is needed to energize such an extended structure. At very high fields, which are of short duration for energetic reasons, the conductor material is compressed in any case, due to the skin effect. Apart from this, the highest fields can be obtained by *Date's* free-standing coaxial solid helices [6.31]. In principle, this system is not limited in the field strength that may be obtained, but in practice the required energy becomes prohibitive at field levels above 1 MG. All high energy systems must be carefully guarded against destructive release of the energy in case of failure.

6.2.2 Facilities and Experiments

The interest in long pulse times has led to the development of the larger facilities required. The first large pulsed magnet was built at the University of Amsterdam [6.27,97]. This well-planned installation is the prototype for magnet systems with long pulse duration. Due to favorable circumstances, 8.2 MW pulsed power can be directly taken from the mains; this is rectified and regulated by 18 thyristors (in place of the originally used ignitrons and thyratrons) at the maximum level of 10 kA, 660 V. Thyristors work very well in this application but they must be carefully protected against overvoltage peaks and current overload. Details on the latest improvements of this installation have not yet been published. The technology of high power solid-state devices is now advancing at such a fast rate that specific information (e. g., on types of thyristors, circuits and protective devices) may become obsolete within a year.

The pulse shape can be programmed by a flexible digital control circuit; the current is regulated by means of a flux comparator with feedback from the magnetic field measurement. The 300 Hz component from the three-phase rectification and its harmonics are removed by passive filters. The field can be kept constant at the programmed value to within 10^{-3}. The pulse shape begins with a rapid rise to peak field to take maximum profit from the precooling, it can be decreased linearly or as a step function while the coil is warming up. The total pulse duration is of the order of one second, the peak field can be maintained for approximately 0.1 s. At the end of the pulse, the polarity can be reversed to

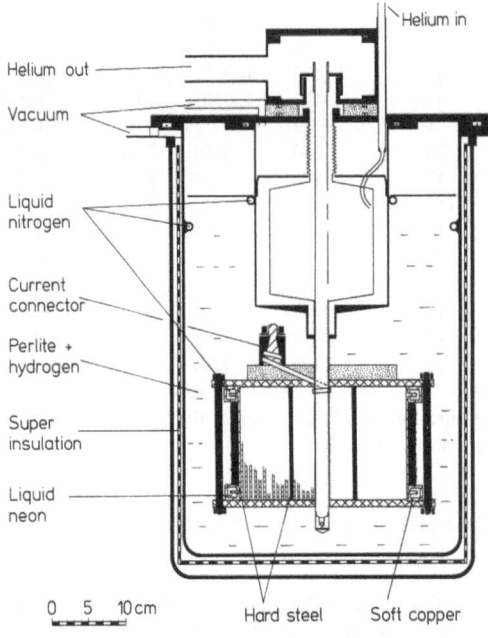

Fig. 6.10. Scale drawing of the Amsterdam magnet with a helium cryostat for the sample space. The i. d. of the helium dewar is 15.6 mm. The magnet is cooled by heat conduction in the outer cylinder made of soft copper. There are 638 windings of hardened copper wire with a cross section of 2.9×7.13 mm; the filling factor is 0.8. The resistance is 235 mΩ at 295 K, 30 mΩ at 77 K and 7 mΩ at 27 K; the inductance is 18 mH. Peak field is 40 T at 9 kA

remove the inductively stored energy from the magnet. The total energy dissipated in one pulse is 1.5 MJ at most. The final temperature of the coil is kept to the conservative value of 150 K. The coil is monitored by measuring its resistance during the pulse and the current is automatically switched off when a critical condition is approached. The coil is precooled by liquid neon to 27 K, which can have a substantial advantage over liquid nitrogen as the function c_p/ϱ rises steeply below 70 K and may go through a maximum above 27 K according to the residual resistance of the copper. While the peak field does not depend directly on the initial temperature, there is a dependence in relation to the power supply; the Amsterdam magnet is reported to give 30 T with liquid nitrogen and 40 T with neon. A disadvantage of neon is the long cooling time, caused not so much by the available refrigeration power but by the reduced heat conductivity. At the Amsterdam facility the cooling time is two to three hours with a 200 W liquefier.

Table 6.4. Summary of presently operational pulsed field installations for nondestructive fields

	Laboratory	max. field [T]	i.d. [mm]	Pulse duration[a] [ms]	Energy [kJ]	Power [MW]	Voltage [kV]	Ref.
Controlled power	Amsterdam	40	20	1000/100	~1500	8.2	0.66	[6.27]
	Sendai	35	18	100/~10	1400	14	0.35	[6.28]
	MIT	40	20	500/100[b]	~1000	4(20[c])	0.2(0.25[c])	[6.29]
Capacitor banks	MIT	50	20	10	100	(30)	4	[6.98]
	Toulouse	40	26	400[d]	400	(~10)	6	[6.30]
	Toulouse	40	30	920[d]	1250	(~10)	10	[6.30]
	Osaka	60	20	0.37	490[e]	(4000)	26.6[f]	[6.31]
	Osaka	39	60	0.41	420	(3000)	26.6[f]	[6.31]
	Tokyo	42	16	4	32	(25)	3.3	[6.99]
	Tokyo	45	16	20	200	(30)	5–10	[6.100]
	Leiden	52[g]	20	17	180	(33)	3.5	[6.26]
	Braunschweig	27	13	150	40	(1)	2.5	[6.33]
	Genève	51[g]	14	15	72	(15)	2.5	[6.32]
	Leuven	45	18	10	70	(20)	3.5	[6.35]
	Moscow (Kurchatov)	50	20	15	190	(40)	5	[6.82]
	Wroclaw	47	10	10	70	(20)		[6.34]
	Nijmegen	40	20	15	100	(20)	5	[Add.]

[a] For controlled dc power, the total pulse duration and the duration of the flat top are given, for capacitor banks the half-period of the (deformed) sine wave.
[b] Computer simulation.
[c] Capability of the power supply.
[d] "Crowbar" waveform: the duration of the peak is smaller than in a sine wave.
[e] Total energy installed 1250+250 kJ.
[f] Convertible to 40 kV.
[g] These coils are reported to last only a few shots under favorable circumstances, typical performance is below 45 T, e.g., at Genève 40 T with 36 ms pulse duration.

The design of the magnet is shown in Fig. 6.10. The coil is wound with hardened, insulated copper strip, not normally available commercially. The layers are interleaved with fiberglass-polyester. The internal reinforcing cylinder is made of polished maraging steel (Ultimar 110). Much attention has been given to the contacts and the places where the wires come out of the winding, since these are always critical points with pulsed field coils. The first coil was in service from 1969 to 1976 without giving any trouble. The research carried out at this facility, mostly magnetization measurements, is described in Chap. 3.

A similar installation has been set up at Sendai [6.28], and at MIT flywheel-assisted dc generators have been used to energize pulsed field coils on a trial basis [6.29]. This involved mainly the development of safe switching techniques. The principal characteristics of all the facilities are summarized in Table 6.4.

The pulsed field laboratory at Toulouse is outstanding for the sustained high level of scientific activity. *Askenazy* [6.30] has pursued the capacitor discharge method to the extreme by setting up a 1.25 MJ capacitor bank. An idea of the size of the installation is given in Fig. 6.11, and Fig. 6.12 shows the effects of a coil blowout. Besides solid engineering, there are two elements that have greatly contributed to make this facility so suitable for experiments: for one, the discharge is crowbarred by power diodes (connected across the capacitors) to obtain an extension of the pulse duration (Fig. 6.13). The most important element is the original high current switch developed at Toulouse. Contact is made by plunging a piston into a pool of mercury at high speed. The piston is driven by compressed air and pre-arcing is reduced by pressurizing the space above the mercury. This switch operates smoothly and reliably and produces a clean pulse that is well suited for sensitive electrical measurements. Currently, the development of high power semiconductor switches is coming to the point where these can provide even smoother and more convenient switching. The coil design is rather conservative although modern materials are used for external reinforcement. Most of the scientific activity has been carried out in fruitful

Fig. 6.11. General view of the 10 kV, 1.25 MJ capacitor bank installed at the Toulouse laboratory. Crowbar diodes are visible on the right side on top of each row of capacitors; the vertical tubes contain fuses. Recently, the facility has been converted to switching with optically triggered thyristors

Fig. 6.12a,b. Damage resulting from a coil failure at an energy level of approximately 100 kJ

collaboration with Stradling's group at the Clarendon Laboratory, Oxford (now at St. Andrews University, Scotland).

The surprisingly high sensitivity of electrical measurements that can be achieved with a well-engineered pulsed field generator is demonstrated by the magnetophonon effect. This resonance effect in the electrical conductivity of semiconductors is similar to the de Haas-Shubnikov effect in that the resonances are periodical in $1/B$. The resonance condition is given by the coincidence of the

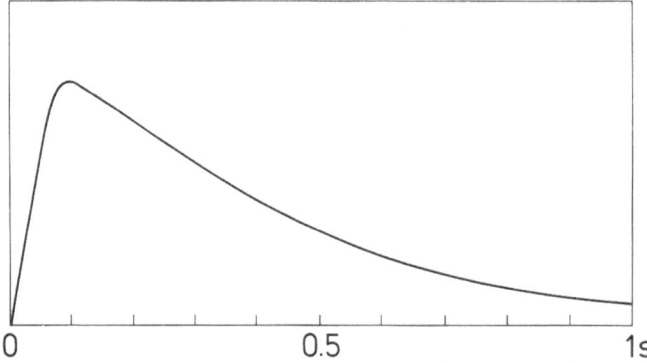

Fig. 6.13. Typical waveform of the Toulouse magnet. The slow decay due to the crowbar is indicative of a large time constant of the coil

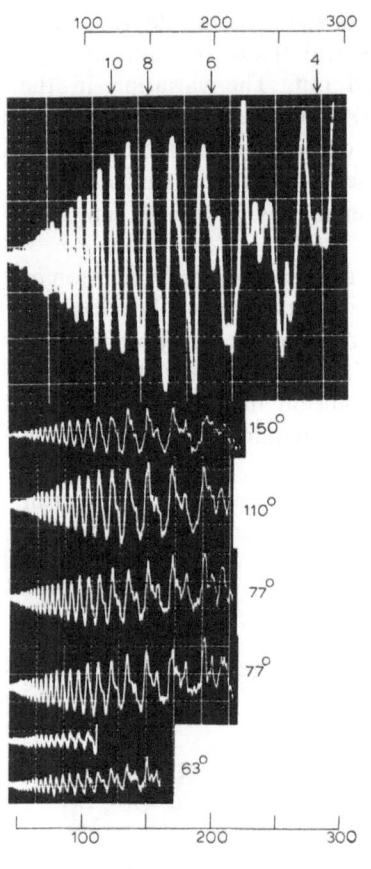

◄ **Fig. 6.14.** The magnetophonon effect in high purity p-Ge with the field orientation $B \| \langle 111 \rangle$ [6.101]. The resistivity change is $\sim 0.1\,\%$ of the zero field resistance. Therefore, the second derivative of the transverse magnetoresistance has been recorded. A triplet structure in the resonances is clearly resolved at the lower temperatures and higher fields. The field axis is 50–300 kG, temperatures 63–150 K

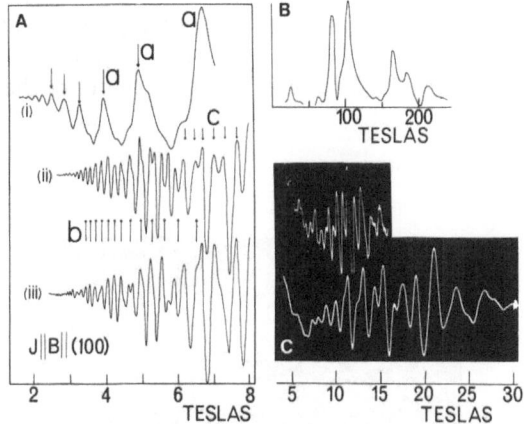

Fig. 6.15 A–C. The magnetophonon effect in n-Si with $B \| \langle 100 \rangle$ [6.102]. **(A)** shows records, with different amplifier time constants, of the longitudinal magnetoresistance at 60 K obtained with a 1/(time) magnetic field sweep. The resistivity is 10,000 Ω cm. The magnetophonon structure is revealed by the Fourier transform in **(B)**. Arrows a, b and c in **(A)** correspond to the peaks at 20.4, 81.1 and 166 T. **(C)** shows recordings of the second derivative of the resistivity at 77 K in a pulsed field

cyclotron frequency ω_c and a well-defined phonon frequency ω_{L_0}, e.g., the longitudinal optical phonon:

$$\omega_{L_0} = N\omega_c = N\,\frac{eB}{m^*}. \tag{6.18}$$

As the case requires, this can be used either to determine the effective mass m^* if the phonon frequency is known from Raman scattering, or to study the phonon spectrum if the band structure is known. Over and above the simple equation (6.18), there are many complications (e.g., "intervalley scattering" and "impurity transitions") which give rise to fine structure in the resonance patterns.

The effect is usually observed at temperatures of the order of 100 K; at low temperatures the carriers or optical phonons will freeze out and at high temperatures the resonances will be smeared out. The variation in the conductivity is quite weak and must be enhanced by double differentiation. It is most remarkable that the oscilloscope traces obtained in pulsed fields are comparable in quality to the best records obtained in continuous fields! Out of the wealth of available records, examples are given in Figs. 6.14 [6.101] and 6.15 [6.102].

A class of experiments where pulsed fields provide an essential extension of measurements is cyclotron resonance. This can be observed only if the condition $\omega_c \tau > 1$ is fulfilled, where the relaxation time τ is the average time interval between collisions of an electron. With this condition fulfilled, the electron can go through a full circle before it is disturbed by a collision. In particular, at higher temperatures and with samples of low mobility, this requires frequencies in the optical range and correspondingly high magnetic fields. A convenient light source is the HCN laser; the 337 µm wavelength results in a free electron resonance at 32 T which is at the center of the range covered by nondestructive

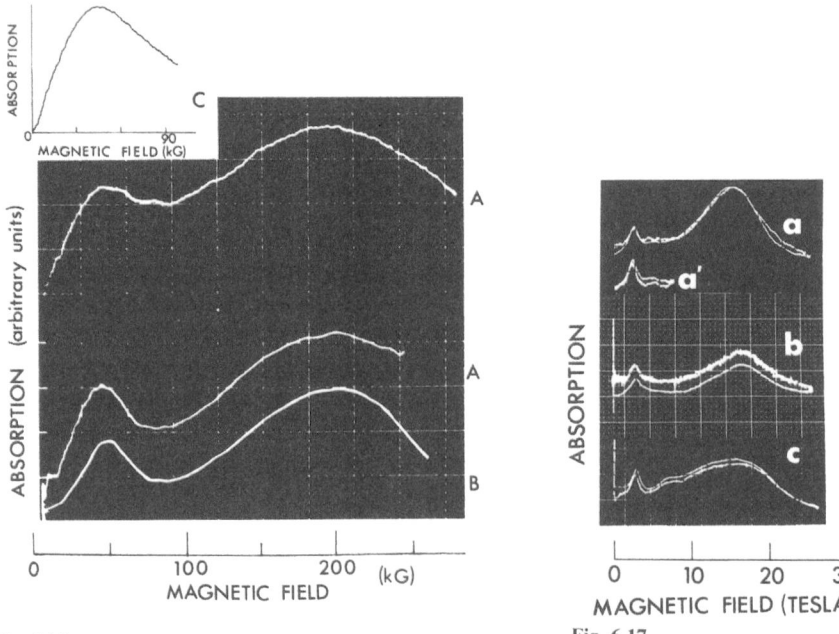

Fig. 6.16

Fig. 6.17

Fig. 6.16. Cyclotron resonance in p-InP at a wavelength of 337 µm (HCN laser) [6.103]. The experimental recordings are labeled (A); (B) is obtained by curve fitting (6.19). (C) is the light hole resonance measured with a dc magnet. The optimum sample temperature is 110 K; at 77 K the resonance cannot be observed because the holes freeze out into the acceptor states

Fig. 6.17. Cyclotron resonance absorption in p-GaAs at 77 K and 337 µm wavelength [6.104]. The change in the transmission is approximately 20% (a) $B \| \langle 100 \rangle$, (b) $B \| \langle 111 \rangle$, (c) $B \| \langle 110 \rangle$ (Table 6.5)

pulsed fields. Selected examples are shown in Figs. 6.16 [6.103] and 6.17 [6.104]. The analysis is made by means of a curve-fitting procedure as illustrated in Fig. 6.16. The absorption is taken to be proportional to the real part of the conductivity,

$$\sigma_r = S \frac{1 + (\omega^2 + \omega_c^2)\tau^2}{[1 + (\omega_c^2 - \omega^2)\tau^2]^2 + 4\omega^2\tau^2},$$

(6.19)

where $S = N(e^2\tau/m^*)$ is the dc conductivity. If there are several types of charge carriers, e.g., light and heavy holes, the expressions (6.19) for the different carriers are added. As the pulsed field technique is so prolific in data taking, it appears impractical to go into much detail with all the results that have been obtained [6.105, 106], so a summary is given in Table 6.5.

Recently, a series of beautiful Shubnikov-de Haas experiments in GaSb, Cd_3As_2 (thin film) and InSe (also cyclotron resonance) has been published [6.108]. An example of a different experimental technique is given by the measurement of the Shubnikov-de Haas effect in a two-dimensional space charge layer [6.109]. In this case, the applied electric field, i.e., the voltage at the MOS gate, is available as additional parameter to sweep the system through the resonances. In this experiment, the electric field is swept so rapidly that the magnetic field remains practically constant during the sweep. It is evident that such a technique could eventually be extended to record multiple sweeps during one field pulse.

Table 6.5. Summary of pulsed field cyclotron resonance experiments

Sample	Tempe-rature [K]	Wave-length [μm]	Orien-tation	m^*/m_0			$\omega\tau$		Ref.
				light hole	heavy hole electron		light	heavy	
p-InP	110	337	⟨100⟩	0.12 ±0.01	0.56±0.02				[6.103]
			⟨111⟩	0.12 ±0.01	0.60 ±0.02				
p-GaP	77	337	⟨111⟩	0.18 ±0.02	0.56 ±0.04				
p-GaAs	77	337	⟨111⟩	0.090±0.001	0.56 ±0.02		2.5±0.3	3.5±0.3	[6.104]
			⟨110⟩	0.090±0.001	0.51 ±0.02		2.5±0.3	3.0±0.3	
			⟨100⟩	0.091±0.001	0.465±0.02		2.7±0.3	3.5±0.3	
p-GaP	77	337	⟨111⟩	0.16 ±0.02	0.54 ±0.05				[6.105]
n-GaP		337	⟨100⟩	0.25±0.01					[6.106]
			⟨110⟩	0.35					
			⟨111⟩	0.40±0.01					
p-GaSb		119	⟨111⟩	0.046±0.02	0.43 ±0.12				[6.95]
			⟨100⟩		0.29 ±0.05				
p-Ge		119	⟨111⟩/⟨100⟩	7/5 lines					[6.95]
n-GaP		119	⟨100⟩	0.254±0.004					[6.107]
			⟨110⟩	0.334±0.005					
			⟨111⟩	0.393±0.006					
n-InSb		contin.	⟨111⟩	0.0116			$g_{eff} = -64$		[6.24]

Yet another experiment that demonstrates the adaptability of pulsed field experiments is the measurement of the anisotropic magnetoresistance in noble metals (Ag, Au) with rare earth impurities (Tb, Dy, Ho, Er, Tm, Yb) [6.110]. The magnetoresistance in metals is fairly weak; with magnetic impurities it can become negative and may go through zero to positive values as the system of magnetic ions becomes saturated. This saturation effect is of interest because of the possibility to distinguish between competitive magnetoresistance mechanisms in the system. It occurs at fields of the order of 30 T, thus perfectly in the range of nondestructive pulsed fields. Refined techniques are needed to detect the weak signals against the background of electrical noise and induced voltages; these experiments demonstrate that it can be done.

The Toulouse facility has been conceived as part of the "Service National des Champs Intenses" but the presently available resources do not allow liberal accommodation of guest experiments, unlike the major magnet laboratories. Potential guests are encouraged to stay for a longer period and to do experiments in cooperation with the local scientific staff. A similar situation, but on a smaller scale, exists at Amsterdam.

A laboratory of comparable size has been set up at the University of Osaka by *Date* and his collaborators [6.31]. As a national service facility, this has already accumulated an impressive list of guest experiments. The uncommon coil design is inspired by the idea to generate a megagauss field nondestructively: the coil consists of free-standing heavy helices machined from solid blocks of maraging steel with an ultimate tensile strength of 2 GPa. Due to the high resistivity of this material ($\sim 50\,\mu\Omega$ cm), precooling with liquid nitrogen would be less effective,

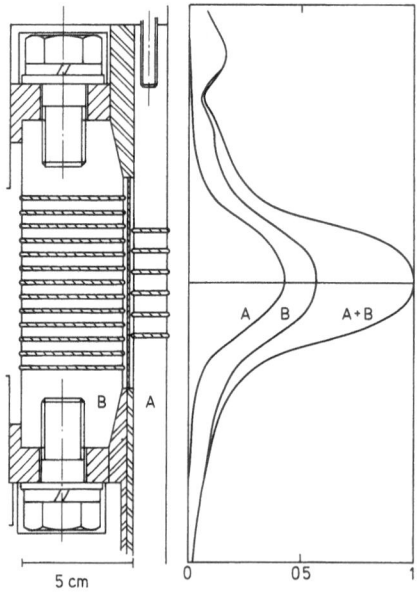

5 cm

0   05   1

Fig. 6.18. Cutaway drawing and field profile of the two-stage coil at Osaka University. The i.d. of coil *A* is 20 mm, that of coil *B* is 60 mm and the o.d. is 150 mm

and thus the coils are used at room temperature with a relatively short pulse duration and with high-temperature insulating materials (mica and glassfiber-polyimide). The two-stage coil (Fig. 6.18) is used conservatively to generate 60 T at 490 kJ (250 kA) with a pulse duration of 370 μs, although it is rated for 70 T at 750 kJ. The outer coil can be used alone to generate 39 T at 420 kJ (300 kA) with a half period of 410 μs; the larger inner diameter is useful for accommodating larger cryostats. A three-stage coil will be built with the goal of generating 100 T at the full energy of the capacitor bank. The main capacitor bank, built by the Nichicon Capacitor Company, consists of 5 identical units with a total energy of 1.25 MJ at 26.6 kV, convertible to 40 kV by series connection. Each 250 kJ unit is switched by a single 1 MA spark gap and can be crowbarred by a second spark gap. The spark gaps are of special patented design with a rotating spark and are reported to work quite well, with some electrical noise at the beginning of the pulse. Below 8 kV, a mechanical switch is used. An older capacitor bank of 250 kJ at 5 or 10 kV (convertible) with ignitron switching is available for experiments with smaller coils.

One of the principal research topics of the resident group is electron spin resonance at ultrahigh frequencies in the far infrared [6.11]. Pulsed HCN (337 μm, 31.8 T at g=2) and H$_2$O (119 μm, 90 T at g=2) lasers with extended pulse duration (~100 μs) are used as light sources. This extension is achieved by adding helium to the gas mixture and inserting a series inductance in the capacitor discharge circuit driving the laser. The magnetic field measurement is conveniently calibrated by electron spin resonance of Cr in ruby which has a g factor of 1.98; the line is sharp but not very strong. In very high magnetic fields, "exchange splitting" of electron paramagnetic resonance can be observed if there are two independent lattice sites with g factors differing by an amount Δg when the condition $\Delta g\mu_B/2J > 1$ is fulfilled, where 2J is the exchange interaction describing the coupling between the two sites and μ_B is the Bohr magneton. At lower fields, the two resonances fuse into a single line. The splitting Δg can be varied from zero to the maximum value $g_\parallel - g_\perp$ by changing the orientation of the crystal with respect to the field, Fig. 6.19. As a fast detector, Ge doped with indium (~10^{17} cm^{-3}, almost metallic) is used. In the available field range, nonlinear effects become discernible, leaving much room for further experiments and discussion.

In the range of visible light, the Zeeman effect of the Na-D lines (Fig. 6.20) and of the R$_1$ and R$_2$ lines in ruby [6.112] has been measured with an optical multichannel analyzer (OMA). A light pulse from a xenon lamp is timed to coincide with peak field, and the OMA gate is closed shortly afterwards. Thus, a single, well-resolved spectrum at peak field is obtained. In these experiments, level mixing and the Paschen-Back effect were observed. As in the infrared, it is anticipated that more interesting nonlinear effects can be seen. As a guest experiment at the OMA setup, the Nishina group from Sendai University has made a study of excitons in the III–VI layer compounds GaSe and InSe [6.114].

The Osaka Laboratory has a long tradition in magnetization measurements [6.113,115], where a number of interesting magnetic phase transitions and

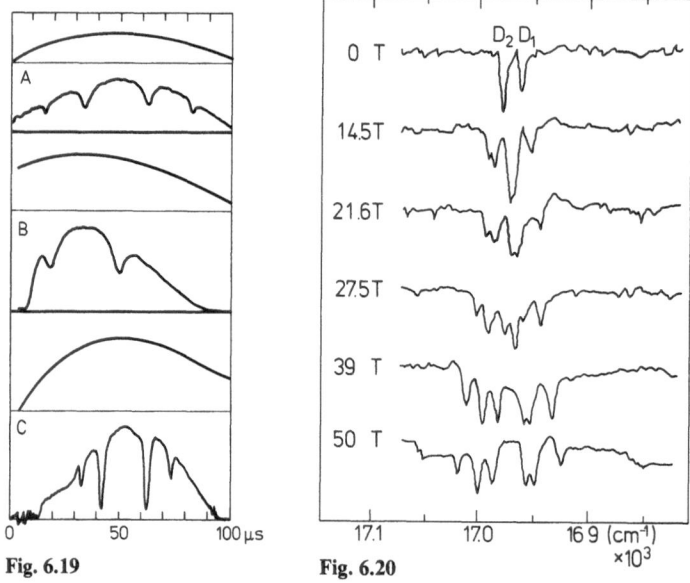

Fig. 6.19   Fig. 6.20

Fig. 6.19. ESR signal of NH_4 – Co-Tutton salt at 337 μm (*A*) and 119 μm (*B*), both at liquid nitrogen temperature. The resonances are superimposed on the waveform of the laser pulse. The waveform of the magnetic field is shown above the recordings of the transmitted light as a function of time. (*C*) shows a resonance of $(NH_4)_2CuCl_4 \cdot 2H_2O$ at 4.2 K and 337 μm [6.111]

Fig. 6.20. High field Zeeman effect of sodium D lines, measured with the gated optical multichannel analyzer [6.112]

saturation effects have been observed in the high fields. It is another demonstration of the adaptability of the pulsed field technique that these delicate experiments have been carried out in pulsed fields of short duration with excellent resolution and precision. They are discussed in more detail in Chap. 3, as well as those of *Ozhogin's* group [6.82].

Magnetic anisotropies in rare earth compounds (e.g., $CeAl_2$, EuB_6) have been studied by magnetoresistance measurements. Localized magnetic moments contribute negatively to the magnetoresistance which can be dominant until the moments become aligned to saturation by a high magnetic field.

Another use of magnetoresistance is to study the upper critical field in high field superconductors. A recent example is shown in Fig. 6.21. In the known materials with high upper critical fields, the transition is rather broad. Some researchers define the upper critical field at the upper edge of the transition curve (insert in Fig. 6.21), others at the transition's midpoint.

High field superconductivity in the ternary molybdenum chalcogenides was discovered and extensively studied at the University of Geneva. As several of these materials have transition points high above the strongest available dc fields, *Fischer* [6.32] set up a pulsed field generator to measure the upper critical

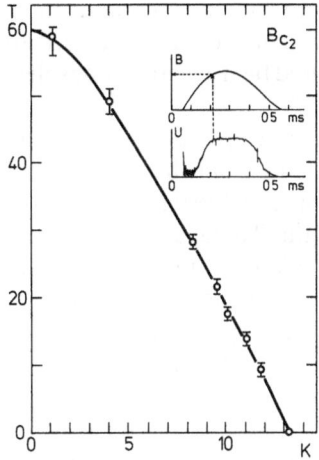

Fig. 6.21. The upper critical field of the superconductor $PbMo_6S_8$ as a function of temperature. The insert shows how these authors [6.116] define and determine the upper critical field from recordings of the magnetic field and the voltage across the sample vs time. This is criticized by *Foner* [6.118] who originally discovered high field superconductivity at 60 T in $PbMo_6S_8$ with a transition temperature of 14.4 K

field. This pulsed field generator with a 72 kJ, 2.5 kV capacitor bank is reported to perform very well with peak fields in excess of 50 T, but no details of the coil construction have been published. The material $Sn_{1.2}Al_{0.6}Mo_{6.35}S_8$ was found to remain superconducting in fields up to 51 T, and an extrapolation of the curve goes to 60 T (which is still the "magic number" for high field superconductivity although this particular result was challenged by *Foner* [6.118] who found only ~ 300 T for $SnAl_{0.5}Mo_5S_6$). This is part of an entire class of materials with high transition fields. Samples are usually prepared by direct reaction of the constituents in powder form at high temperatures. The resulting inhomogeneities, together with an anisotropy that is smeared out in the polycrystalline sample, can account for the observed broad transitions. A critical problem is the transient response of the superconductors; the flux movement induced by the pulsed field will indeed result in some heating of the material. The transient response was investigated by testing the same sample in different coils with a pulse duration in the range 15–300 ms. This had to be done at fields below 17 T which is the limiting field of the 300 ms coil (6.3). In all these tests, the transition fields were found to be independent of pulse duration. However, a recent review [6.117] reports that some deviations at smaller pulse times have been found.

The pulsed field facility at MIT [6.98] where much pioneering work was done [6.24, 119] is still operational but it has been used only occasionally in recent years, mainly by *Foner* et al. to study high field superconductors [6.118]. The change in resistivity was detected by means of a contactless radio-frequency technique, monitoring the losses in a small *rf* coil placed against the sample surface. The results were found to agree with conventional resistance measurements. A comparison was made between the transition behavior in pulsed (half-period 10 ms) and continuous fields; no difference was observed. Similar results were found by *Schneider* at Braunschweig [6.120] and by *van der Sluis* et al. at Leiden [6.121]. *Van der Sluis* [6.122] has reviewed earlier work where in some experiments differences had been observed. In any case, if a material stays superconducting in a pulsed field, it will certainly do so in a continuous field. There is hope that one or other of these materials can eventually be used to generate very high dc fields, thus rendering those pulsed field generators obsolete which were used to investigate them!

Meanwhile, at MIT it has been considered to generate up to 50 T for a period of one second or longer, using the flywheel alternator from the new Alcator experiment (Chap. 5) at a power level of 150 MW [6.123]. Design estimates for the magnet are 10 cm i.d., 1 m o.d. and 60 cm length. To cope with the stress problem, the use of composite conductors, e.g., copper wires with internal reinforcement or laminates of copper and high strength materials, is proposed. The laminates could be either plates from which the spirals are sawed, or cylinders· that are machined into helices. A number of tough technological problems has to be solved before such a coil could become fully operational.

At other laboratories, the use of inductive storage as an alternative energy source for a pulsed magnet has been investigated. Switching of inductive circuits is difficult because it involves interrupting high current which induces a very high voltage. Inductive storage is attractive because of the high energy density and greater versatility of the output waveform. In the pilot experiment at Braunschweig [6.33], fields up to 40 T were generated with pulse times in the range 5 μs–100 ms. With ordinary conductor materials, even large inductors cannot have time constants (6.9) of more than just a few seconds. To be charged up efficiently, they need very high power. While such inductors can be useful to provide pulses of very short duration, all other applications will be taken over by superconducting storage coils. The required technology is now under intensive development. The feasibility of driving a pulsed field coil with a superconducting storage inductor has been demonstrated by *Hairie* et al. [6.124]. A pilot experiment was done with a 12.2 kJ, 700 A inductor: 20 T were generated with a rise time of 2 ms and a decay time of 20 ms.

The experimental work at Braunschweig is still done with a capacitor bank. The main research topic is an extensive study of semiconducting bismuth-antimony alloys, mostly with Shubnikov-de Haas and de Haas-van Alphen measurements at temperatures as high as 77 K [6.125].

An elaborate pulsed field installation has been built at the University of Leiden under the guidance of D. de Klerk. The 180 kJ, 3.5 kV capacitor bank is switched and crowbarred by ignitrons, and is designed for fully automatic operation. The design and operation are described to the last detail in [6.122], with a thorough discussion of all related problems, in particular the coil design and calculations of the mechanical deformation of the coils. This thesis contains a most complete review of all pulsed field work up to 1967. In one experiment, a record field of 52 T with a half-period of 17 ms was achieved. However, it turned out later that this had just been a lucky shot and could not be repeated at will. As in most other laboratories, 45 T are now considered as an ultimate limit for the repeated operation of copper coils with outer reinforcement. Besides the work on superconductivity, the Leiden group has performed a large number of fine magnetization measurements which are reviewed in Chap. 3.

There are many potential applications of very high magnetic fields in the physics of elementary particles, such as the determination of momentum and charge of short-lived particles, orientation of spins, critical beam transport problems, and improvement of the efficiency of colliding beams. To obtain reasonable statistics, a large number of experiments is usually required. This calls for a high repetition rate and is a severe condition for a pulsed field generator. It appears that the pioneering work of *Furth* and *Waniek* [6.22] was much inspired by this application. They generated 17.5 T with a repetition interval of 5 s, using a 7.8 kJ, 3 kV capacitor bank. The coils were Bitter-type stacks of discs in a solid housing. A large volume 22.5 T magnet for a bubble chamber (7 cm i.d., 10 cm length) was described by *Bergmann* in 1966 [6.126]. Recently, an elaborate pulsed magnet system for elementary particle analysis was built at the Rutherford Laboratory [6.87]. The coils are water-cooled helices machined from either CrCu or 0.2%BeCu and have the following characteristics: 5 cm i.d., 12 cm o.d., 15 cm length, 30 turns, inductance 16 μH, resistance dc 1–2 mΩ, ac 12 mΩ, crowbar resistor 8 mΩ (800 μs), peak field 40 T at 480 kJ (8 kV, 180 kA), rise time to peak field 750 μs, pulse duration 3 ms, interval between pulses 5.2 s, water flow rate 3 l/s, axial prestress 1000 kg/cm^2. It was found that glassfiber-polyimide insulation was much more durable than glassfiber-epoxy. The coils have been used for very large numbers of shots (from 10^3 at 33 to 2×10^5 at 25 T); failure has mostly occurred by flashover due to deformation caused by metal fatigue. A typical fatigue crack at the inner radius is shown in Fig. 6.7. Experiments with this instrument and further improvements are still in progress.

At the Institute for Solid State Physics (ISSP) of the University of Tokyo, nondestructive pulsed field generators have been set up to complement the large megagauss facility. Together with superconducting coils, this gives complete coverage of the entire magnetic field range. Two types of pulse magnets are employed. The small solid helix magnet, adapted from the design of *Foner* and *Kolm* [6.23], is shown in Fig. 6.9. The coil is machined from a rod of Cu-Be with a yield strength of more than 120 kg/mm^2 after heat treatment. The windings are insulated with sheets of Kapton and Mylar. With a 32 kJ, 3.3 kV capacitor bank, pulsed fields up to 450 kG have been produced [6.95]. This type of magnet has the advantage of large mechanical strength and easy precooling in a liquid nitrogen bath. On the other hand, it requires elaborate machining to make a strong magnet which can really produce high fields. Moreover, because the inductance of the coil is relatively small, the duration of the pulsed field generated by this type of magnet is fairly short, typically a few hundred microseconds.

For obtaining a longer pulse duration, wire-wound coils are used. These work best with a larger energy source, ideally of the order 100 kJ. To this end, a new capacitor bank has been set up; with 200 kJ at 5 kV or 10 kV (convertible) this provides the required energy with a comfortable margin. The coils are impregnated with epoxy resin under high pressure; protected by a plastic bag, they are subjected to circa 50 kg/cm^2 in an oil bath. With coils wound from copper wire, pulsed fields higher than 40 T were produced with a rise time of

7–10 ms [6.100]; experiments with stronger alloys are in progress. These long duration pulsed fields are particularly convenient for magnetotransport measurements; they have been applied to measurements of the cyclotron resonance [6.95, 107, 127–129], far-infrared magnetoreflection and magnetotransmission in semimetals [6.129–133], and quantum transport phenomena [6.99, 134].

In very high magnetic fields, a quantum effect in the cyclotron resonance is enhanced. The quantum peaks of the cyclotron resonance in p-type GaSb were well resolved in the far infrared range, and the valence band parameters were accurately determined for GaSb [6.95]. In n-type GaP, the cyclotron resonance can be observed only in the pulsed field range, because of the low mobility of this substance. A very peculiar nonparabolicity near the conduction band edge was studied and it was confirmed that the conduction band has a so-called camel's back shape with a double minimum just as the valence band in p-type Te [6.107, 128], based on experiments in both nondestructive and megagauss fields [6.107, 128, 135, 136].

Electronic energy levels of low quantum number in Bi, and Bi-Sb alloys were studied by measuring the Shubnikov-de Haas effect [6.99, 134], magnetoreflection [6.129] and magnetoabsorption [6.130]. Since Bi has small effective masses of carriers and a small band overlap, the influence of the high magnetic fields on the electronic energy levels is large. Particularly when the magnetic fields are applied parallel to the binary axis or the bisectrix axis, the lowest Landau levels of the light electrons both in the conduction and valence bands at the L point show an unusual field dependence including level repulsion. At high magnetic fields, Bi transforms into a semiconductor. From the investigation of the field dependences of the Landau levels in the nondestructive pulsed field range, the critical field for the magnetic field induced semimetal-semiconductor transition was estimated to be 85–100 T [6.99, 138]. This transition was observed at 88 T in the far infrared magnetotransmission [6.137]. By doping with a small amount of Sb, the critical field was reduced, and the magnetic field induced transition could be directly observed in $Bi_{95.6}Sb_{4.4}$ [6.134].

The band parameters were determined [6.138] on the basis of Vecchi's model [6.139] from the data of the Shubnikov-de Haas experiments and the magnetic field induced semimetal to semiconductor transition. The magnetostriction in Bi was measured in fields up to 40 T to investigate the carrier densities in pulsed high magnetic fields [6.140, 141]. The result of the magnetostriction experiments supports the validity of the newly determined band parameters. To measure the magnetostriction, the sample dilatation is transmitted to a parallel plate capacitor and the capacitance variation is measured by means of a sensitive capacitance bridge. The mechanical parts must be well designed to avoid spurious vibrations which may be caused by eddy currents or by direct transmission from the pulsed field coil. In a recent design, the sensitivity has been greatly improved by using the surface of the sample directly as one plate of the capacitor (G. Kido, private communication).

Besides the electronic energy levels, the dielectric responses to the far infrared radiation, i.e., magnetoreflection and magnetotransmission, were investigated in

Bi and Bi-Sb alloys. In magnetoreflection, the cyclotron resonances of electrons and holes were observed together with the plasma resonance and interband transitions [6.129]. In magnetotransmission at a wavelength of 337 μm, the Fabry-Perot interferograms due to the Alfven wave propagation were observed [6.130]. The mass density was obtained from the periods of the interferograms. The magnetic field was applied in the direction of the binary axis, and the electric field vector of the radiation was chosen in the directions of either the bisectrix axis $(E\|y)$ or the trigonal axis $(E\|z)$. It was found that although the agreement between theory and experiment is reasonably good regarding the mass density for $E\|z$, there is a large discrepancy for $E\|y$ in fields higher than 14 T. The discrepancy was interpreted in terms of the increase of the dielectric constant with the magnetic field, caused by the interband transition of electrons between the lowest Landau levels in the light electron pockets at the L point. The Shubnikov-de Haas oscillation was observed in the magnetotransmission spectra as an envelope of the Fabry-Perot interference fringes. The correspondence between magnetotransmission and magnetoresistance is convenient for observing the Shubnikov-de Haas effect optically in the megagauss range where transport measurements are extremely difficult.

In graphite, a striking anomaly was observed in the high field magnetoresistance at low temperatures [6.142]. A typical example of the experimental recording of the transverse magnetoresistance is shown in Fig. 6.22. After the last minimum of the Shubnikov-de Haas oscillations at 7.4 T the magnetoresistance has a maximum at about 12 T. Then it increases suddenly at a critical field around 30–37 T. The critical field increases with temperature as shown in the inset. Because of the large temperature dependence, this anomaly cannot be explained by a one-electron model. *Yoshioka* and *Fukuyama* proposed a model

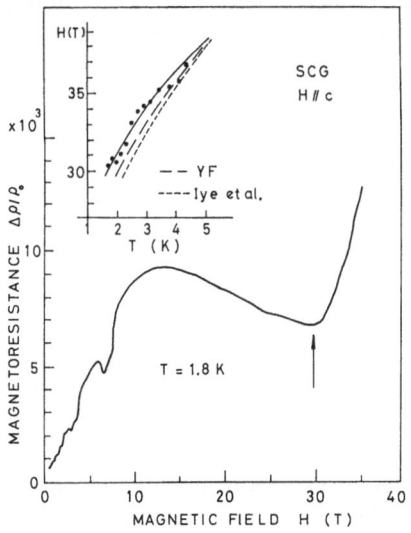

Fig. 6.22. The transversal magnetoresistance in a single crystal of graphite (SCG). At a critical field (indicated by the arrow) the magnetoresistance increases abruptly. This critical field depends on temperature as shown in the inset. YF is the theoretical curve from [6.143] and *Iye* et al. the extrapolation of experimental values obtained in [6.144]

involving a phase transition to a charge density wave (CDW) state to explain the anomaly [6.143]. This effect was extensively investigated by both dc and ac transport measurements up to 45 T [6.133]. In the dc measurement, a large decrease of the magnetoresistance was found before the onset of the anomaly, considered to be a precursor of the phase transition. In the ac measurement in the frequency range 1–150 MHz, the anomaly was found to be remarkably suppressed.

For further investigation of this anomaly, the far infrared magnetoreflection of graphite was measured [6.131]. Since the temperature of the sample was not sufficiently low, the anomaly has not been seen in the spectra as yet. Instead, a new structure was observed and interpreted as originating from the electron-hole coupled plasma mode. A quantum mechanical calculation of the dielectric function revealed that the theoretical line shape of the magnetoreflection is in excellent agreement with the experimental results. The magnetic field dependence of the energy of the new structure was explained by this calculation as well [6.132].

In two-dimensional electronic systems, i.e., Si-MOSFET and GaAs-AlGaAs heterojunction interfaces, quantum transport phenomena were investigated. For the Si-MOSFET, the Shubnikov-de Haas effect has been observed by sweeping the gate voltage at constant magnetic field. Taking advantage of the relatively long duration of the field pulse, the gate voltage was swept rapidly at the top of the pulse where the magnetic field can be regarded as nearly constant. The Shubnikov-de Haas oscillations were observed up to 37 T [6.145]. It was found that the first Shubnikov-de Haas peak is associated with the lowest Landau level. This is usually very small at low temperatures in moderate magnetic fields but grows as the field is increased. The results were interpreted in terms of the mobility edge model of the Anderson localization. The magneto-phonon resonance was observed in the GaAs-Al$_x$Ga$_{1-x}$As heterojunction interface in fields up to 35 T [6.146]. The fundamental peak with the harmonic number N = 1 was observed for the first time. The magnetophonon mass of electrons which is larger than in bulk GaAs was explained by the nonparabolic character of the two-dimensional electronic state. The larger damping factor of the oscillation was also attributed to the two-dimensionality.

6.3 Megagauss Generators

An ultrastrong magnetic field is best characterized by the inevitable destruction of the structures in which it is generated and confined. This is a direct consequence of the high energy density resulting in heating and compression of the conductor material. As an example, take 500 T (5 MG) with an energy density of 100 kJ/cm^3 corresponding to a pressure of 1 Mb = 100 GPa. The energy density of pulsed energy sources (Table 6.6) is much lower. Therefore it is necessary to use energy concentrating systems. These can be low inductance transmission lines or an imploding cylinder which concentrates energy as well as

Table 6.6. The energy density of fast storage systems

High explosives	10 kJ/cm^3
Lead accumulator	500 J/cm^3
Flywheel	100 J/cm^3
Compressed gas	20 J/cm^3
Magnetic field	20 J/cm^3
Electric field	0.2 J/cm^3 (electrolytic capacitor)
	0.08 J/cm^3 (pulse capacitor)

Note: The given energy densities refer to the "bare" storage volumes, i.e., excluding peripheral equipment and switchgear. In particular, the energy density of a contained explosion relates to the size of the containment chamber, which is three orders of magnitude larger than the charge itself. This is comparable to inductive storage, where the peripheral equipment may occupy a volume about equal to that of the inductor.

power. The destruction originates at the conductor surface facing the magnetic field, where the current is concentrated by the skin effect. The surface is pushed away from the field region, resulting in a compression usually accompanied by a shock wave traveling into the conductor material. Simultaneously, magnetic flux penetrates into the conductor, where about one-half of the magnetic energy is converted into Joule heat; this leads to melting and vaporization. All of these effects can be related to a speed of energy flow. The net result is a rapid enlargement of the volume which must be filled with magnetic energy for building up the field. The Poynting vectors set up by the energy source must be strong enough to overcome these losses, requiring both high voltage and high current. The strength of an implosion system is in the local generation of a Poynting vector field converging from all sides onto the experimental volume, in opposition to the speed vectors that would carry the energy away.

6.3.1 Interaction of a Pulsed Magnetic Field with Conducting Surfaces

If a pulsed magnetic field is applied to a conducting surface, an opposing current is induced in the skin layer. As this current is attenuated by the resistivity, magnetic flux diffuses into the conductor with a speed v_f. The concept of a flux diffusion speed can be visualized with a Gedankenexperiment as shown in Fig. 6.23a. A box with infinitely conducting walls is closed on one side with a movable sheet of constant resistivity ϱ. This results in the exponential decay of any magnetic flux initially present in the cavity. The magnetic field can be kept constant by moving the sheet as a "piston" into the cavity at a constant speed v, such that the induced voltage

$$U = Eh = -d\phi/dt = -Bh\,d\xi/dt = -Bhv \tag{6.20}$$

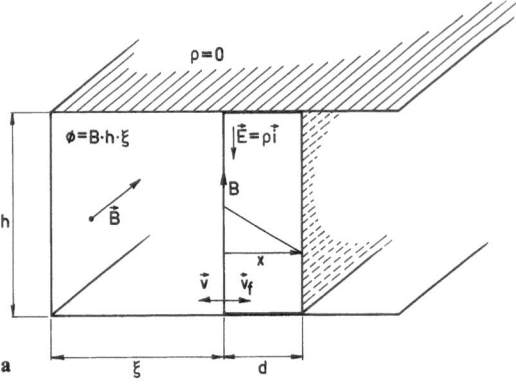

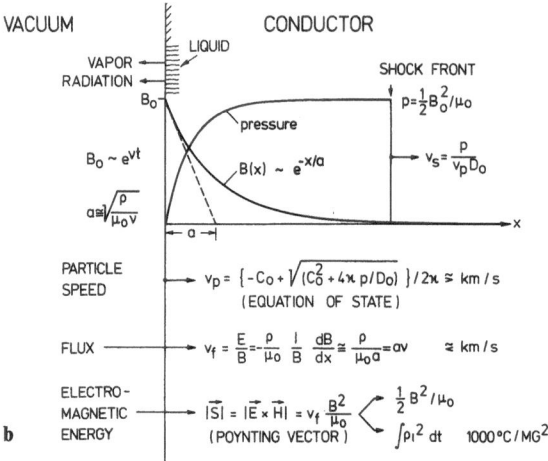

Fig. 6.23. (a) Gedankenexperiment to illustrate magnetic flux diffusion and flux compression: a conducting slab is pushed into a superconducting cavity containing magnetic flux
(b) Interaction of a magnetic field B_0 with a conducting wall: p: pressure, v_s: shock speed, v_p: particle speed, D: density, c_0, κ: constants related to the equation of state. a: skin depth, E: electric field, ϱ: resistivity, S: Poynting vector, i: current density, t: time. The illustration refers to an exponential field rise and constant resistivity

is equal to the voltage drop across the resistance of the sheet. As the flux density in the cavity remains constant, all the flux encountered by the sheet in its movement penetrates into the sheet; thus the flux diffusion speed v_f is equal to the speed of movement:

$$v_f = \frac{E}{B} = \frac{\varrho i}{B} = -\frac{\varrho}{\mu_0}\frac{1}{B}\frac{\partial B}{\partial x}. \tag{6.21}$$

The flux diffusion is related to the spatial distribution of magnetic field, which in the eddy current approximation can be calculated from

$$\frac{\partial B}{\partial t} = \frac{1}{\mu_0}\frac{\partial}{\partial x}\left(\varrho\frac{\partial B}{\partial x}\right). \tag{6.22}$$

If the piston is accelerated to increase the magnetic field exponentially, we have a situation that is closely approximated in many flux compression experiments. Assuming $\partial\varrho/\partial x = 0$ we find a simple particular solution of (6.22):

$$B(x,t) = B_0 e^{vt} \sinh\left[(d-x)/a\right] \sinh(d/a) \qquad a = \sqrt{\varrho/(\mu_0 v)}$$

$$d \gg a: \quad B(x,t) = B_0 e^{(vt-x/a)} \qquad v_{\mathrm{f}} = \varrho/(\mu_0 a) = av = \sqrt{\varrho v/\mu_0} \qquad (6.23)$$

$$d \ll a: \quad v_{\mathrm{f}} = \varrho/(\mu_0 d) \qquad \text{at} \qquad x = 0 \text{ (inner surface)},$$

where a is the skin depth. As the megagauss limit is approached, it will be more realistic to take resistivity variation into account. *Bryant* [6.14] found a useful particular solution for this case. He assumed a resistivity function $\varrho = \varrho_0[1 + (B/B_c)^2]$ [6.148] which is a reasonable approximation of the overall behavior of metals in the case of equipartition between electromagnetic and Joule energy densities. Here B_c is a constant depending on the conductor material, for copper this is 40 T. A self-similar solution for $d \gg a$ is given by

$$Be^{(B/B_c)^2/2} = B_0 e^{(v_0 t - x/a_0)} \qquad a_0 = \sqrt{\varrho_0/(\mu_0 v_0)} \qquad v_{\mathrm{f}} = a_0 v_0. \qquad (6.24)$$

This was actually derived from the postulate that the flux diffusion speed is constant in space and time. In some theoretical work, a "skin layer" approximation is used where the current is assumed to flow with constant density within the skin depth. This is not advisable because the magnetic energy, the Joule heating and the flux loss will be off by factors of the order of two. Solution (6.23) and its adaptation to cylindrical symmetry are better suited for an elementary discussion of flux loss and energy balance in flux compression experiments [6.56, 149, 150]. If the flow of electromagnetic energy (Poynting vector) is expressed in terms of the flux diffusion speed, it becomes apparent that there will be approximate equipartition between the energy densities of the magnetic field and the Joule heat (Fig. 6.23b). The particular solutions (6.23, 24) both result in exact equipartition, thus justifying the assumption made by Bryant. The specific heat of metals is in the range 2.5–4 kJ/cm^3 corresponding to a temperature increase of 1600–1000 °C/MG^2. The heating is independent of the resistivity as long as the conductor dimension d is large compared to the skin depth. The equipartition is disturbed when the energy flow encounters a discontinuity, e. g., when this condition is not fulfilled. The heating will then be larger if the field is applied on only one side of the conducting sheet as in the case of a field coil, and it will be smaller if the conductor is immersed into the field as in the case of a conducting sample. Examples calculated for an exponential field rise are given in Table 6.7. The boiling point of any metal will be reached at a few megagauss. However, boiling takes place at the surface only and the speed of the vapor-liquid interface depends on the energy input; in practice this turns out to be of the same order as other critical velocities in the system. In addition, some of the heat at the surface is carried away as radiation. Eventually, the inner layer

Table 6.7. The ratio (Joule heat)/(magnetic energy density) for an exponential field rise where a is the skin depth

d/a (r/a)	Plane (thickness d)		Cylindrical[c] radius r
	one side[a]	immersed[b]	
0.01	10,000	0.000006	
0.02	2,500	0.000025	
0.05	400	0.000625	
0.1	100.7	0.0025	0.0025
0.2	25.7	0.01	0.01
0.5	4.68	0.08	0.059
1	1.72	0.214	0.199
2	1.08	0.58	0.487
5	1.002	0.973	0.798
10	1	0.9998	0.9
20	1	1	0.95

[a] "One side" refers to a sheet which carries the conduction current needed to support the magnetic field on one side, with no field on the other side.
[b] "Immersed" refers to a plane sample heated by eddy currents only.
[c] "Cylindrical" refers to a solid cylinder parallel to the magnetic field.

will be transformed into a fully ionized plasma whose resistivity decreases with the 3/2 power of temperature.

It is a fairly straightforward procedure to solve (6.22) or its counterpart in cylindrical geometry,

$$\frac{\partial B}{\partial t} = \frac{1}{\mu_0 r} \frac{\partial}{\partial r} \varrho r \frac{\partial B}{\partial r} \qquad (6.25)$$

with numerical methods. This involves a system of equations with a "tridiagonal" matrix which can be solved on a modest computer [6.151]. While this is fun and gives useful insight into the flux diffusion process, a rigorous treatment must include the effects due to the mechanical compression of the conductor material. Elaborate computer codes that solve the complete magnetohydrodynamic equations are now available [6.152]. These are needed for the detailed analysis of a specific flux compression experiment, e. g., an experiment aimed at the isentropic compression of a sample. For a general understanding and an estimate of the peak performance of a megagauss generator, the following arguments will be adequate.

Magnetic stress is converted into mechanical stress within the skin layer. High pressure results in a substantial reduction of the resistivity [6.149, 152] and thus in improved flux trapping. The resistivity of metals as a function of pressure has been measured by *Bridgman* [6.153]. The experimental data fit the law $(d \ln \varrho)/(d \ln V) = y$, where V is the specific volume. The exponent y is temperature dependent; specific values for copper are 2.67 at 20 °C and 2.55 at 75 °C. An extrapolation to very high temperatures will therefore give only a rough estimate of the effect. As the field increases, pressure waves run into the conducting material at a speed that increases with pressure. The later disturbances therefore catch up with the earlier ones and a shock wave is formed in the material [6.57]. The conservation of mass, energy and momentum across the shock front can be expressed by three simple relations between pressure p, density D, the speed of the shock front v_s and the particle speed v_p behind it:

$$v_s D_0 = (v_s - v_p) D$$

$$v_s v_p D_0 = p - p_0 \qquad\qquad (6.26)$$

$$W - W_0 = \tfrac{1}{2} (p + p_0)(V_0 - V),$$

where W is the internal energy and $V = 1/D$ is the specific volume.

These Rankine-Hugoniot relations are useful in shock wave physics as it is easier to measure speeds than pressure and volume dynamically. The particle speed is of particular interest because the conducting surface recedes from the magnetic field at approximately this speed [6.154]. An empirical relationship between the pressure and the particle speed is included in Fig. 6.23b; by virtue of (6.26) this is equivalent to the equation of state of the solid. Examples are shown in Fig. 6.24 [6.155]. The particle speed related to ultrastrong fields is of the order

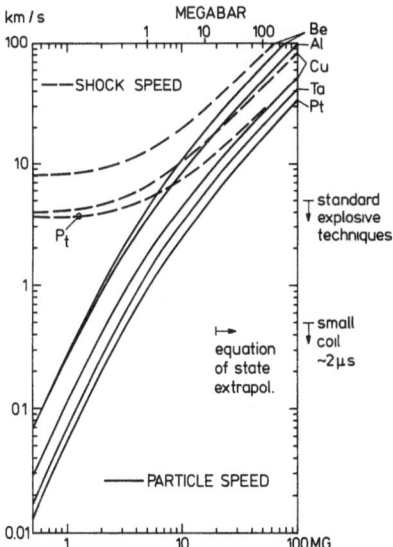

Fig. 6.24. Shock speed and particle speed behind the shock front for a number of metals as a function of the applied magnetic field

Table 6.8. Properties of conductor materials related to interaction with pulsed magnetic fields

Material		Al[d]	steel[e]	Cu	Ta	W	Au	Pt
Density[a]	D_0 [g/cm³]	2.785	7.896	8.930	16.65	19.22	19.24	21.45
Heat capacity[a,b]	S [J/cm³ deg]	2.4	3.9	3.4	2.3	2.5	2.5	2.9
Vaporization heat[b,e]	[kJ/cm³]	30	60	40		80	30	53
Resistivity[a]	ϱ_0 [μΩcm]	2.8	73	1.7	14	5.5	2.4	10
Resistivity at boiling point[c]	ϱ_b [μΩcm]	50	<200	34		>100	32	
Boiling point	θ_b [°C]	2400	2500	2582	5300	5900	2660	4000
$\sqrt{2S\theta_b\mu_0}$	[T]	130	160	150	180	200	130	170
Shock wave	c_0 [km/s]	5.328	4.569	3.940	3.414	4.029	3.056	3.598
parameters	κ	1.338	1.490	1.489	1.201	1.237	1.572	1.544

[a] at 20 °C.
[b] per unit volume.
[c] approximate.
[d] 2024 aluminum.
[e] AISI 304 stainless steel.

of several kilometers per second. This is the primary factor limiting the peak field and restricting the pulse time; the flux diffusion speed is usually somewhat lower. The relevant properties of a number of metals are compiled in Table 6.8.

6.3.2 Exploding Single-Turn Coils

To generate a magnetic field in a single-turn coil, the electromagnetic energy must be squeezed through the narrow feed gap. In the megagauss range, this requires very high voltage and current; a capacitor bank with low internal inductance and resistance is most suitable and convenient for this purpose. The stored energy and consequently the coils are relatively small. The lifetime of the field is determined by the expansion speed in relation of the size of the coil; the first quarter period of the discharge is of the order of 2 μs. An optimized system may produce several megagauss in a volume of 0.1–1 cm^3 with a total stored energy of no more than 10–100 kJ. The capacitor voltage will be in the range 10–40 kV at currents of a few MA. Because of the distributed inductance in the pulsed power supply, only a fraction of this appears at the coil terminals, where the voltage is given by the coil dimensions and the rate of rise of the field. In a multimegagauss system, the voltage easily becomes high enough for breakdown or flashover to occur at the surface of the coil or in places where metal vapors have been deposited. The coil can be immersed into oil or water contained in a small plastic bag, but this is likely to mess up experiments. It was found that carefully applied insulating tape and brush-on insulating materials (e. g., "liquid-tape" and "corona dope") provide adequate protection.

Figure 6.25 shows the principal design features of a practical megagauss generator built with relatively modest resources at the Illinois Institute of Technology [6.40]. The lightweight coils were made from 2 mm sheet metal, mostly copper, and a solid dielectric switch was directly incorporated in the coil clamping mechanism with a hydraulic press. The simplified operation results in a repetition rate of up to one shot in ten minutes. With a 55 kJ, 20 kV capacitor bank, 100 T were obtained in 10 mm diameter, 150 T in 5 mm diameter and 200 T in 2.5 mm diameter. This generator was used as a target for the electron beam of the Stanford accelerator [6.156,157], and for the first experiments on cyclotron resonance in semiconductors [6.158]. The capacitor bank was assembled from surplus capacitors; with optimal specifications, in particular with a lower internal resistance, about one-half of the energy would have been sufficient to achieve similar results. *Andrianov* et al. in fact obtained 310 T in 2.1 mm diameter with 25 kJ, using tantalum as the best suitable coil material [6.54], found as the result of a systematic study using different metals. The superiority of tantalum may be attributed to a combination of factors: the shock wave characteristics result in a small particle speed and the resistivity in a relatively large skin depth; the ductility ensures continuity of the coil during expansion, apart from the ease of manufacturing the coils.

It was a long way to arrive at this point. Initially, progress with the single-coil technique had been rather slow. The development is traced in Table 6.9. After

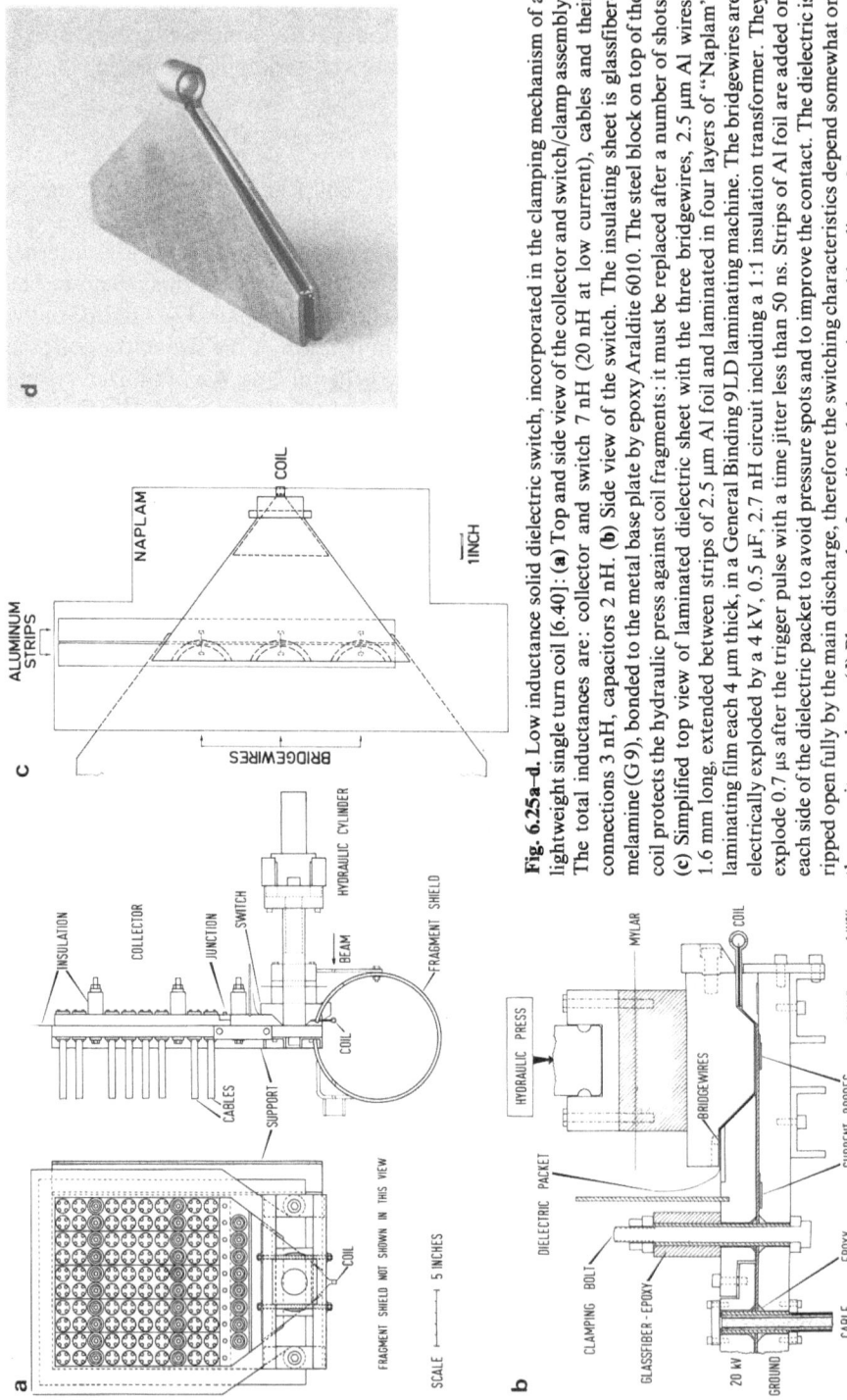

Fig. 6.25a–d. Low inductance solid dielectric switch, incorporated in the clamping mechanism of a lightweight single turn coil [6.40]: (**a**) Top and side view of the collector and switch/clamp assembly. The total inductances are: collector and switch 7 nH (20 nH at low current), cables and their connections 3 nH, capacitors 2 nH. (**b**) Side view of the switch. The insulating sheet is glassfiber-melamine (G 9), bonded to the metal base plate by epoxy Araldite 6010. The steel block on top of the coil protects the hydraulic press against coil fragments: it must be replaced after a number of shots. (**c**) Simplified top view of laminated dielectric sheet with the three bridgewires, 2.5 μm Al wires, 1.6 mm long, extended between strips of 2.5 μm Al foil and laminated in four layers of "Naplam" laminating film each 4 μm thick, in a General Binding 9LD laminating machine. The bridgewires are electrically exploded by a 4 kV, 0.5 μF, 2.7 nH circuit including a 1:1 insulation transformer. They explode 0.7 μs after the trigger pulse with a time jitter less than 50 ns. Strips of Al foil are added on each side of the dielectric packet to avoid pressure spots and to improve the contact. The dielectric is ripped open fully by the main discharge, therefore the switching characteristics depend somewhat on the capacitor voltage. (**d**) Photograph of a coil made by cutting and bending a 2.1 mm copper sheet

Table 6.9. Experiments with capacitor discharges into single-turn coils

Authors	*Furth* et al.	*Shneerson*	*Forster Martin*	*Shearer*	*Andrianov* et al.	*Knoepfel Luppi*	*Herlach McBroom*	*Gennadiev* et al.	*Botcharov* et al.	*Nakao* et al.
Ref.	[6.36]	[6.43]	[6.49]	[6.52]	[6.54]	[6.159]	[6.40]	[6.64]	[6.67]	[Add.]
Year	1957	1962	1967	1969	1970	1972	1973	1979	1979	1984
capacitor bank { Voltage [kV]	4	125	15	70	15	15	20	25	30	40
capacitor bank { Energy [kJ]	24	222	63	820	26	200	55	38	20	100
capacitor bank { Inductance [nH]	10	} 90[a]	0.9	15	14	6	14	5 (20)[c]	(25)[c]	18
capacitor bank { Resistance [mΩ]	2		0.5	1.5	2.8	0.6	6	(5)[c]	(5)[c]	3
Peak current [MA]	1.1		3.5	8.8	1.5	2.6	1.3	1.4	1	2.1–2.1–2
Quarter period[d] [μs]	6.4	3.9	1	3.3	2.4	5	1.6	1.7	1.6	2.5–2–1.9
Peak field [T]	160	150	260	355	310[b]	180[b]	200	225	350	150–200–240
Coil inner dia [mm]	3.2	3.4	5	3	2.1	5	2.5	2.4	2	10–6–4
Coil length [mm]	(6)	(8)	(15)		2.5	10	5	5	2	10–7–6

From top to bottom: capacitor bank peak voltage and energy, total circuit inductance and resistance, peak current, rise time (quarter period) of the field, peak field, initial inner coil diameter and length.

[a] Total impedance; [b] with a tantalum coil; [c] estimated from the discharge characteristics; [d] rise time to peak field.

the initial work of *Furth* et al. [6.36], *Shneerson* [6.43] was the first to use very high voltage but the results were still modest. A breakthrough was made by *Forster* and *Martin* [6.49] who developed a capacitor bank with moderately high voltage and energy but with very low internal inductance and resistance, using a multiple-channel solid dielectric switch. They introduced lightweight coils and demonstrated that these are most efficient; in the heavy coils that were used previously, the current is allowed to spread to regions where it does not contribute efficiently to the magnetic field in the bore, while the bulk of the conductor material does not contribute to the confinement. *Shearer* [6.52] obtained higher fields with heavy coils, using a very large capacitor bank. This is the most impressive example for the destructive power of megagauss fields. All recent work has been done with capacitor banks of modest size and with lightweight coils [6.64, 67]. Over a period of nearly 20 years, *Shneerson* and his collaborators have made a thorough experimental and theoretical study of the performance of single-turn coils. This work has been rewarded with the achievement of the highest fields obtained with moderately sized equipment [6.53, 67].

At the base of this exceptional performance there is a new and unexpected effect: under certain conditions of a very fast field rise and particular coil dimensions, a delay in the expansion of the inner surface is observed, both by measuring the field/current ratio and by flash x-ray photography [6.160]. The coil inductance is usually small compared to the inductance of the discharge circuit. The peak field therefore depends only on the peak current sustained by the discharge circuit and on the inner diameter of the coil at peak field [6.161]. The delay of the expansion, as compared to the calculated hydrodynamics of one-dimensional compressible flow, is sufficient to almost double the peak field from the expected values. The mechanism of this surprising effect is not yet well understood. Since it occurs in coils with a small inner diameter (~ 2 mm) and with an axial length and wall thickness of similar dimensions, it is speculated that axial compressible flow due to magnetic stress at the side walls develops a tendency to squeeze the inner wall towards the center. It would be interesting to analyze this with a two-dimensional computer code and to study it experimentally with refined diagnostics. A real breakthrough would be achieved if this effect could be exploited in coils with a larger volume, sufficient for experimental applications.

Among all methods for generating megagauss fields, the capacitor discharge into a small single-turn coil is unique in allowing the survival of even delicate samples. Although the destruction of the coil is quite violent, it is directed mainly towards the exterior, and destructive effects are minimized by using lightweight coils with optimal wall thickness. This is empirically determined, keeping in mind that the reflected rarefaction wave should preferably not arrive at the inner surface before the peak field is reached. The expansion of the coil can be monitored by recording the magnetic field and current. Figure 6.26 illustrates a well-optimized coil where the expansion becomes noticeable in the difference between the field and current traces only around peak field. Flash x-ray pictures as in

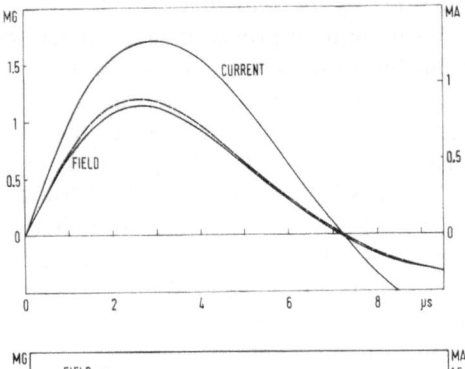

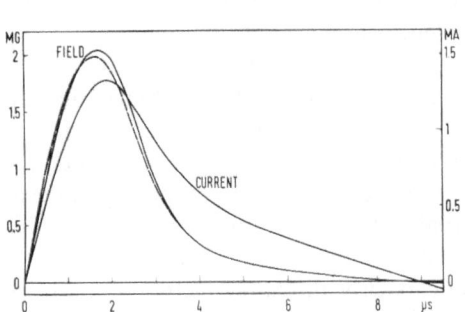

Fig. 6.26. Field and current waveforms from capacitor discharges into small single-turn coils made from 2.1 mm copper sheet [6.40]. (*Top*) coil with stainless steel collar, 5 mm i.d., 10 mm axial length, at 15 kV (31 kJ). (*Bottom*) coil 2.4 mm i.d., 4 mm axial length, at 20 kV (56 kJ). (−−−) are from field probes located 2 mm off the midplane to show the field homogeneity

Fig. 6.27. Flash x-ray shadowgraphs of single-turn coils. (*Lower row*) before shot. (*Upper row*) at peak field. From *left* to *right*: copper with stainless steel collar at 193 T, copper at 204 T, tantalum at 218 T

Fig. 6.27 are quite helpful in analyzing the coil performance. The current distribution in a single-turn coil is modified by the skin effect which tends to concentrate the current at the edges, resulting in improved field homogeneity [6.158]. This is a bonus for applications but in some cases it may be a mixed blessing because the axial current distribution becomes time dependent.

Quite recently, an improved megagauss generator of this type, built by the Nichicon company, became operational at ISSP. The 100 kJ capacitor bank is switched by 20 pressurized spark gaps that work in the range from 40 kV down to 2 kV (convenient for testing before the shot). This method of switching generates much less electrical noise than the solid dielectric switch in the IIT generator. Apart from the fact that the peak field is limited (eventually, it may become possible to generate about 500 T with this method), this noise has been the only drawback of the direct capacitor discharge. In an implosion system there is plenty of time for the switching noise to die out before the field rises to its peak. However, in the single-turn coil the field pulse is of much longer duration and measurements can be extended over the entire half-cycle of the damped capacitor discharge. It is remarkable that the pulse shape of the magetic field does not show any discontinuity due to the disintegration of the coil, it decays smoothly towards zero in a time interval up to twice as long as the rise time. The first experiments with the new instrument (Table 6.9) have confirmed the excellent reproducibility of the direct capacitor discharge method and the fact that the sample is not disturbed if it is somewhat protected against the blast effects that accompany the explosion of the coil.

If a single-turn coil is powered by an explosive-driven generator as described in Sect. 6.3.4, both the volume and the pulse duration can be larger [6.45, 48, 51, 59, 62, 65]. However, peak fields have been restricted to about 250 T, due to the limited liner speed in planar compression combined with the expansion of the coil and its transmission line. Wherever facilities exist for experimenting with explosive-driven devices, unquestionably this type of device is preferable for all applications in this field range that require a larger field volume than available in a direct capacitor discharge. At Los Alamos, simplified "miniature" devices were developed especially for this purpose [6.63]. These could easily be fired in a containment vessel under laboratory conditions. On the other hand, their performance is close to that of a direct discharge; they could be most useful for modeling larger systems. In the USSR, and in particular at Novosibirsk, there has been a long tradition in the design and use of bellows generators. Optimization has resulted in efficient and reliable devices well suited for experimental applications [6.65]. Typical waveforms, characterized by a slow increase and decay, are shown in Fig. 6.28 and some results are given in Table 6.10.

6.3.3 Magnetic Flux Compression

The most efficient way to concentrate electromagnetic energy is flux compression by an imploding cylinder. Magnetic flux is compressed as long as the

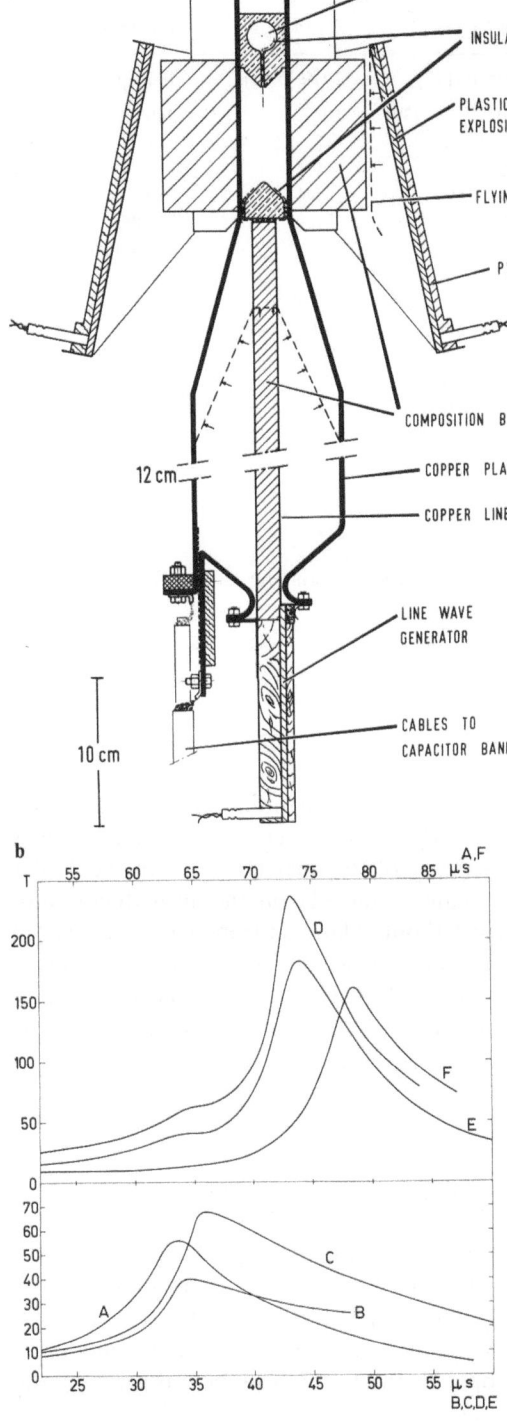

Fig. 6.28. (a) Scale drawing of the "bellows" pulsed field generator [6.45]. Note the flying plate plane wave generator which initiates the explosion of the second stage by impact. The axial length of the field coil is 15 cm. In the single-stage devices, a big coil is connected directly in place of the second stage. **(b)** Field waveforms for different coil diameters. Single stage: (*A*) 26 mm, (*B*) 30 mm, (*C*) 40 mm. Two stages: (*D*) 9 mm, (*E*) 15 mm, (*F*) 8 mm. Note the different time scales for *A* and *F* which refer to devices of an older design, with an axial length of 7 cm [6.44]

Table 6.10. "Bellows" or "strip" flux compression devices

Authors		*Herlach* et al.		*Caird* et al.	*Bichenkov* et al.
Year, reference		1965 [6.177]		1979 [6.63]	1979 [6.65]
Peak field	[T]	70	230[a]	108	240–290
Initial field	[T]	6		8	7
Coil i.d.	[mm]	40	9	9.5	14
length	[mm]	150		25.4	50
Compression volume length	[mm]	380		178	2 × 500
width	[mm]	150		44	20–200 (tapered)
Explosive thickness	[mm]	15	+2 × 50[a]		6
quantity	[kg]	1	2.5	0.087	~1
Capacitor bank energy	[kJ]	180		300	33[b]
voltage	[kV]	15		20	
Compression time	[μs]	(Fig. 6.28b)		25	~50
Peak duration	[μs]	(Fig. 6.28b)		~3	

[a] Second stage (Fig. 6.28)
[b] Initial inductive energy in the generator

implosion speed $v_i = -dr/dt$ if the inner wall exceeds the flux diffusion speed:

$$B/B_0 = (r_0/r)^2 \Phi/\Phi_0 \tag{6.27}$$

$$\frac{1}{B}\frac{dB}{dt} = -\frac{2}{r}\frac{dr}{dt} + \frac{1}{\Phi}\frac{d\Phi}{dt} = \frac{2}{r}(v_i - v_f), \tag{6.28}$$

where r is the inner radius and the index 0 denotes the initial condition. The field rises very sharply at the end of the compression, when the energy transfer is strongly dependent on the compressibility of the liner material. In the earlier stages of the implosion, energy is gradually extracted from the entire thickness of the liner; in many experiments this just about balances the increase of the speed of the inner wall due to cylindrical convergence [6.56, 149], i.e., this speed remains approximately constant. As the field begins to rise rapidly, the inner surface is suddenly decelerated by shock compression. The peak field is determined by the "field turnaround" condition

$$v_{it} = v_p + v_f, \tag{6.29}$$

where v_{it} is the speed of the inner wall before the onset of the final deceleration and v_p is the particle speed related to the peak field (Figs. 6.23b, 24). At peak field, the inner surface still moves at a speed of the order v_f which accounts for the observed probe destruction shortly afterwards. Field turnaround was first observed with devices of the type shown in Fig. 6.29a, where this phenomenon occurs at a relatively large radius, due to the large initial field and the moderate liner velocity [6.44, 149, 162]. An upper bound to the peak field is given by the

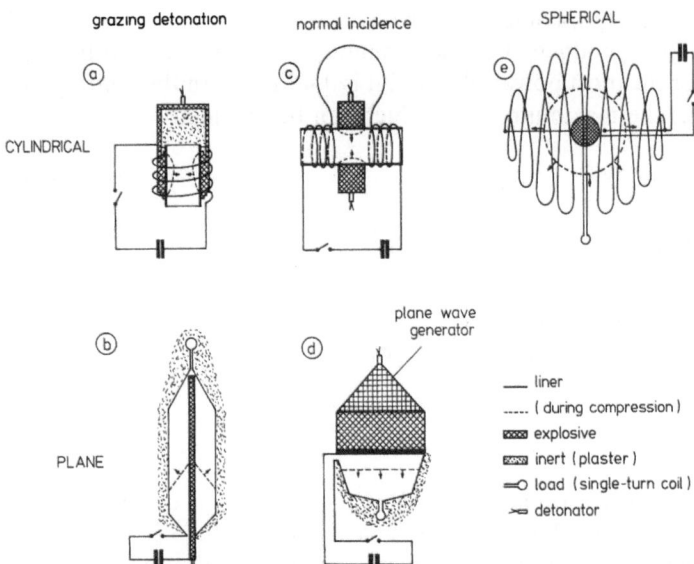

Fig. 6.29a–e. Flux compression configurations. (**a**) single detonator "top hat" [6.56], (**b**) single-stage "bellows" [6.45], (**c**) cylindrical implosion with a coil pair [6.37]; (**d**) high-speed flat plate generator [6.51] which may be used as a second stage for the bellows device, (**e**) spherical explosion which would be suitable for a nuclear blast. A spherical device compressing an azimuthal field was proposed by Fowler et al. [6.163]

implosion speed and the equation of state (Fig. 6.24). Since the other loss mechanisms of flux diffusion and vaporization also translate into speeds at the inner surface, for a given material it is always the implosion speed which determines the final field that can be reached.

a) Liners Driven by High Explosives

The acceleration of metal plates by high explosives is limited by the momentum and energy carried away by the detonation products [4.56, 164]. While the numerical analysis of the shock acceleration is somewhat cumbersome, a reasonable estimate of the final velocity can be obtained by the Gurney method which is based on the momentum and energy balances. The final speed v_L of the liner, i.e., the metal plate which is a lining on one side of the explosive charge, is given by

$$v_L^2 = \frac{6w\alpha}{5 + \alpha + 4/\alpha},$$

(6.30)

where w is the "Gurney energy" of the explosive, and α is the ratio (explosive mass)/(metal mass). The Gurney energy is the part of the internal energy of the explosive (approximately 3/4 or less) which is converted to kinetic energy of both metal and detonation products. For the commonly used composition B, this is 3.7 MJ/kg and for the most powerful HMX it is 4.4 MJ/kg. In the limiting case $\alpha \to \infty$ (6.30) gives the maximum speed which can be obtained:

$$v_{Lmax} = \sqrt{6w} \sim 5 \text{ km/s}. \tag{6.31}$$

The efficiency of energy transfer is given by

$$\varepsilon = \frac{m_L v_L^2}{2m_x w} = \frac{v_L^2}{2\alpha_L w} = \left[\frac{\alpha_L}{3} (\chi^2 - \chi + 1) + \chi^2 \frac{\alpha_L}{\alpha_T} + 1 \right]^{-1} \tag{6.32}$$

$$\frac{v_T}{v_L} = \chi = \frac{1 + 2/\alpha_L}{1 + 2/\alpha_T}.$$

Considering practical aspects of explosive charge design, an optimal value of the liner speed is about one-third of the possible maximum. In (6.32) the index T refers to a "tamper", i.e., a second metal plate on the back side of the explosive. Without a tamper the equations are valid with $\alpha_T \to \infty$ and v_T is then the speed of the most advanced detonation products. Although the assumption of a linear speed distribution is a crude estimate, the Gurney method gives values for the final speed that are within the experimental uncertainties of most experiments. This method cannot describe the process of acceleration which is characterized by shock waves traveling back and forth in the liner, resulting in stepwise increase of the speed at its free surface [6.165]. An unwanted effect that sometimes occurs is "spalling": at the intersection of two reflected rarefaction waves, a substantial layer of the liner can be split off and will move ahead at higher speed. A smooth acceleration can be achieved by placing a soft material, e.g., lucite, between the liner and the explosive. Another effect to be avoided is the shaped charge effect which results in the projection of a powerful jet from a pit or groove in the surface.

Even in a cylindrically imploding system where the speed at the inner surface is somewhat increased due to the convergence, it is in practice difficult to exceed 5 km/s. This accounts for the fact that explosive-driven flux compression is limited to about 1000 T. Apart from this fundamental limitation, there are two technical problems which are related to each other: the introduction of sufficient initial flux and the achievement of a regular implosion. Both determine the volume in which the final field is obtained. It is almost a miracle that a cylinder of 100 mm diameter and 1 mm wall thickness, for example, can be regularly imploded to an inner diameter of less than 5 mm when the wall thickness is 10 mm! The stability of cylindrical shells imploding against a magnetic field has been discussed theoretically [6.166, 167]; so far there has been no direct experimental evidence of instabilities. The common configurations for the arrangement of the field coils and the explosive charge are shown in Fig. 6.29. The coil pair leaves plenty of space for an elaborate explosive system but is relatively inefficient in generating the initial field. Coils inside the liner and between the liner and the explosive [6.37, 55, 168] were tried but did not give fully reproducible results.

This difficulty was finally overcome in the most advanced reproducible high field system developed at the Kurchatov Institute [6.69]. In place of the liner, a multilayer coil is used that transforms into a conducting cylinder upon the impact of the detonation wave. This induces some gradually increasing instabilities at the inner wall of the imploding cylinder; it was, however, possible to eliminate these by means of so-called cascades. These are cylinders made of many insulated copper wires, strung in the axial direction and embedded in epoxy. When the imploding liner hits such a cylinder, this becomes conducting and unites with the liner. Elaborate flash x-ray diagnostics has revealed that this indeed reduces the development of instabilities; in the present 10 MG system three cascades were found to be sufficient. These are very big and complicated devices, but they generate 10 MG reproducibly in a fairly large volume of more than 8 mm diameter, thus they are well suited for experimental applications and they are now in use for a number of solid-state experiments [6.169].

For applications at more modest field levels, a practical and uncomplicated device is given by a minimized explosive charge with grazing detonation where the coil for the initial field surrounds the charge [6.44, 56, 162] (Figs. 6.29a, 30). A slightly more sophisticated version which is open at both ends was developed later [6.170]. *Hawke* et al. [6.57] used a coil system that is half way between (a) and (b). Typical performance data of flux compression devices are given in Table 6.11 and field waveforms are shown in Fig. 6.31.

A reproducible, regular implosion can be obtained only with a seamless liner. This requires introduction of the initial flux by diffusion, which is governed by the time constant

$$\tau = L/R \cong \mu_0 rd/(2\varrho) = -\Phi/(d\Phi/dt) = r/(2v_f), \tag{6.33}$$

where r is the radius and d the wall thickness of the liner [6.171]. Typical values for stainless steel are $\gtrsim 100\,\mu$s, for copper $\gtrsim 1$ ms. Copper would be preferable from the point of view of better flux trapping during the implosion [6.168, 172]. *Sakharov* et al. [6.47] used a thin copper coating at the inside of a stainless steel liner. For massive copper liners, very heavy coils with reinforced leads are needed to generate the initial field with a slow rise time; otherwise the device may be disturbed by deformation of the coil or whiplash movements of the leads. If the current rises too fast in relation to the liner time constant, the magnetic stress due to the difference between the inner and outer fields will buckle the liner and cause an imperfect implosion. A radical solution would consist in using a containment vessel for the explosive device, surrounded by a large superconducting coil. This would allow complete freedom for the design of the explosive charge, and bring megagauss fields to a laboratory environment with the versatility inherent in explosively driven devices. Techniques for the containment of experimental explosions have been developed on a large scale at Novosibirsk [6.173]. An

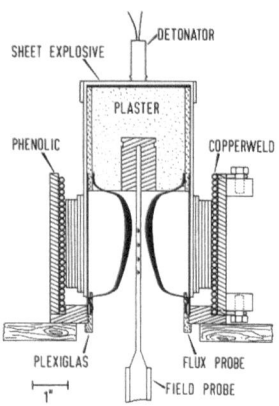

Fig. 6.30. Scale drawing of the top hat flux compression device [6.56]. The liner (in solid black) is shown in the initial position and at one time during the implosion, as determined from a flash x-ray photograph

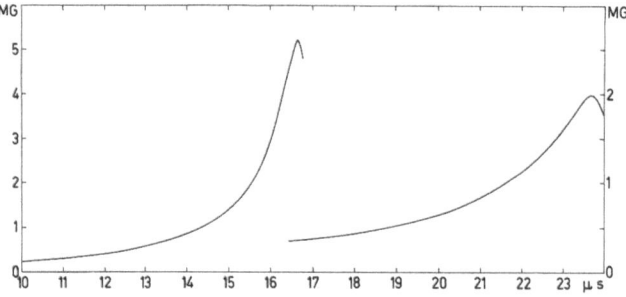

Fig. 6.31. Representative field waveforms from implosion devices. The time scale starts at the beginning of the implosion. The *left curve* is from experiment "Frascati 3", the *right curve* from experiment "Frascati 1", Table 6.11

impressive example of the technical effort needed to generate 1000 T reproducibly has recently been given by *Pavlovskii* et al. [6.69].

A new method of explosive-driven magnetic flux compression has been invented and tested recently by *Nagayama* and *Mashimo* [6.169, 174], and also by *Bichenkov* et al. [6.169]. Flux is compressed by a shock front that transforms a semiconductor into a conductor. Preliminary experiments with powdered silicon have yielded magnetic fields up to 100 T with compression factors of the order of 30. These results agree with a theory predicting the possibility of obtaining fields up to 1000 T. It was later found that powdered aluminum works as well, even without being imbedded in an insulating material. In this method it is easier to introduce the initial flux, and for the application to experiments it has the advantage of a good sample space protection. However, in contrast to compression by moving liners, there is practically no amplification of magnetic energy.

Table 6.11. Cylindrical flux compression systems

Laboratory		Los Alamos	Limeil	Moscow	Frascati 1	Frascati 3	Chicago	Livermore 1	Livermore 2
Explosive		comp. B	comp. B		plastic[a]	plastic[b]	Detasheet[c]	9404/TNT	9404/TNT
outer diameter	[mm]	235	200	300	87	140	104.6	211	406
radial thickness	[mm]	64	50	74	6	20	15	50	100
ignition system		det. ring	expl. lens		hat	hat	hat	4 det. rings	4 det. rings
Liner		AISI 304	copper	copper	AISI 304	AISI 304	AISI 304	AISI 304+Cu	AISI 304+Cu
inner diameter	[mm]	105	92	140	73	97	72.6	106.2	200
wall thickness	[mm]	1.59	2	6[d]	1	1.5	1	1.43+0.038	1.52+0.0254
Capacitor bank	[kJ]	300	200	16	100	200	50	1000	1000
Initial field	[T]	3.1	4.2		8.2	5.7	6.5	6.5–7	3.5
Peak field	[T]	610	1165	950±50[e]	200	540	350	~1000	>350
Final i.d.	[mm]	~2.5	4.4	8.5±0.5	8.6	5.6	5.2	~6	~12
Compression time	[µs]	13	10	17	24	16	18	11/12	~17/23

[a] Dynamit Nobel AG, Troisdorf; [b] Bombrini Parodi-Delfino, Colleferro; [c] E.I. duPont deNemours & Co; [d] The multiwire initial field coil is transformed into the liner when the implosion starts. The mass is 9 g/mm; [e] In ~100 shots.

b) Electromagnetically Driven Cylinders

The imploding cylinder is an amplification mechanism for the Poynting vector field. It can take up kinetic energy gradually over a large surface and deliver it to a small volume in a short time interval. Electromagnetic forces are well suited for the acceleration of a conducting cylinder to very high speed [6.41]. The commonly used arrangement is a single-turn coil closely surrounding the cylinder to be imploded. From plasma physics, we borrow the term "theta pinch" for this configuration; the imploded cylinder is usually called the "liner" or, in small systems, "the foil".

A first large series of experiments aimed at the development of a megagauss generator for practical applications was carried out at the Illinois Institute of Technology [6.175]. Besides coils made from copper sheet which are destroyed in every shot, a permanent solid steel coil was used. It was found that the close electrical coupling results in a more homogeneous current distribution in the axial direction and pulls the current away from the side walls of the solid coil. Due to the effects of current redistribution, the best results were obtained with coils that were 1.5 to 2 times as wide as the imploded cylinder. This is effective only for sufficiently fast rise times of the primary current; in these particular experiments this was 5 µs. In this small system, the heat treatment (temper) of the foil had a strong effect on the implosion.

The first electromagnetic implosion apparatus for practical use in solid state experiments was developed at the ISSP (Fig. 6.32, Table 6.12). It has provided magnetic fields in the 2 MG range for a great number of solid-state experiments, Sect. 6.5. Magnetic flux diffuses into the liner in the early stages of the

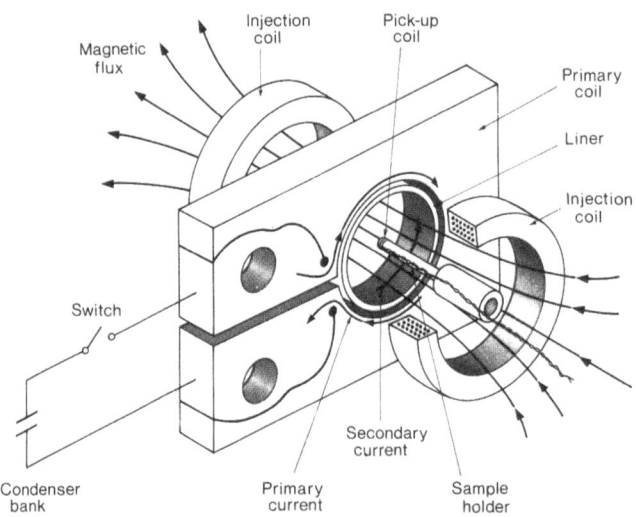

Fig. 6.32. Scheme of an electromagnetic implosion system with a solid steel single-turn coil as used at ISSP. Optional injection coils for an additional 'seed field" are included

Table 6.12. Electromagnetic flux compression systems

Authors		Cnare	Alikhanov et al.	Kachilla et al.	Mikhkelsoo et al.	Andrianov et al.	Miura et al.	Miura et al.
Year, reference		1966 [6.41]	1968 [6.50]	1970 [6.175]	1974 [6.60]	1979 [6.66]	1979 [6.42, 58]	1983 [6.176]
Capacitor bank								
energy	[kJ]	136	570	27	92	315	285	1000 5000
voltage	[kV]	20	4.8	20	120[b]	30	30	40 40
inductance	[nH]	160[a]	13	60	75	7	220	179 30
resistance	[mΩ]	3.5[a]	0.3	7			6.2	3.6 0.760
Imploding cylinder		Al	Cu	Al	Cu	Al	Cu	Cu Cu
initial thickness	[mm]	0.41	2	0.3	0.47	0.6	1.0	
initial i.d.	[mm]	25	60	31	<45	30	64	
final i.d.	[mm]	~2	6	~2	~2	~2	~8	
Peak field[c]	[T]	210	310	130	320	320	280	
Max. field rise	[MG/μs]	2.3	2	1.5		~6	~0.8	
Total time to peak	[μs]	18	56	12		10	50	

[a] Private communication.
[b] Coupled to the single-turn driving coil by means of a 14:1 step-down transformer.
[c] These are the highest values in a series of experiments that are generally about 2/3 of this value.

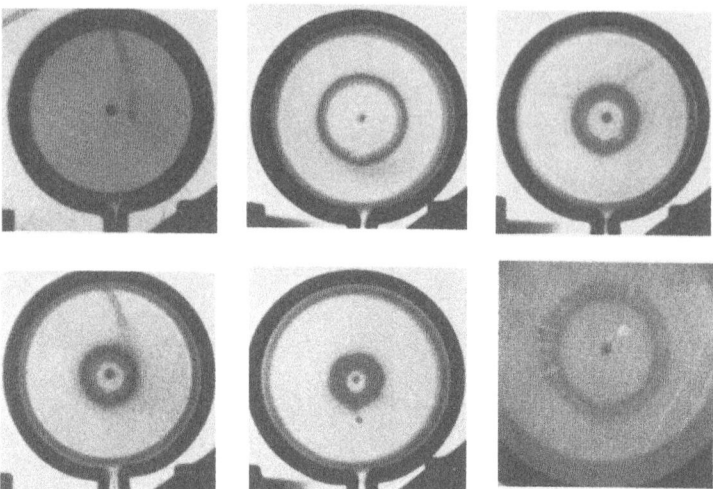

Fig. 6.33. Flash x-ray shadowgraphs from electromagnetic implosions. Consecutive stages of the implosion are shown but each picture is from a different experiment. The still picture (*upper left*) belongs to the implosion picture at *lower left*, which shows the deformation of the driving coil. (*Lower right*), the foil (actually a ring cut from an aluminium beer can) has been exploded by acceleration to a speed exceeding the limit given by (6.38)

implosion, when the implosion speed is smaller than the flux diffusion speed. This is sufficient for compression to a very high field in a small volume, but for applications where the magnetic field is needed in a larger volume, additional initial flux must be introduced by an external coil pair with a slow rise time.

The peak field depends on a delicate balance between the size and thickness of the liner and the waveform of the driving field which is influenced by the inductance changes caused by the movement of the liner [6.58]. Due to the limited energy available from the power supply, most systems are axially short and thus the field distribution is essentially three-dimensional. The length/radius ratio of the liner is usually in the range 0.5–1 (e. g., 0.61 at ISSP, 1 at IIT). A problem arises since the field inhomogeneity in the vicinity of the feed gap causes a bulge in the imploding cylinder and a subsequent radial displacement of the center of the implosion (Fig. 6.33). This can be much improved by suitably shaped slits in the solid coil [6.42] or by inserting a copper sheet between the cylinder and the feed gap [6.66].

Alikhanov et al. [6.50] have circumvented this problem by driving the cylinder in *z*-pinch geometry, passing the driving current in axial direction through the cylinder itself; contact is made by means of slightly conical end plates. In this system, all the initial flux must be introduced from external coils. To obtain an estimate of the driving efficiency in the two configurations, we compare the total force on the circumference of the cylinder, neglecting edge effects:

Table 6.13. Comparison of electromagnetic implosion systems with 5 cm primary radius and length (edge effects neglected)

r/r_p	Inductance [nH]		$F/(\frac{1}{2} LI^2)$ [N/J]		F/I^2 [N/kA^2]	
	θ	z	θ	z	θ	z
0.99	3.9	0.1	1990	2010	3.91	0.101
0.98	7.8	0.2	990	1010	3.87	0.102
0.95	19	0.5	390	410	3.75	0.105
0.9	38	1	189	211	3.55	0.111
0.5	148	6.9	27	58	1.97	0.2
0.2	189	16	8	62	0.79	0.5
0.1	195	23	4	87	0.39	1.0
0.05	197	30	2	134	0.2	2.0

$$\frac{F}{\frac{1}{2} LI^2} = \overset{\theta \text{ pinch}}{\frac{2r}{r_p^2 - r^2}} \qquad \overset{z \text{ pinch}}{\frac{1}{r \ln \frac{r_p}{r}}}. \tag{6.34}$$

Here L is the inductance of the system at cylinder radius r, I is the primary current and r_p the radius of the primary. As an illustration, a few numerical examples are given in Table 6.13. The z pinch has a slightly higher driving efficiency but this is largely offset by the much smaller inductance which has to share the electrical energy with the residual inductance of the power supply.

For a more detailed analysis of the flux compression process, a computer simulation is needed. In a first approximation, the cylinder is represented by a single average radius r. The system is then described by the following equations where the edge effects can be included in the parameters L and M:

θ pinch:

$$L_0 \frac{dI}{dt} + R_0 I - \left(M \frac{dI}{dt} + \frac{dM}{dt} I \right) = V_0 - \frac{1}{C} \int I \, dt, \tag{6.35}$$

$$M \frac{dI}{dt} + \frac{dM}{dt} I - \left(L_\theta \frac{dI_\theta}{dt} + \frac{dL_\theta}{dt} I_\theta \right) + \mu_0 H_0 \frac{dS}{dt} - R I_\theta = 0, \tag{6.36}$$

$$-m \frac{d^2 r}{dt^2} = 2 \pi r l \frac{\mu_0}{2} [(H_e + H_0)^2 - (H_i + H_0)^2]; \tag{6.37}$$

z pinch:

$$(L_z + L_0) \frac{dI_z}{dt} + \frac{dL_z}{dt} I_z + (R_0 + R_z) I_z = V_0 - \frac{1}{C} \int I_z dt, \tag{6.35a}$$

$$L_\theta \frac{dI_\theta}{dt} + \frac{dL_\theta}{dt} I_\theta + R_\theta I_\theta = 0, \tag{6.36a}$$

$$-m \frac{d^2r}{dt^2} = 2\pi r l \frac{\mu_0}{2} \left[\left(\frac{I_z}{2\pi r}\right)^2 + H_0^2 - (H_0 + H_z)^2 \right]. \tag{6.37a}$$

Currents: I in the primary coil, I_z axial, I_θ azimuthal in the liner.

Liner: r radius, l length, m mass, $S(r)$ enclosed cross section, $L_z(r)$, $L_\theta(r)$, $R_z(r)$, $R_\theta(r)$ inductances and resistances (axial and radial), $M(r)$ mutual inductance between liner and primary coil.

Capacitor bank: C capacitance, L_0 internal inductance, R_0 resistance, V_0 initial voltage.

Magnetic fields: H_0 initial, H_θ azimuthal and H_z axial: H_i at the inner, H_e at the outer surface of the liner.

The force related to elastic-plastic deformation can be introduced in (6.37), and a cylinder of finite thickness can be described by taking the derivative of the kinetic energy:

$$\frac{dv_i}{dt} = \frac{1}{r_i} \left[\left(\frac{p}{D} + \frac{1}{2} v_i^2 \frac{r_{e0}^2 - r_{i0}^2}{r_e^2} \right) \Big/ \ln \frac{r_e}{r_i} \right], \tag{6.37b}$$

where p is the difference in pressure between the inner and the outer surface, D is the density and v_i is the speed of the inner surface.

These equations have been used by *Miura* and *Chikazumi* [6.58] for a numerical analysis of their experiments in θ-pinch geometry (Fig. 6.34). Such a straightforward simulation with ordinary differential equations will always give a fair representation of the acceleration phase. It cannot directly be used to predict peak fields in excess of a few megagauss because these are mainly determined by the compressibility of the liner material and by the maximum speed of the inner surface. However, this type of simulation is quite useful for parameter studies to optimize the conditions for obtaining a high implosion speed. An estimate for the limiting peak field can then be derived from the simple criterion given in (6.29) and the related discussion.

In practice, it is by no means trivial to obtain a high implosion speed with a large capacitor bank. Even if the internal inductance of the capacitor bank is small, the large capacitance results in a relatively long half period of discharge. To drive the implosion efficiently, the half period ought to be of the order of the implosion time. One would anticipate that this can be improved by increasing the initial diameter of the imploding cylinder. However, the simulations demonstrate that this is balanced by the increase of the inductance in the larger coil-liner system. The theta pinch initially develops a good driving force, but the inductance increases strongly after the liner starts moving. One possible remedy consists in incorporating additional elements into the system, such as fast-acting transfer switches that shorten the rise time of the current pulse. Another solution

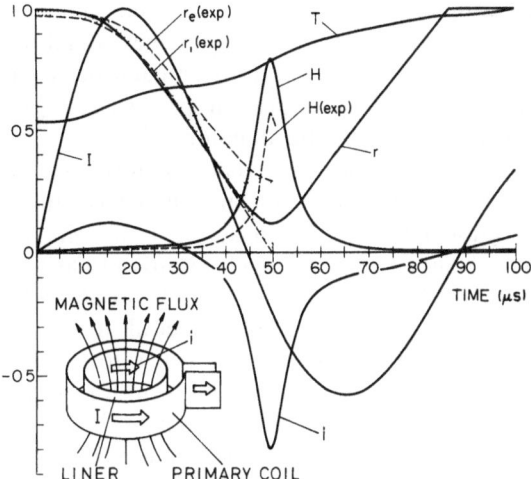

Fig. 6.34. Computer simulation of electromagnetic flux compression [6.58]. I denotes the current in the primary coil, i the secondary current in the liner, r the radius of the liner, H the magnetic field, and T the temperature of the liner. r_e (exp) and r_i (exp) stand for the experimentally observed outer and inner radius of the liner, respectively, and H (exp) stands for the experimentally observed field. The inset shows a schematic diagram of the coil system of the electromagnetic flux compression

would be a two-stage system which combines the initial driving efficiency of the theta pinch with the high force developed by the z pinch in the later stages of the implosion. In this arrangement, the liner itself can possibly be used as the transfer switch when it hits the electrodes of the z pinch. As illustrated in Table 6.13, the z pinch alone needs a very high current to develop a large force in the beginning. This can in principle be improved by a step-down transformer. Each of these or other possible arrangements for optimization involve large technical efforts and it is thus important to determine first the most promising arrangement via a computer simulation.

The first electromagnetic flux compression systems were fairly small and limited in field strength and final volume (Table 6.12). However, there are good prospects for future development because there is no fundamental speed limit such as with high explosives. When a metal sheet is accelerated by electromagnetic forces, it will eventually be vaporized by the induced current. Assuming a homogeneous current distribution, *Cnare* [6.41] has shown that the speed is directly related to the temperature increase of the sheet, in proportion to the sheet thickness d:

$$v = (\mu_0/2)(d/D) \int (I/A)^2 dt = (\mu_0/2)d \int (c_p/\varrho)dT = d \cdot f(T). \tag{6.38}$$

This is related to (6.2); A is the cross section of the conducting sheet, I the current, c_p the specific heat and T the temperature; the integral includes the heat of melting and vaporization. For example, aluminum will theoretically burst at

25 km/s per mm thickness, and copper at 14 km/s. This has been confirmed experimentally; however, the available data base is limited and does not allow a prediction on how closely one may approach these limits in a reliable device. In *Cnare's* extended series of experiments [6.41], the implosion speed was 0.6 ± 0.1 of the theoretical speed for copper and 0.3 ± 0.1 for aluminum.

Very large energies are needed to drive an electromagnetic implosion system to really high fields because of the required liner thickness and large initial radius. A first step in this direction has been taken at the ISSP with the installation of a 5 MJ capacitor bank aimed at generating fields up to 1000 T for laboratory use [6.100, 177]. With a view to high-pressure research, proposals for large-scale implosion systems have been made by *Alikhanov* et al. [6.70] and *Fowler* et al. [6.178], including the use of explosive-driven generators. Eventually, the large power supplies which are now under development for nuclear fusion research [6.179] may be used for driving electromagnetic implosion systems with peak fields up to 10^4 T, e. g., rotating machines, inductive storage, superconducting coils, compressed gas and internal combustion devices. Eventually, this will make explosive-driven systems obsolete.

A new branch of the development involves devices with plasma cylinders imploding at very high speed [6.71, 180–182]. The power sources are large sophisticated capacitor banks with low inductance and resistance. The field rise is so fast in these devices that diagnostics becomes a major problem, and only preliminary results are available at present.

A three-dimensional plasma implosion can be obtained with a fairly simple device called the "plasma focus", a coaxial plasma gun with the open end shaped so that the ejected plasma ring concentrates in a "focal point". There is some indirect evidence for the formation of small threads containing fields of the order of 100 MG in the turbulent plasma created at the focal point [6.61].

Alikhanov et al. [6.71] proposed a "magnetopressed discharge", whereby a highly conductive plasma cylinder is driven ahead of a heavy imploding shell with high resistivity. Calculations indicate that fields as high as 200 MG might be obtained if the cylinder implodes at sufficient speed. The problem of accelerating the cylinder is taken for granted by these authors.

In all implosion systems, a problem is posed by the shock wave running ahead of the imploding cylinder. This can severely disturb delicate samples, and it was observed that even sturdy inductive field probes were substantially compressed [6.56]. For most applied experiments, it is therefore necessary to evacuate the volume inside the imploding cylinder. Even in a vacuum, the probe may be disturbed by material ejected from the surface of the imploding cylinder. The "cascades" used in the big implosion device at the Kurchatov Institute provide a good protection of the sample space [6.69].

6.3.4 Pulsed Power from Flux Compression Devices

A flux compression device is a generator of electricity. This becomes more evident when the design provides for two terminals where the energy is delivered

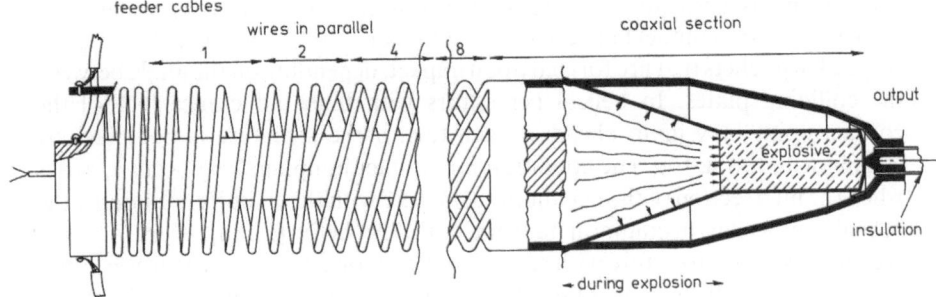

Fig. 6.35. Flux compression generator with a helical and coaxial stage, showing the principle of varying pitch by splitting turns. For simplification, the helical section has been shortened. For proper matching to the coaxial stage, it would have to be extended with further splitting of turns

to a load. In principle, any device in which an inductance is decreased by mechanical deformation can be used as a generator for a giant pulse of electromagnetic energy. These generators are usually described in engineering terms such as lumped circuit parameters. In most generators, a distinction can be made between the "stator", i.e., a stationary conductor in which the initial flux is generated, and an "armature", which is a conductor driven by high explosives or another primary propulsion system into the stator to reduce the inductance; in most devices there is a moving point of contact where the stator is gradually short-circuited by the armature. The variety of configurations in which an inductance can be decreased by mechanical deformation is virtually unlimited. Practical solutions are based on the geometry imposed by the explosive system, Figs. 6.29b,d,e, and 35. In the common configurations, the explosive is used to drive either an imploding cylinder or a flat plate, each with a normally incident [6.37, 183] or a grazing detonation wave [6.44], or a cylinder which is expanded in the shape of a cone [6.47, 154, 184]. While generators with a coaxial cavity appear to be most efficient, high inductance and flexibility in the inductance variation are obtained by using helical windings for the stationary part of the generator (Fig. 6.35). More examples with different configurations can be found in the ingenious work presented by Russian research groups at the second and third megagauss conferences [6.169, 185].

The current increase can be calculated from

$$(1/I)\,(dI/dt) = -(1/L)\,(dL/dt + R) = -(1/L)dL/dt + (1/\Phi)d\Phi/dt, \qquad (6.39)$$

where L is the total inductance including the load and $\Phi = LI$ is the total flux in the circuit. The term dL/dt is a negative resistance, which must exceed the circuit resistance R to give a net current increase. The total inductively stored energy increases according to

$$W/W_0 = (L_0/L)\,(\Phi/\Phi_0)^2 = [(L_{g0} + L_1)/(L_g + L_1)]\,(\Phi/\Phi_0)^2, \quad L_g \to 0. \qquad (6.40)$$

The flux loss is a critical factor in the energy amplification. In most generators, there is a moving contact between the armature and the stator, where flux may be trapped in pockets that are formed upon impact, depending on the angle between the colliding plates. In helical generators with high inductance [6.154] the contact point may jump ahead at each turn if the armature is not precisely coaxial with the stator. The condition for contact jumping, $2\pi\varepsilon p \geq \tan\alpha$, arises when the intersection of the expanding armature with the cylinder containing the helical stator has a common tangent with the helix [6.186]. Here ε is the eccentricity of the armature with respect to the stator, p is the pitch (turns per unit length) and α is the half-opening angle of the cone-shaped armature. To obtain good energy efficiency, each part of the armature must have delivered most of its kinetic energy before it collides with the stator, but the speed at impact must still be sufficient to assure a good contact. This requires matching of the electrical power output to the power developed by the explosive throughout the compression, by means of a suitable progression of inductance along the detonation path. For constant power output, the inductance will vary approximately as $1/t$. In the helical device, the progression of inductance is affected by a stator coil with variable pitch. The highest permissible pitch is determined by the condition for contact jumping. In the bellows device, the variation of the width both decreases the explosive power and increases the inductance towards the detonator end [6.48]. An important point regarding efficiency is the complete enclosure of the explosive by the liner, e.g., in a cylinder or between flat plates. The performance of a flux compression generator depends on the ratio of the load inductance L_l to the generator inductance L_{go} (6.40). A survey of published results is given in Table 6.14. Ironless pulse transformers can be used for adapting larger load inductances, and for staging several generators [6.163, 186, 189–191]. The rise time of the output pulse can be shortened, with a corresponding increase in output voltage, by a fast opening switch, usually a fuse [6.62, 184, 191].

These generators are the most compact and powerful self-contained packages for delivering pulsed electromagnetic energy. Pulses with similar characteristics can always be delivered by a nondestructive generator, but only at huge investment cost for a large installation. Explosive generators are most useful for testing advanced concepts, to find out whether it is worthwhile to invest in the engineering of a nondestructive power supply. Due to their inherent simplicity and versatility, they can easily be adapted to a given experiment. A recent example is the work on rail guns which excellently demonstrates the usefulness of explosive generators. This work was initiated by *Hawke* [6.192] and later carried out in collaboration with the Los Alamos group [6.193]. It is now also pursued at Novosibirsk [6.169]. For experiments in outer space or other remote locations, there is no alternative to explosive-driven generators when a strong burst of pulsed power is needed. *Fowler* et al. [6.194] have built a rocket-borne generator which reliably delivered 0.3 MJ to a plasma gun that was fired in space. For more stationary devices employing the principle of magnetic flux compression, compressed gases [6.195] and gaseous explosives [6.196] have

Table 6.14. Performance of explosive-driven megajoule generators

Type	Ref.	Inductance		Inductive energy		Efficiency		Compression time
		generator [μH]	load [nH]	initial [kJ]	final [kJ]	$(\Phi/\Phi_0)^2$ [%]	magn./expl. [%]	[μs]
Helix, machined	[6.154]	350	800	6.3	144	5.2	0.2	62
Helix, machined	[6.154]	725	110	17.8	5500	4.7	3.9	121
Helix, wound	[6.154]	46	85	57.5	4250	13.7	1.0	76
Helix, twin	[6.184]	10	70	72	1715	16.7	3.7	45
Helix	[6.187]	2.4	168	195	1278	45.9	5.4	75
Coaxial	[6.187]	0.168	32	1278	5390	79.1	17.8	85
Helix, optimized	[6.187]	1.68	30	170	7260	76.2	28.3	100
Bellows	[6.188]	1.16	15	84	1920	29.6	1.3	150
Bellows	[6.48]	0.3	10	38	980	72	15	~120

been used as propellants, and *Cowan* et al. [6.197] proposed a "pulsar" device with superconducting coils for the initial flux.

6.4 Measurement of Pulsed Magnetic Fields

Precise and reliable measurement of the magnetic field is a prerequisite for all applications. The most common method consists in measuring the voltage induced in a calibrated pick-up coil [6.37, 44, 198]. This may amount to a few kilovolt even for small coils, and the signal will contain very high frequency components. Originally, there was much concern about the precision of this measurement. However, comparison with the Faraday [6.21] and Zeeman effects [6.199] proved that it can be surprisingly accurate, i.e., ±1%, if the proper precautions are taken. Probes are wound with thin wire (∼50 μm) for better definition and are potted in epoxy for stability. They are best individually calibrated by comparison to a known standard probe in an alternating magnetic field at a frequency within the frequency spectrum of the field pulse. In the long run, it would be best to establish a number of well-defined fixed points on the magnetic field scale. The problem consists in finding sufficiently sharp transitions that can be measured with ease. As an example, the transition observed in the magnetization of *Pr* metal at 32 T (Fig. 3.2) might be used. In a laboratory equipped for measuring resonance absorption with lasers, certain resonances could be used for this purpose. At both Tokyo and Osaka, the paramagnetic resonance of Cr in ruby has been used; the g factor is 1.98. At Leuven, the paramagnetic resonance in a thin layer of DPPH, recrystallized from benzene, was found to be most practical for field calibrations, with a g factor of 2.0037 and a linewidth of 30 mT at helium temperature. This also works at higher temperatures but the calibration is best done in the vicinity of the temperature where measurements are carried out, to avoid errors due to the thermal expansion of the pick-up coil. The cabling on the pick-up coils must be terminated with the characteristic impedance if the cables are long and the pulse times are short. Other factors to be considered are the voltage division between the terminating resistor and the combined resistance of the probe and its cabling, the finite integration time of an RC integrator, and its disturbance by the input resistance R' and capacitance C' of the recording instrument [6.200]. The errors related to the integrator can be removed by partial numerical integration:

$$\int U dt = R(C+C')U_1 + (1+R/R') \int U_1 dt, \qquad (6.41)$$

where U is the probe voltage and U_1 the RC integrated voltage at the input of the recording instrument. The field measurement by RC integration alone can be deceptive because irregularities in the induced signals due to malfunctions (e.g., spurious voltages coupled into the probe circuit) are smoothed out. It is always advisable to run a direct measurement of the induced voltage in parallel. In a flux compression experiment, this rises very steeply in the final phase of the

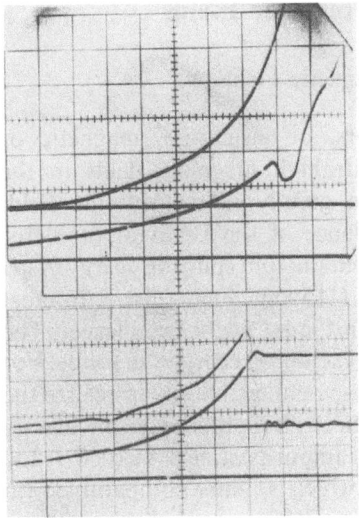

Fig. 6.36. Oscilloscope records from a top hat device where the liner was not evacuated [6.56]. From *top* to *bottom*: initial dB/dt, 0–20 µs, 0.92 T/µs div; intermediate dB/dt, 17–22 µs, 9.2 T/µs div; final dB/dt, 19–24 µs, 91 T/µs div; final B, 19–24 µs, 90 T/div. The two pictures are from the same experiment but different probes (Fig. 6.30). Therefore the dip in dB/dt which is caused by a precursor shock wave does not occur at the same time. Note the time correlation by the 1 µs blanking markers

compression; to obtain adequate accuracy the measurement must be divided into two or three recordings with different sensitivities. As an example, Fig. 6.36 shows a dB/dt recording from an explosive-driven flux compression experiment [6.56] which reveals the effect of a slight compression of the magnetic probe due to a precursor shock wave. This would not be visible in a recording of the integrated voltage.

With stable nondestructive coils, instead of measuring the field directly one can measure the current. The precise measurement of strong current pulses is a nontrivial problem. Current transformers with a saturable core work well but become quite bulky for permitting the passage of the total charge involved in a pulsed field experiment. In a current transformer, the secondary current will be exactly proportional to the primary current if the resistance of the secondary circuit is zero. In practice, a resistance is needed for measuring the secondary current. One may then consider the current transformer as a pick-up coil with L/R integration. An inductive current pick-up without a ferromagnetic core is called a Rogovski belt; in its elementary form it is a toroidal winding encircling the current to be measured. The inductance is usually too small for L/R integration, therefore external integration is needed, with precaution for the inductive effects. For both types of inductive current monitors, careful shielding is required. If a shunt is used, it must be compensated for inductive pick up and guarded against ground loops. An elegant method is the measurement of the Faraday effect in an optical fiber wound many times around the conductor [6.201]. Altogether, the inference of the field from a current measurement will never be quite satisfactory because the field coils undergo deformations during the pulse.

6.5 Solid-State Physics in Ultrahigh Magnetic Fields

6.5.1 Electrons in Solids Under Ultrahigh Magnetic Fields

When solids such as metals, semiconductors, or insulators, magnetic or nonmagnetic substances are subjected to ultrahigh magnetic fields in the megagauss range, their electronic states are greatly altered by the field. Figure 6.37 shows the magnetic field dependence of the behavior of a free electron. The cyclotron energy $\hbar\omega_c = \hbar e B/m^*$ and the spin splitting energy $g\mu_B H$ are 134.0 K (11.58 meV) at 100 T, and 1340 K (115.8 meV) at 1000 T. In most typical semiconductors with an effective mass $m^* < m_0$, $\hbar\omega_c$ is even larger. The values of $\hbar\omega_c$ and $g\mu_B B$ may exceed the various excitation energies in solids. For example, in most semiconductors the megagauss field enables us to realize the conditions whereby $\hbar\omega_c$ is larger than the optical phonon energy, the energy band gap or the binding energies of excitons or impurity states. At 1000 T, the radius of the cyclotron motion is reduced to 8 Å which is almost comparable with the lattice constants in crystals. In such situations, various new phenomena are expected to occur. In particular, this will result in all kinds of nonlinear effects which can give a much more profound insight into solid-state properties.

6.5.2 Experimental Techniques for Solid-State Physics in Megagauss Fields

The effort needed to perform a nontrivial experiment in a megagauss field is comparable to that of setting up a reliable pulsed field generator. This reflects on the time scale of the development. The first experiments were nothing more than feasibility demonstrations and a calibration and cross-check of the magnetic field measurement. They were all done with optical methods, where the detection

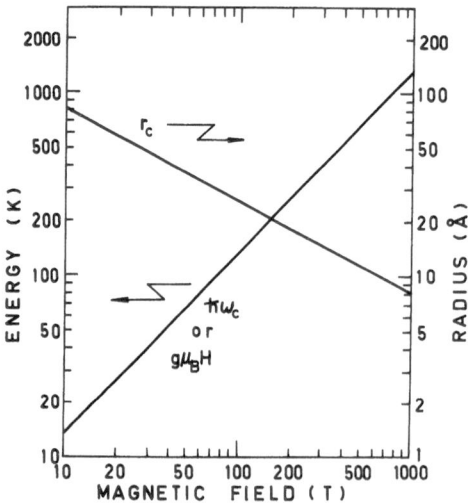

Fig. 6.37. Magnetic field dependence of quantities concerning a free electron. $r_c = (\hbar/eH)^{1/2}$ is the radius of the cyclotron motion, $\hbar\omega_c = \hbar e H/m_0$ is the energy splitting between Landau levels and $g\mu_B H$ is the spin splitting

equipment can be kept at a respectable distance using either lasers and/or fiber optics, and where no electrical wiring to the sample is needed. For a straightforward measurement of the Faraday effect, only a monochromatic light source, a detector and two polarizers are needed [6.183] (one polarizer is sufficient if the field volume is accessible from one side only, with a mirror behind the sample [6.202]).

Measuring the Zeeman effect requires more sophisticated instrumentation, e.g., a streak spectrograph for time-resolved spectroscopy. The first experiments were made by the pioneering Los Alamos group with a rotating mirror spectrograph [6.200]. The short duration of the experiment requires light sources of extreme brightness. This is not too much of a problem; an exploding wire is an excellent spectral source, and plenty of white light is obtained from an explosive-driven argon flashbomb (e.g., a white balloon filled with argon and taped to a small brick of high explosive). At Los Alamos, the streak spectrograph was used to measure the Faraday effect simultaneously over the entire visible spectrum in a single shot. This is a good example for the art of extracting much information from a single shot which can make experimentation with pulsed fields so efficient.

The design of the grating spectrograph is shown in Fig. 6.38 [6.200]. The wavelength scale is calibrated with the spectrum obtained from an exploding gold wire. The magnetic fields were produced by explosive-driven flux compression, mostly with "strip" flux compression devices. *Fowler* et al. obtained fine magnetooptical spectra such as Faraday rotation and the magnetooptical absorption spectra in MnF_2, CdS, GaSe and zinc chalcogenides, etc. [6.39, 203–205].

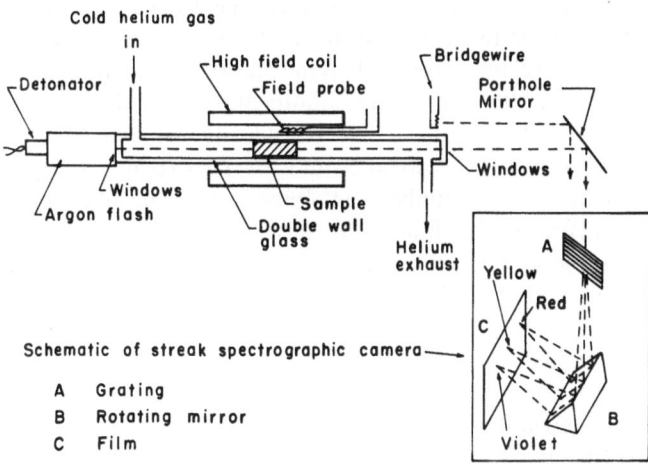

Fig. 6.38. Schematic of the experimental setup to obtain high field magnetooptical spectra using a rotating mirror [6.203]

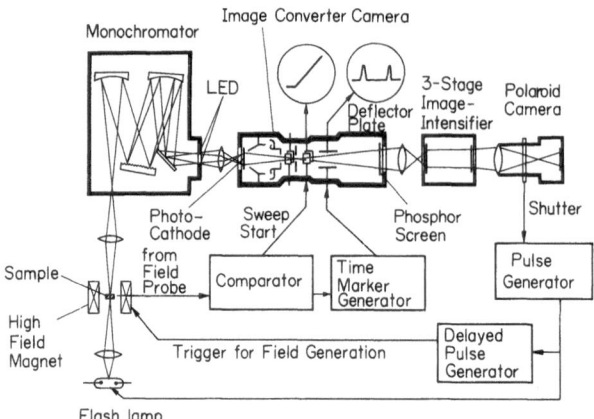

Fig. 6.39. Scheme of a streak spectrometer using an image converter camera [6.42]

The rotating mirror spectrograph has the advantage of high resolution, but the sensitivity of direct recording on photographic film is low and has to be made up for by using extreme light sources. For indoor experiments, the convenience of explosive-driven light sources is not readily available and an alternative system must be developed. This can be done by using an image-converter camera (ICC) [6.42, 206], which converts an optical image to an electron beam and thus sweeps the image. The ICC has a very high optical gain which can be further increased by combination with an image intensifier (II). Thus it is possible to obtain time-resolved optical spectra with a light source of moderate brightness. In addition, we can more easily synchronize the sweep of the image with the magnet system.

Figure 6.39 shows schematically the block-diagram of the streak spectrometer using an ICC and an II developed by the ISSP group [6.42, 206]. The image of the optical spectrum on a horizontal line is focused on the photocathode in the ICC. Then photoelectrons are emitted from the photocathode, and they are accelerated and focussed on the phosphor screen where the optical image is reproduced. During this process, the electron beam is swept perpendicularly to the spectral line (vertical direction) by applying a voltage ramp to one of the pairs of electrodes. Consequently a streaked image is formed on the phosphor screen, displaying a time-resolved spectrum. This image is then intensified by the II and finally recorded on a film in a polaroid camera. To calibrate the time scale as a function of magnetic field, a short double pulse voltage produced by the time marker generator is applied to the deflector plates to modulate the beam horizontally. The timing of the double pulse is adjusted so as to be recorded on both sides of the swept spectrum. The time marker signal is made more clearly visible by putting two tiny LED's (light emitting diode) at both sides of the spectra at the exit of the monochromator. As there is a small time jitter in the rise of the magnetic field every time, the sweep of the ICC is triggered when the magnetic field reaches some adjustable threshold so that the recording time may coincide with the period when the magnetic field rises to its maximum. For this purpose, a part of the signal from the pick-up probe is led to a comparator which compares this signal with a preset threshold. This generates a trigger pulse for starting the sweep of the ICC. By using the streak spectrometer the ISSP group obtained a variety of magnetooptical spectra in the megagauss fields which were produced by the electromagnetic flux-compression technique [6.42].

Another powerful technique for solid-state experiments in megagauss fields is laser spectroscopy. By using various lasers with many different wavelengths in the visible, infrared or far-infrared region, magnetospectroscopy can be performed in megagauss fields. Measurements of the Faraday rotation using lasers (especially a He–Ne laser with 6328 Å wavelength) or other monochromatic light sources have been made by many authors [6.44, 50, 202, 207–210] as one of the easiest solid-state experiments in the megagauss fields. Experiments of the cyclotron resonance were first performed by *Herlach* and co-workers on InSb and GaAs using a CO_2 laser and a CO laser [6.158, 211, 212]. More extensive experiments have been performed by the ISSP group on a variety of semiconductor crystals, using many different lines from a CO_2 laser and an H_2O laser in the infrared region [6.213–217]. Figure 6.40 shows a block diagram of the system for the infrared cyclotron resonance measurements in megagauss fields [6.209, 210]. The infrared light is detected by the extrinsic photoconductivity of doped Ge cooled down to liquid helium temperature. The response time of such a detector in combination with a fast preamplifier is well matched to the megagauss pulse.

Magnetization measurements in megagauss fields are difficult because the intensity of the samples' magnetization is generally much smaller than the flux density in the megagauss range. In the usual induction method (Sect. 3.10), compensation of the pick-up signal induced by the applied field becomes almost impossible. It is more practical to measure the reversible susceptibility by modulating the magnetization with a high-frequency magnetic field. Figure 6.41 shows a block-diagram of the system for measuring the reversible susceptibility,

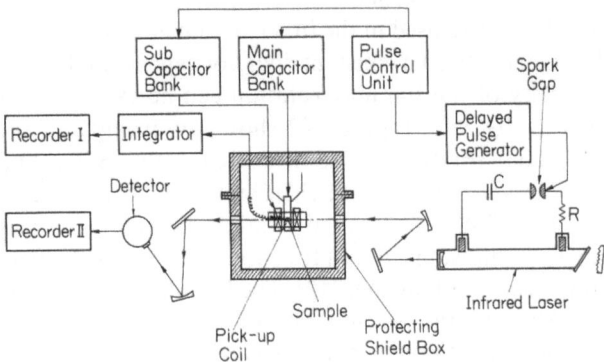

Fig. 6.40. Block diagram of the experimental setup for measuring the cyclotron resonance in megagauss fields [6.209]

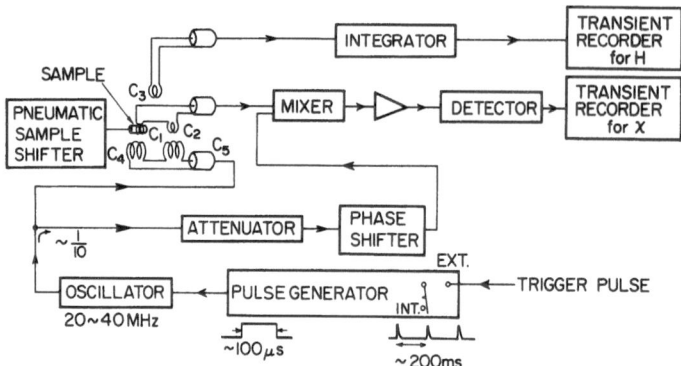

Fig. 6.41. Block diagram of the system to measure high-frequency reversible susceptibility in megagauss fields [6.177]

developed by the ISSP group [6.177]. In this system, a high-power oscillator generates a high frequency (20–40 MHz) current pulse 100 μs in width with 150 W peak power in intervals of 200 ms. The current is supplied to exciting coils C_4 and C_5. Underneath are two search coils C_1 and C_2 which are wound in opposite directions to each other; the sample is contained in C_1. Taking out only the exciting frequency component of the signal from C_1 and C_2, we can measure the reversible susceptibility, avoiding the effect of induced voltage on the search coils by the field. The spin-flip transition was observed by this system in iron garnet crystals [6.177].

6.5.3 Faraday Rotation

When linearly polarized light passes through a substance in the direction of the applied magnetic field, the plane of the polarization rotates. The angle of rotation is given by VlB, where l is the length of the light path in the sample and V is the "Verdet constant" which is a function of the wavelength. In most materials, such as glasses and quartz which were used in the first experiments, the effect is linear up to very high fields, i.e., the Verdet constant does not depend on the field [6.44, 50, 202, 207]. The linearity of the Faraday rotation can be exploited for precise field measurements [6.209, 210]. Figure 6.42 shows an example of the experimental trace of the Faraday rotation in CdS measured by the ISSP group with a He–Ne laser at 6328 Å wavelength. As the field is high and the wavelength is close to the absorption edge of CdS, the observed rotation is very large. It was found that the rotation angle is almost linear as a function of the magnetic field up to the highest field at this wavelength [6.209, 210]. The Faraday rotation was conveniently utilized for calibrating the pick-up coil for field measurement.

The wavelength dispersion of the Faraday rotation can also be measured by using a streak spectrometer, as mentioned in the last section. The Los Alamos group observed the Faraday rotation in CdS, ZnO, etc., using the system shown

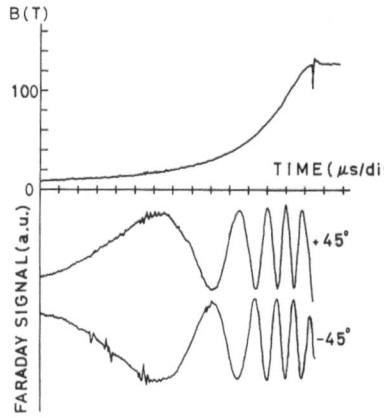

Fig. 6.42. Experimental trace of the Faraday rotation in CdS [6.215]. A He–Ne laser was used as the light source at 6328 Å wavelength. Two signals for the Faraday rotation were measured simultaneously corresponding to the angle of analyzers +45° and −45° relative to that of the polarizer. Sample thickness was 1.05 mm

in Fig. 6.38. Figure 6.43 shows streak spectra of the Faraday rotation in ZnO up to 104 T [6.03]. The interval between each adjacent stripe corresponds to a rotation of 180°. It was found that the Verdet constant decreased in ZnS in very high magnetic fields. However, similar behavior was not observed for ZnO at fields up to 180 T [6.203].

The ISSP group also measured the interband Faraday rotation in CdS, GaSe and GaP near the absorption edge using the streak spectrometer [6.216].

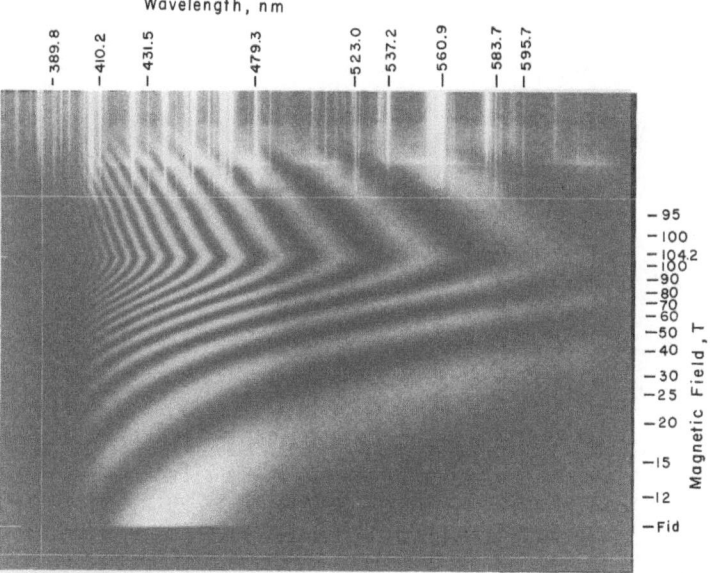

Fig. 6.43. Faraday rotation spectrum for ZnO observed by the system shown in Fig. 6.38 [6.203]. A few selected wavelengths from a bridgewire spectrum are labeled on the horizontal axis

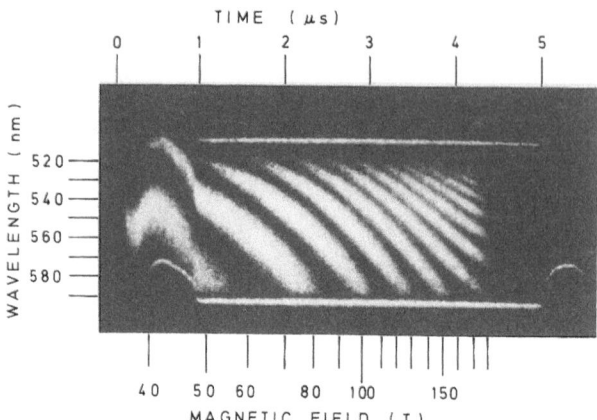

Fig. 6.44. Faraday rotation spectrum for CdS observed by the system shown in Fig. 6.39. $T = 180$ K. The sample was 0.781 mm thick [6.216]

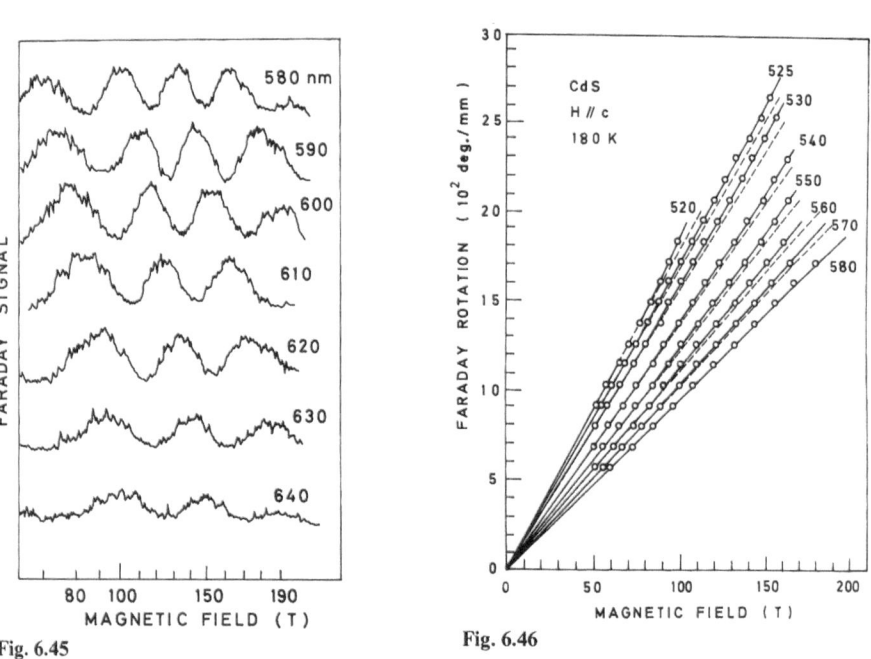

Fig. 6.45

Fig. 6.46

Fig. 6.45. Microphotodensitometer traces of the streak photograph for the Faraday rotation in CdS [6.216]. $T = 298$ K. The wavelength is indicated for each curve in nm

Fig. 6.46. Faraday rotation angle vs magnetic field for CdS. The wavelength is indicated for each curve in nm [6.216]

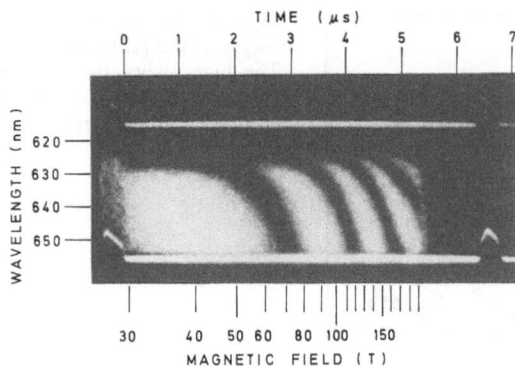

Fig. 6.47. Faraday rotation spectrum in GaSe [6.216]. $T = 300$ K. The sample was 0.570 mm thick

Figure 6.44 shows the streak photograph of Faraday rotation in CdS, in which a large dispersion is observed. The Faraday rotation signal can be plotted as a function of magnetic field via the optical density recorded on the photograph by a microphotodensitometer. An example of the trace of the microphotodensitometer is shown in Fig. 6.45. The graphs are almost sinusoidal, as expected. From these curves we obtain the angle of the Faraday rotation as a function of magnetic field for each wavelength. Figure 6.46 is a plot of the rotation angle versus magnetic field for CdS [6.216]. It was found that the rotation increases almost linearly with increasing magnetic field at the long wavelength far from the absorption edge. At shorter wavelengths in the vicinity of the absorption edge, the rotation deviates from the linear relationship at higher fields, with the slope increasing as a function of the field. As the wavelength becomes shorter, the deviation starts at lower fields. A similar nonlinearity was observed by the Los Alamos group [6.39]. It was found that the rotation at 6328 Å was almost linear up to the highest field with a Verdet constant of $7.2 \deg \cdot \mathrm{T}^{-1} \mathrm{mm}^{-1}$ at room temperature, coinciding with the value measured with a He–Ne laser.

Compared with CdS, GaSe shows quite different behavior in the Faraday rotation. Figure 6.47 shows a streak spectrum of the Faraday rotation in GaSe in the megagauss range [6.216]. The rotation exhibits a rather small wavelength dependence at longer wavelengths, but at wavelengths shorter than a given threshold, the wavelength dependence suddenly becomes large. Figure 6.48 shows the magnetic field dependence of the rotation angle and also its photon energy dependence at 150 T in the inset. It is clearly seen in the inset that the rotation starts to increase abruptly above 2.02 eV with increasing photon energy. The sudden increase of the Faraday rotation (inset in Fig. 6.48) shifts to higher photon energy as the field increases. At photon energies above this position, the field dependence of the rotation angle exhibits anomalous behavior. As shown in Fig. 6.48, at wavelengths shorter than 6125 Å, the rotation versus field curves do not tend to the origin by a linear extrapolation, indicating that these curves should bend downwards at lower fields where the experimental data are missing. At longer wavelengths, on the other hand, the rotation is a linear function of magnetic field.

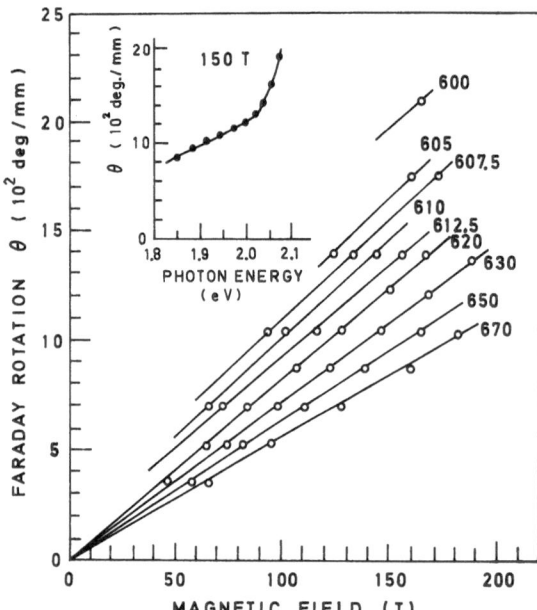

Fig. 6.48. Faraday rotation vs magnetic field for GaSe [6.216]. The inset shows the rotation angle as a a function of photon energy at 150 T. $T = 180$ K

Interband Faraday rotation has been theoretically studied by several authors. *Boswarva* et al. derived the expression for interband Faraday rotation for a direct allowed transition [6.218, 219],

$$
\begin{aligned}
\theta \propto \frac{\gamma H}{\omega} \sum_{n=0}^{\infty} \Big\{ & [\omega_g - \omega + (n + \tfrac{1}{2})\omega_c - \gamma H]^{-1/2} \\
& - [\omega_g - \omega + (n + \tfrac{1}{2})\omega_c + \gamma H]^{-1/2} \\
& - [\omega_g + \omega + (n + \tfrac{1}{2})\omega_c - \gamma H]^{-1/2} \\
& + [\omega_g + \omega + (n + \tfrac{1}{2})\omega_c + \gamma H]^{-1/2} \\
& - \omega [\omega_g + (n + \tfrac{1}{2})\omega_c - \gamma H]^{-3/2} \\
& + \omega [\omega_g + (n + \tfrac{1}{2})\omega_c + \gamma H]^{-3/2} \Big\}.
\end{aligned}
\tag{6.42}
$$

where ω is the frequency of light, $\hbar\omega_g$ is the gap, $\hbar\omega_c$ is the combined cyclotron frequency for electrons and holes and γH is the combined spin splitting of electrons and holes. When the magnetic field is low so that γH or ω_c is much smaller than ω or ω_g, it can readily be verified that (6.42) is reduced to a form proportional to the magnetic field. However, as the magnetic field is increased, higher-order terms of the field in the expansion of the parentheses of (6.42) become important. The nonlinear Faraday rotation observed in CdS can be explained as arising from these higher-order terms. In fact, the first higher-order terms are those proportional to $(\gamma H)^3$ and they tend to bend the rotation vs field curve in the sense of increasing slope. As the field is increased and the photon energy approaches the absorption edge, the higher-order terms are no longer negligible in comparison to other terms.

The nonlinearity observed in GaSe was attributed to the exciton effect [6.216]. In GaSe, this effect is large in the absorption spectra up to high temperatures. The sudden increase of the rotation with increasing photon energy shown in Fig. 6.48 corresponds to the onset of exciton absorption. Since the diamagnetic shift of the ground-state exciton line is fairly large in GaSe, the exciton absorption line shifts to shorter wavelength as the field increases. Therefore, if we measure the Faraday rotation at a wavelength longer than the exciton line, a part of the rotation originating from the exciton effect should decrease, which should lead to sublinear dependence of the rotation on the field.

Nonlinear Faraday rotation was also observed by *Pavlovskii* and co-workers in NiO, EuO and $Er(PO_3)_3$ [6.59, 220, 221], who produced the high magnetic fields by the explosive method. A He−Ne laser beam with a wavelength of 6328 Å was used as a light source. They found in NiO that the rotation reached a maximum at a magnetic field of 90 T and then it decreased, reversing sign at 150 T. The nonlinearity was ascribed to the electronic transition in the divalent nickel ion [6.59]. In EuO, an abrupt increase of the Faraday rotation was observed at about 45 T. *Pavlovskii* et al. suggested that this abrupt increase may be connected with the formation of tremendous ferromagnetic molecules – "ferrons" – around the oxygen vacancy [6.220]. They also investigated the nonlinear Faraday rotation in $Er(PO_3)_3$ glass in fields up to 100 T. A complex field dependence of the rotation was observed: the rotation saturated at 20–40 T, then increased again to a maximum at 80 T and it changed sign at 90 T. An electron paramagnetic resonance was observed at 30 T as an absorption peak for a wavelength of 6328 Å. The sudden increase of the Faraday rotation at 80 T was analyzed in terms of resonant enhancement of the rotation [6.21].

6.5.4 Cyclotron Resonance

Cyclotron resonance is a powerful means to study the energy states of conduction electrons and holes in megagauss fields. First, it is possible to study the energy band structure far away from the band extrema. As the field is increased, the nonparabolic nature of the energy band becomes more prominent. Second, we can investigate the energy band structure of substances which have carriers with low mobility or large effective mass, because $\omega_c \tau$ becomes large enough even for small τ if ω_c becomes large in the megagauss field. In this respect, many new interesting materials including molecular solids, various metals and amorphous materials will become objects of research. Third, if the cyclotron resonance energy becomes equal to or larger than the optical phonon energy, the polaron pinning effect can be investigated. The cyclotron resonance will become a useful probe in studies of possible new effects which may occur in ultrahigh magnetic fields.

Table 6.15. Cyclotron resonance fields, in tesla, for n-InSb and n-GaAs at 10.6 and 5.56 μm radiation [6.158]

Material	InSb	InSb	InSb	InSb	InSb	InSb	GaAs	GaAs	GaAs
Laser	CO_2	CO_2	CO_2	CO_2	CO_2	CO	CO_2	CO_2	CO_2
Transition	0+→1+	0-→1-	1+→2+	1-→2-	2+→3+	0+→1+	0±→1±	1±→2±	2±→3±
Resonance field									
Theoretical[a]	28	38	57	68	87	87	67	80	95
Theoretical[b]	29	39	58	69	88	90	69	81	94
Experimental[c]	27±1	39±3	61±3	78±4		85±1[d]	68±1		94±1
Peak field			100			140		100	
No. of experiments	11	5	5	3		4	2		2

[a] InSb [6.222]: $E_g = 0.18$ eV, $m_0^* = 0.0116\,m_e$, $g_{eff} = -64$; GaAs: $E_g = 1.25$ eV, $m_0^* = 0.0553\,m_e$ (hypothetical).

[b] InSb [6.223]: $E_g = 0.20$ eV, $m_0^* = 0.0130\,m_e$, $g_{eff} = -55$; GaAs: $E_g = 1.42$ eV, $m_0^* = 0.058\,m_e$.

[c] Error limits refer to statistical variations between different experiments and observers only. Add calibration errors for field probe (1%), oscilloscope (1-2%) and integrating network (0.3%).

[d] This is for two experiments with a peak field of 90 T. The mean value for two experiments with peak fields of 120 and 140 T is 80±2 T, and for all four experiments 83±3 T.

The first cyclotron resonance experiment at megagauss fields was performed by *Herlach* and co-workers [6.158, 211, 212]. By using a CO_2 laser and a CO laser, the cyclotron resonance in InSb and GaAs was measured. Table 6.15 lists the cyclotron resonant fields obtained in these experiments [6.158]. A large nonparabolicity in both substances was observed at ultrahigh fields. The ISSP group investigated the cyclotron resonance in a variety of semiconductors such as GaP, Ge, InSb, GaAs, Te, CdS and CdSe [6.213–217]. Figure 6.49 shows typical examples of the experimental traces. In n-type InSb, two absorption peaks were resolved near 9.5 μm wavelength (photon energy 130 meV) at room temperature, corresponding to the transitions of electrons between the Landau levels, $(0^+ \rightarrow 1^+)$ and $(0^- \rightarrow 1^-)$ [6.213]. The apparent cyclotron masses were found to be $0.0290 m_0$ and $0.0415 m_0$ (m_0 is the free electron mass) for the two peaks, respectively. These values are much larger than the band edge effective mass $m_0^* = 0.0139 \, m_0$ which was measured at 10–60 K [6.223]. Such a large nonparabolicity is due to the small energy band gap and small effective mass of electrons in InSb. In fact, it should be noted that the photon energy for the measurement is almost comparable with the energy band gap $\varepsilon_g = 170$ meV. *Lax* et al. [6.222] gave an expression for the energy of the nth Landau level at $K_H = 0$ based upon *Kane's* model [6.224]:

$$
\varepsilon_n^{\pm} = \left(n + \frac{1}{2}\right) \cdot 2\mu_B H \frac{m_0}{m^*} \frac{\varepsilon_g(\varepsilon_g + \Delta)}{3\varepsilon_g + 2\Delta} \times \left(\frac{2}{\varepsilon_n^{\pm} + \varepsilon_g} + \frac{1}{\varepsilon_n^{\pm} + \varepsilon_g + \Delta}\right)
$$
$$
\pm \frac{1}{2} g^* \mu_B H \frac{\varepsilon_g(\varepsilon_g + \Delta)}{\Delta} \times \left(\frac{1}{\varepsilon_n^{\pm} + \varepsilon_g} - \frac{1}{\varepsilon_n^{\pm} + \varepsilon_g + \Delta}\right),
$$

$$(6.43)$$

where ε_g is the energy gap, Δ is the spin-orbit splitting, μ_B is the Bohr magneton, and m_0^* and g^* are the effective mass and g factor at the bottom of the band, respectively. Figure 6.50 shows the plot of the resonant photon energy calculated from (6.43) together with experimental data for InSb and GaAs. Using the known band parameters of InSb at room temperature, values of m_0^* and g^* were determined by fitting the calculation with the experimental data. The best fit values were $m_0^* = 0.0127 \, m_0$ and $g^* = -70$. An interesting point is that with these values *Roth's* relation [6.225] between m_0^* and g^* does not hold, indicating the necessity of a more rigorous formulation including the effect of remote bands [6.226] in the megagauss fields.

A large nonparabolicity was found in the conduction band of Te, but in this case it was very anisotropic [6.227]. Many investigations have been made on the valence band of Te and its unusual structure has been revealed [6.228]. As for the conduction band, however, much less information has been obtained, since Te is always p-type at the extrinsic region at low temperature. At temperatures higher than about 200 K, Te containing a usual impurity concentration becomes n-type as the intrinsic carriers are excited across the band gap and the mobility of electrons is higher than that of holes. Therefore, the cyclotron resonance in the conduction band can be observed at room temperature if $\omega_c \tau$ becomes large

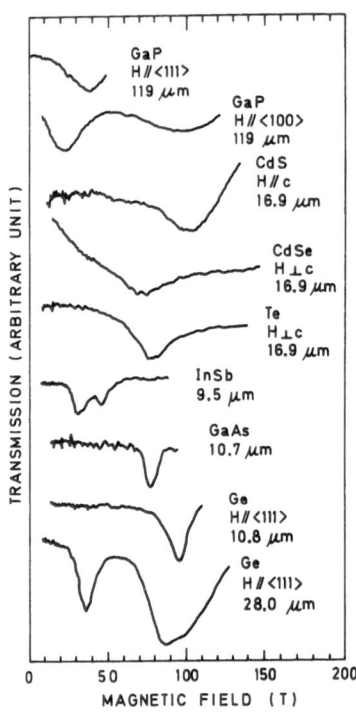

Fig. 6.49. Experimental traces of the cyclotron resonance absorption spectra for various crystals [6.42]. The wavelength of the incident light is indicated for each curve

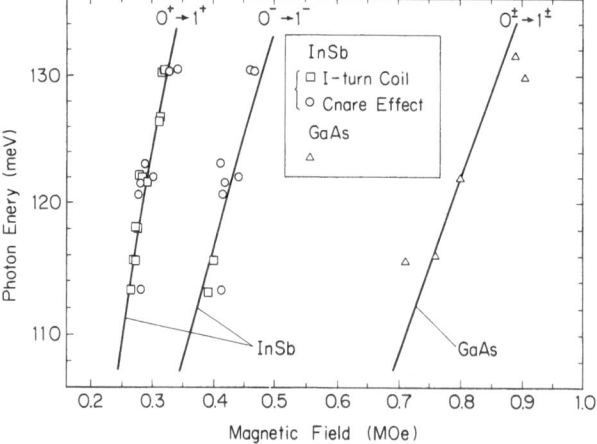

Fig. 6.50. Resonant photon energy as a function of magnetic field for n-InSb and n-GaAs [6.213]. (——) were calculated from (6.43). The following band parameters were used: InSb; $\varepsilon = 170$ meV, $\Delta = 810$ meV, $m_0^* = 0.0127\, m_0$, $g^* = -70$. GaAs; $\varepsilon_g = 1.43$ eV, $\Delta = 340$ meV, $m_0^* = 0.065\, m_0$

enough at high magnetic fields. The observed peak shown in Fig. 6.49 is an electron peak and the cyclotron masses of electrons were determined from these measurements at different wavelengths. Reflecting a small energy gap of 320 meV, the nonparabolicity was found to be very large for the motion of electrons perpendicular to the c axis. However, for the parallel direction, it was found to be fairly small. This remarkably anisotropic nonparabolicity was discussed in terms of the anisotropic coupling between the conduction and the valence bands [6.227]. The nonparabolicity of the conduction band was also observed in GaAs and Ge [6.213–215], but it is relatively small.

A rather unusual type of nonparabolicity was observed in the conduction band of GaP. With the magnetic field parallel to the $\langle 100 \rangle$ direction, two resonance peaks were observed at far infrared wavelengths of 337 µm and 119 µm in megagauss fields [6.135]. The apparent longitudinal and transverse masses were determined for the first time directly by the cyclotron resonance based on the ellipsoidal band model. The apparent anisotropy factor of the effective masses, $K = m_l^*/m_t^*$, was found to be much larger at $\lambda = 337$ µm ($K = 28 \pm 7$) than at $\lambda = 119$ µm ($K = 19 \pm 2$). The results were explained by the "camel's back" structure model of the conduction band in GaP. The conduction band minima in GaP are located near the X point of the Brillouin zone. Because the double degeneracy of the two bands is lifted at the X point due to the lack of inversion symmetry, the conduction band has a peculiar structure with a double minimum in the $\langle 100 \rangle$ direction, similar to that of the valence band of Te. The Landau level energies in such a band depend on the magnetic field in a complicated way. Moreover, the electronic transitions between the Landau levels do not necessarily satisfy the condition $\Delta N = 1$, so that many harmonic resonances are not clearly resolved, as the mobility is low. By comparing the experimental results with the calculated resonance positions and the intensity of the absorption peaks, the observed peaks were assigned and the band parameters were determined [6.136] within the framework of the "camel's back" model.

In the cyclotron resonance experiment, significant information can be obtained not only from the peak positions, but also from the linewidth of the resonance. In particular, it is interesting to study the linewidth in the quantum condition, i.e., $\omega_c \tau \gg kT$, which can be satisfied at high fields even if the temperature is not so low. The relaxation time $\langle \tau \rangle$ of the carrier scattering was investigated for Ge from the measurement of the cyclotron resonance linewidth near room temperature [6.214]. It was found that $\langle \tau \rangle$ decreased in proportion to $\omega_c^{-1/2}$, but $\langle \tau \rangle$ did not show any significant temperature dependence near room temperature. The relaxation time $\langle \tau \rangle$ is predominantly determined by the acoustic phonon scattering near room temperature. According to Meyer's theory [6.229] for the acoustic phonon scattering, $\langle \tau \rangle$ should be proportional to $\omega^{1/2}T^{-1}$ in the quantum condition. The experimentally observed ω_c dependence agrees well with the theory but temperature dependence does not, indicating that further theoretical work is required. The ISSP group also found that the peak height of the absorption coefficient α_m decreased as the magnetic field was

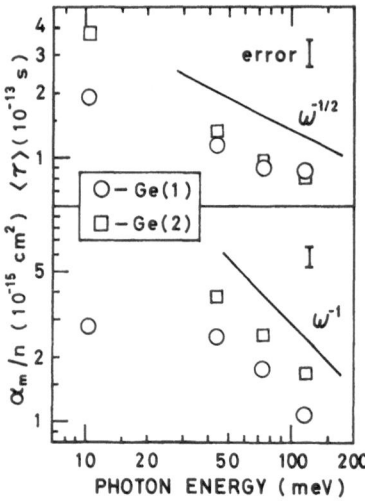

Fig. 6.51. (*Top*) dependence of the relaxation time $\langle \tau \rangle$ on the resonant photon energy for two samples of n-Ge at 300 K. (*Bottom*) dependence of α_m/n on the resonant photon energy [6.214]. α_m is the peak height of the absorption coefficient and n is the density of conduction electrons. (——) indicate merely the slope of the ω dependence

increased. The dependence of α_m on the resonant photon energy was almost ω^{-1}, as is shown in Fig. 6.51.

In CdS and CdSe, large polaron effects were observed [6.216, 217]. These substances are relatively polar, and have large electron-phonon coupling constants among semiconductors, i.e., $\alpha = 0.6$ for CdS and $\alpha = 0.45$ for CdSe. When we sweep the cyclotron resonant photon energy across the longitudinal optical (LO) phonon energy $\hbar\omega_0$, a large polaron effect should be observed in the vicinity of $\hbar\omega = \hbar\omega_0$. As is shown in Fig. 6.51, when $\hbar\omega_c = \hbar\omega_0$ is satisfied, the energy of an electron of the Landau quantum number $N = 0$ plus one phonon becomes equal to that of an electron at $N = 1$ with no LO phonon. Near the crossover point, the admixture of the two states due to the electron-phonon interaction is enhanced, and the levels are split into two [6.230]. This phenomenon is known as the polaron pinning effect. It was first discovered in n-InSb for which $\alpha \simeq 0.02$, and a number of studies have been made on InSb [6.230–233]. The pinning effect should be much larger for CdS and CdSe, because the effect is almost proportional to α and $(\hbar\omega_0)^2$ [6.232], and both of these factors of CdS and CdSe are larger than those of InSb. On the other hand, we need high magnetic fields to attain the condition: i.e., the condition holds at about 55 T for CdS and at about 27 T for CdSe, which are much higher fields than that for InSb, 3.3 T. Moreover, the piezoelectric coefficients are large in CdS and CdSe, and the piezoelectric polaron effect was observed in the cyclotron resonance in CdS at lower fields [6.234–237]. Therefore, the piezoelectric polaron effect may have an influence on the high field cyclotron resonance.

Tables 6.16, 17 list the apparent cyclotron mass of electrons observed by the ISSP group in the experiments on CdS and CdSe, respectively, against the resonant photon energy [6.216, 217]. At $\hbar\omega = 44.3$ meV, which is just above the LO phonon energy $\hbar\omega_0 = 37.8$ meV, the cyclotron mass is much smaller than the

Table 6.16. Variation of the apparent effective mass of electrons in CdS with photon energy ($\hbar\omega_0 = 37.8$ meV)

$\hbar\omega$ [meV]	$\hbar\omega/\hbar\omega_0$	T [K]	$m_c^*/m_0(H \parallel c)$	$m_c^*/m_0(H \perp c)$	
0.206	0.00545	1.7		0.17	*Sawamoto* (1963) [6.238]
0.289	0.00765	1.3	0.171	0.162	*Baer, Dexter* (1964) [6.234]
3.68	0.0974	3	0.16		
6.36	0.168	3	0.17		*Button* et al. (1970) [6.235]
4.96	0.131	19–38	0.174		*Nagasaka* (1977) [6.237]
10.5	0.278	35	0.188		*Nagasaka* et al. (1973) [6.236]
44.3	1.172	100~300	0.131	0.107	*Miura* et al. (1978) [6.216, 217]
73.4	1.941	100~300	0.164	0.158	

Table 6.17. Variation of the apparent effective mass of electrons in CdSe with photon energy ($\hbar\omega_0 = 26.3$ meV)

$\hbar\omega$ [meV]	$\hbar\omega/\hbar\omega_0$	T [K]	$m_c^*/m_0(H \parallel c)$	$m_c^*/m_0(H \perp c)$	
3.68	0.140		0.120		*Button, Lax* (1970) [6.239]
10.5	0.399	21	0.127		
44.3	1.68	150~300	0.115	0.107	*Miura* et al. (1978) [6.216, 217]
73.4	2.79	150~300	0.118	0.114	

masses at lower photon energies. Moreover, the cyclotron mass at 73.4 meV is smaller. A similar feature is found in CdSe. This dependence of the cyclotron mass on the photon energy is interpreted in terms of the polaron pinning effect.

The polaron pinning effect was theoretically considered by *Larsen* and colleagues, based on the Wigner-Brillouin perturbation theory and the variational treatment [6.232, 233, 240, 241]. According to the Wigner-Brillouin perturbation theory

$$\varepsilon_0 = \frac{1}{2}\hbar\omega_c(0) - \frac{(\hbar\omega_0)^2}{2\pi^2}\int dk \sum_n \frac{|H_{n0}(k)|^2}{\left(n+\frac{1}{2}\right)\hbar\omega_c(0) + \hbar\omega_0 + \frac{\hbar^2 k_z^2}{2m_z} - \varepsilon_0}$$

$$\varepsilon_1 = \frac{3}{2}\hbar\omega_c(0) - \frac{(\hbar\omega_0)^2}{2\pi^2} P\int dk \sum_n \frac{|H_{n1}(k)|^2}{\varepsilon_0 + n\hbar\omega_c(0) + \hbar\omega_0 + \frac{\hbar^2 k_z^2}{2m_z} - \varepsilon_1}$$

(6.44)

are the perturbed energies of the $N=0$ and $N=1$ Landau levels, where $\hbar\omega_c(0) = \hbar e B/(m_b^*)$ and m_b^* is the effective mass of bare electrons at the band edge (Fig. 6.52). The matrix element of the perturbing Hamiltonian for the

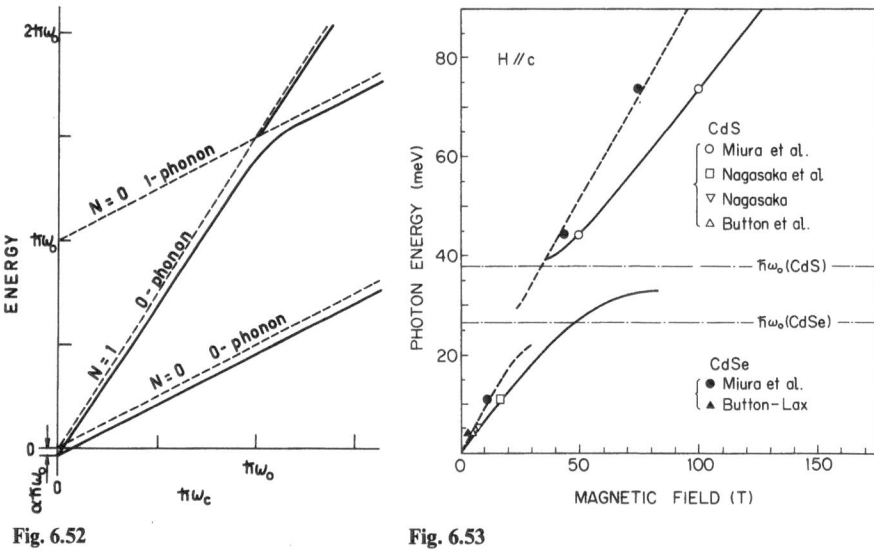

Fig. 6.52

Fig. 6.53

Fig. 6.52. Schematic diagram of Landau levels in the presence of the polaron pinning effect

Fig. 6.53. Resonant photon energy vs magnetic field for the polaron cyclotron resonance in n-CdS and n-CdSe for $H \parallel c$ [6.217]. (——, ———) were calculated from (6.44) for CdS and CdSe, respectively, taking into account the band nonparabolicity

electron-phonon interaction is $H_{nN}(k)$. The resonant photon energy in the cyclotron resonance can be calculated as the difference between ε_1 and ε_0, if the bare band edge mass m_b^* is given. Experimental points are compared in Fig. 6.53 with calculated results assuming $m_b^* = 0.165\, m_0$ for CdS and $m_b^* = 0.116\, m_0$ for CdSe, which were the best fit values of m_b^* for the data of the ultrahigh field experiment. The effect of nonparabolicity of the conduction band was taken into account in the calculation by assuming a two-band model. The values of m_b^* determined from the ultrahigh field experiment are larger than those obtained from the experimental results at lower fields. The experimental data at lower fields predict $m_b^* = 0.155\, m_0$ for CdS and $m_b^* = 0.103\, m_0$ for CdSe [6.217]. The differences between the values of m_b^* may be due to the piezoelectric polaron effect which was not taken into account in the calculation. According to the theory of the piezoelectric polaron effect [6.240], it should make the apparent cyclotron mass smaller, and the effect should be more pronounced at lower fields. Thus the general feature of the difference in the cyclotron mass can be qualitatively explained in terms of the piezoelectric polaron effect, as far as the magnetic field dependence is concerned. The temperature dependence, however, remains an open question to be solved by future experimental and theoretical work.

6.5.5 Excitons and Impurity States

Excitonic and shallow impurity states in semiconductors under ultrahigh magnetic fields are important in conjunction with a fundamental problem regarding the atomic energy levels of hydrogen in high magnetic fields. The binding energy of excitons or impurity states is given in terms of the hydrogenic model by $R_y^* = m^* e^4 / 2\hbar^2 \varepsilon^2$, where ε is the dielectric constant and m^* is the effective reduced mass of excitons or, for impurity states, the effective mass of electrons. To express the intensity of the magnetic field acting on these states, a dimensionless parameter $\gamma = \hbar\omega_c / 2R_y^*$ is used, where $\hbar\omega_c$ is the quantized energy of the cyclotron motion. The value of γ indicates the relative importance of the effect to that of the coulomb interaction. For a hydrogen atom, γ reaches unity at an extremely high magnetic field of 2.4×10^5 T. For excitons or impurity states in semiconductors, however, the condition $\gamma > 1$ can be realized at much lower fields, because γ is proportional to ε^2 / m^{*2}, and this factor is much larger than that for a hydrogen atom, ε_0^2 / m^2. For instance, γ becomes 1 at about 0.2 T for the donor states in InSb, and at 44 T for the excitons in GaSe. Therefore, by means of ultrahigh magnetic fields we can sweep the parameter γ across the vicinity of 1 in many substances.

For $\gamma \ll 1$ at low magnetic fields, the energy of exciton states (hereafter we describe only the exciton states, but the same argument holds for impurity states) is expressed by hydrogen-like states such as $1s$, $2s$, ..., i.e., quantum numbers (n, l, m) which are perturbed by the magnetic fields. For $\gamma \gg 1$ in the high magnetic field limit, on the other hand, the exciton states are those associated with each Landau level. The states are usually labeled by quantum numbers $(N, M, v^\pm)$, where N is the Landau quantum number, M is the quantum number of L_z and $v^\pm$ specifies the exciton substate associated with the Landau level whose quantum numbers are (N, M) [6.242,243]. An interesting question arises as to how each hydrogen-like state (n, m, l) for $\gamma \ll 1$ is connected to the state $(N, M, v^\pm)$ for $\gamma \gg 1$, when γ is varied. There have been many theoretical studies of this problem [6.242,243]. *Shinada* et al. gave a connection scheme between the (n, m, l) and $(N, M, v^\pm)$ states, postulating that the total number of the nodal surfaces of any hydrogen eigenfunction in the magnetic field should be conserved [6.242]. According to their theory, when any two energy states with the same total number of nodal surfaces approach each other, mixing of the states causes a repulsion of the two levels without crossing. For example, the states $2s$ and $3d_0$ mix with each other as the magnetic field is increased, and they tend to (001^+) and (100^+) states without crossing in the high field limit. *Lee* et al., on the other hand, insisted that crossing of the two $m = 0$ levels should not occur, and thus all the $m = 0$ hydrogen-like states at zero field in the high field limit should lie below the lowest free electron Landau level [6.243]. Quantitative calculations of the quantum levels of excitons were made by many authors. *Yafet* et al. obtained the energy of the ground state by a variational method for $\gamma \gg 1$ [6.244]. Using a similar method, *Wallis* and *Bowlden* calculated several excited states as well [6.245]. Calculations including the range in the

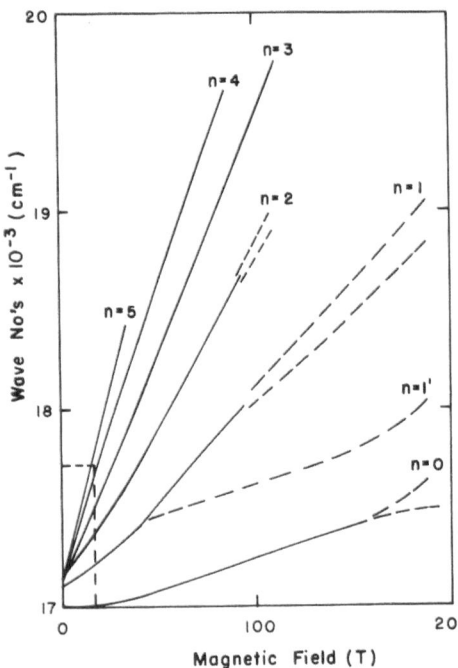

Fig. 6.54. Magnetic field dependence of the photon energy of the exciton absorption lines in GaSe at 7 K [6.62]. (---) indicate weak or poorly resolved structure

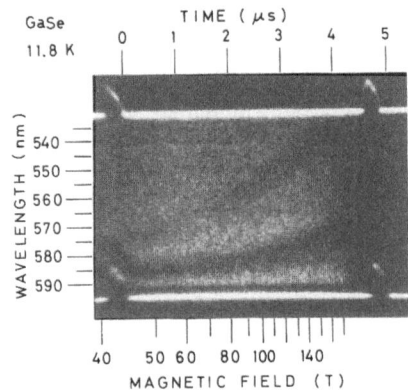

Fig. 6.54

Fig. 6.55

Fig. 6.55. Streak photograph of the magnetooptical spectra of excitons in GaSe [6.248]. $H \parallel c$. $T = 11.8$ K

vicinity of $\gamma = 1$ were made by *Cabib* et al., *Lee* et al., and *Baldereschi* and *Bassani* [6.243, 246, 247].

The magnetooptical spectra of excitons were first observed for GaSe by the Los Alamos group [6.39, 62, 205]. They observed the absorption lines of the exciton up to the $N = 5$ line by a streak spectrometer. The photon energy of the observed absorption lines is plotted in Fig. 6.54 as a function of magnetic field. Similar spectra for GaSe were also observed by the ISSP group, as shown in Fig. 6.55 [6.216, 217, 248]. The crystal structure of GaSe is layer-type. In this respect, it aroused considerable attention that an exciton with 2-dimensional character might exist in GaSe [6.249]. However, a recent experiment has revealed that the exciton in GaSe is more 3-dimensional, because of the moderate anisotropy of the band structure [6.250]. It so happens that GaSe is one of the most suitable substances for studying the magnetic field dependence of excitons, because the exciton in GaSe is easy to observe and also because it does not involve a complexity arising from the degenerate valence bands as in zinc-blende type crystals. *Mooser* and *Schlüter* studied the magnetic field dependence of the exciton lines in magnetic fields up to 9 T [6.250]. The ISSP group [6.248] found that the shift of the ground-state line agrees well with the theories by *Yafet* et al.

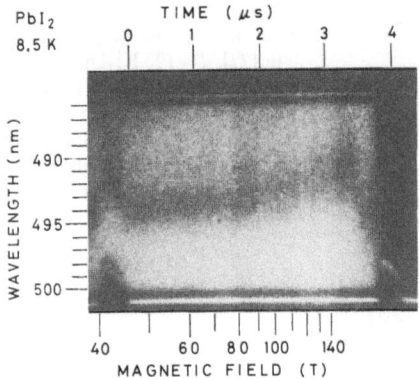

PbI$_2$
8.5 K

TIME (μs)

WAVELENGTH (nm)

490 —
495 —
500 —

40   60   80 100   140
MAGNETIC FIELD (T)

Fig. 6.56. Streak photograph of the magneto-optical spectra of excitons in PbI$_2$ [6.248]. The sample is an evaporated film 1400 Å thick. $T = 8.5$ K

and *Cabib* et al. [6.244, 246] in the ultrahigh field range. However, it turned out that the line next to ground state does not agree with the theoretical $2s$ line, but it looks like a line which tends to (100^+) in the high field limit. More theoretical and experimental work will be required to solve the problem of the connection of the quantum states mentioned above.

The ISSP group observed exciton absorption in PbI$_2$ under ultrahigh magnetic fields [6.216, 217, 248]. As for the exciton in PbI$_2$, there has been considerable controversy concerning the binding energy of the ground states [6.251]. One of the difficulties in investigating the exciton in PbI$_2$ stems from the "insensitivity" of its absorption lines to magnetic field. *Skolnick* et al. reported that they observed no diamagnetic shift of the exciton absorption lines in PbI$_2$ up to $n = 3$ [6.252]. The ISSP group measured the magnetooptical spectra in PbI$_2$ and observed a diamagnetic shift of 5.4 meV at 100 T for the ground state, as shown in Fig. 6.56. From this value of the diamagnetic shift, they evaluated the reduced mass of excitons in PbI$_2$ as $\mu^* = 0.21\ m_0$. Moreover, a sudden increase of the slope of the diamagnetic shift was observed at 100 T. The origin of this sudden increase is not presently clear [6.248]. The diamagnetic shift of the ground state of the A exciton was also observed for CdS. Reflecting a large electron-phonon interaction, a considerably large quenching of the diamagnetic shift was observed [6.248].

6.5.6 Zeeman Splitting

The Zeeman effect of atomic energy levels or ionic energy levels in crystals can be observed in absorption lines due to the optical transition between these levels in the visible range. The Zeeman effect in ultrahigh magnetic fields is of interest in many ways. First, it may be a good means for measuring the field intensity. If the splitting is linear with respect to the fields, it will be convenient for confirming the reliability of measurement by pick-up probes. Second, we can expect that the various nonlinearities of Zeeman splitting may be observed in ultrahigh

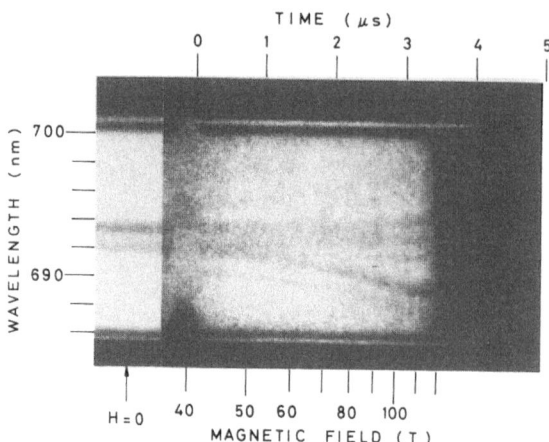

Fig. 6.57. Zeeman splitting of the R lines in ruby for $H \| c$ [6.177]. $T = 190$ K. The two absorption lines R_1 and R_2 which are separated by 1.4 nm at zero applied field are split into more than seven lines at high fields

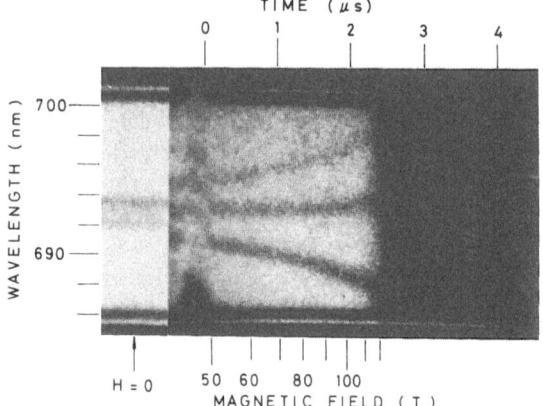

Fig. 6.58. Zeeman splitting of the R lines in ruby for $H \perp c$ [6.42]. $T = 150$ K

magnetic fields, and it will give us new information on the spin Hamiltonian. The measurement of the Zeeman effect in megagauss fields was first carried out by the Los Alamos group on several atomic levels in In, Na, Pb, Au and Ag [6.199], who measured the spectra using a sweeping image spectrographic camera described in Sect. 6.5.2. As the spectral source, they employed an exploding wire light source of Au or Ag. Spectral lines were obtained from this type of light source not only for Au and Ag, but also for elements contained in solder components (In, Pb and Sn) and an appropriate salt placed on a portion of the wire as a contaminant. The Zeeman splitting of these atomic spectral lines was used to measure magnetic fields up to 500 T.

The ISSP group measured the Zeeman splitting of the *R* lines in ruby in ultrahigh magnetic fields [6.42, 177]. In ruby there are many absorption lines due to the electronic transitions between $3d$ crystalline field levels of Cr^{3+} impurities

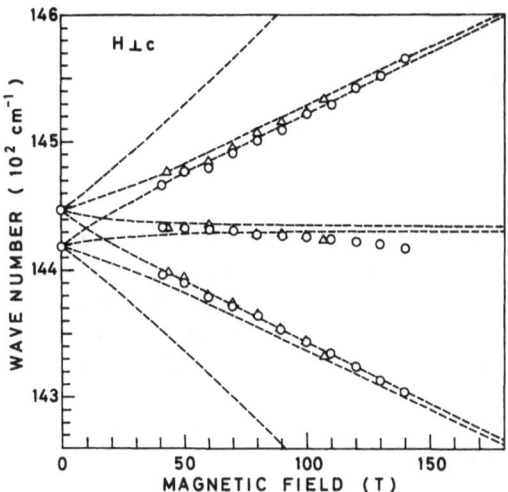

Fig. 6.59. Plot of the Zeeman splitting of the R lines in ruby for $H \perp c$. (– – –) are theoretical curves taking into account the Paschen-Back effect. (O, △) represent data for two different samples. $T = 150$–160 K

[6.253]. The two R lines R_1 and R_2 correspond to the transitions $^4A_2 \rightarrow \bar{E}$ and $^4A_2 \rightarrow 2\bar{A}$, respectively, and they appear at 693.4 nm and 692.0 nm [6.254]. The ground state and the excited states for each of the R lines are fourfold and twofold degenerate levels, respectively, and are split in magnetic fields.

Figures 6.57, 58 show the streak spectra of the R lines in ruby for the magnetic field directions parallel and perpendicular to the c axis. When the magnetic field was applied parallel to the c axis, seven lines were resolved in total, some of which had not been resolved in earlier measurements at lower fields. Moreover, it was found that the Zeeman splitting showed a nonlinear dependence on the magnetic field in the high field region, while the theory predicted a linear dependence in the range of fields used in the experiments.

When the magnetic field is perpendicular to the c axis, the lines are theoretically expected to show a nonlinearity due to the Paschen-Back effect [6.255]. Figure 6.59 shows the theoretically calculated Zeeman splitting of the R lines in the Paschen-Back region [6.255]. The splittings of the experimentally observed three strong lines in Fig. 6.57 agree fairly well with the theoretical prediction, except for the small deviation of the central line from the theoretically expected horizontal line at high fields.

The ISSP group measured the electron spin resonance in ruby at the wavelengths 337 μm and 119 μm in the far infrared [6.256]. At 337 μm, a line split into three absorption peaks was observed in the range of nondestructive pulsed fields. This corresponds to the electronic transitions between the four split levels of the ground state of Cr^{3+} ions in ruby. At the 119 μm wavelength a single peak was observed at 91 T. Splitting was not observed because it is too small in comparison to this high magnetic field. The measured resonance fields are in good agreement with the theoretically predicted positions assuming a constant g

factor. It can be concluded that the *g* factor remains constant up into the megagauss range. As the resonance peaks are quite sharp, this provides a good means for calibrating the field intensity.

6.5.7 Spin-Flip Transition

In ordered magnetic substances such as ferromagnetic, antiferromagnetic or ferrimagnetic substances, the spin direction at each lattice site is regularly oriented by the exchange interaction. In Weiss's molecular field theory, the strength of the exchange interaction is represented by the internal molecular field. When we apply an external magnetic field comparable with the molecular field, the magnetic order in these substances should be greatly altered. Particularly in antiferromagnetic and ferrimagnetic materials, the spin-flip transition takes place at certain critical fields.

Let us first consider antiferromagnetic substances. The magnetic moments M^+ and M^- of two sublattice sites are oriented in the opposite directions at zero field as shown in Fig. 6.60 I. The absolute values of M^+ and M^- are identical so that there is no net magnetization for $H_0 = 0$. The direction of the moments with respect to the crystalline axes is determined by the anisotropy field H_A. When we apply the external field H_0 along the anisotropy axis, the spin configuration remains the same as in Fig. 6.60 I, as long as H_0 is low enough. In this region the crystal shows only a small susceptibility. However, the moment M^- is energetically unstable against the interaction with H_0 as H_0 is increased. Consequently, when H_0 exceeds some critical value H_{c1}, the spin configuration suddenly jumps to that shown in Fig. 6.60 II, i.e., the direction of M^+ and M^- becomes perpendicular to H_0. The angle θ between M^+ and M^- is first π at the transition field, but it decreases as H_0 is increased. Here the susceptibility of the crystal is larger than that for I. This configuration is called the spin-flop phase. When H_0 becomes higher than the second critical field H_{c2}, θ becomes 0, and both M^+ and M^- become parallel to H_0, Fig. 6.60 III. The phase diagram of the spin configuration of an antiferromagnet in the field-temperature plane is shown

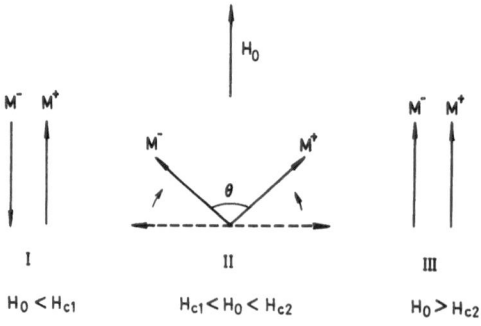

Fig. 6.60. Spin-flop transition in antiferromagnetic substances. M^+ and M^- denote the two sublattice magnetizations. The external field H_0 is assumed to be applied along the easy axis of anisotropy. $I (H < H_{c1})$: antiferromagnetic phase. $II (H_{c1} < H < H_{c2})$: spinflop phase. $III (H > H_{c2})$: paramagnetic phase

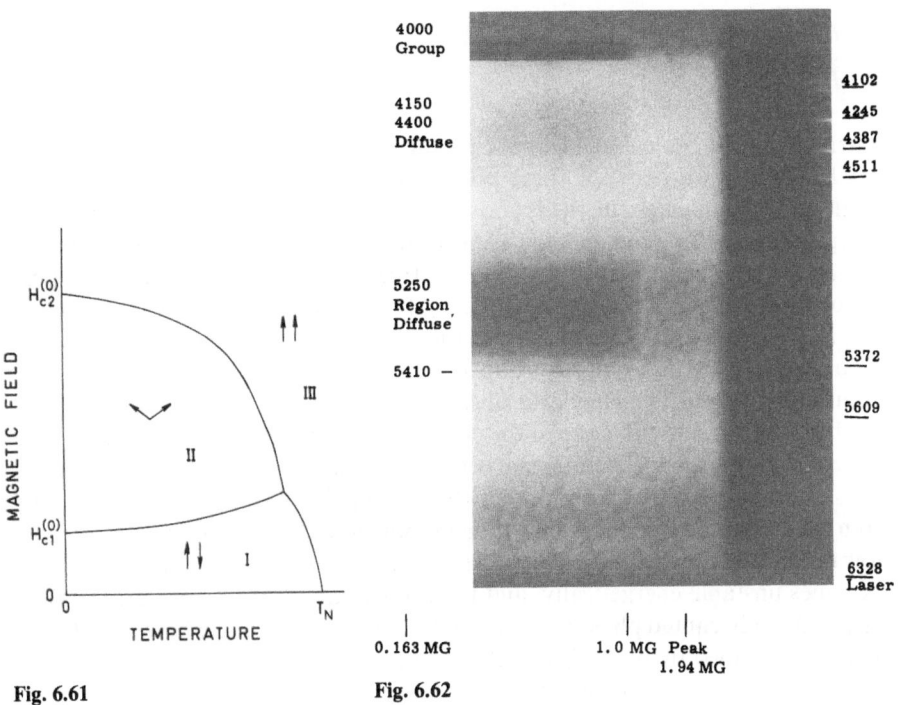

Fig. 6.61

Fig. 6.62

Fig. 6.61. Magnetic phase diagram for an antiferromagnetic substance. Regions *I*, *II*, *III* correspond to the three phases shown in Fig. 6.60

Fig. 6.62. Streak photograph of the absorption lines in MnF_2 at 7 K [6.258]. Much of the structure disappears at about 100 T

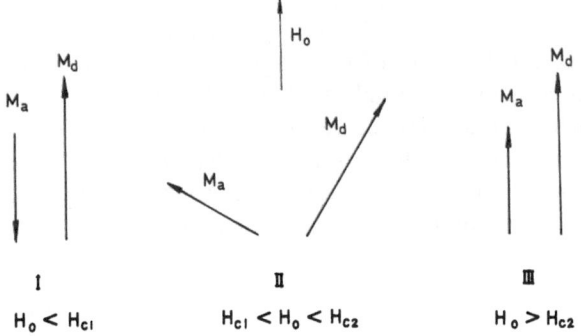

Fig. 6.63. Transition of the spin configuration in a two-sublattice ferrimagnet [6.260]. M_a and M_d ($|M_d| > |M_a|$) represent the sublattice moments. $I (H_0 < H_{c1})$: ferrimagnetic phase. $II (H_{c1} < H_0 < H_{c2})$: spin cant phase. $III (H_0 > H_{c2})$: paramagnetic phase

in Fig. 6.61 [6.57]. The critical fields $H_{c1}^{(0)}$ and $H_{c2}^{(0)}$ at $T=0\,\mathrm{K}$ are given by

$$H_{c1}^{(0)} = (2H_A \cdot H_E)^{1/2}$$
$$H_{c2}^{(0)} = 2H_E \tag{6.45}$$

where H_E is the molecular field due to the exchange interaction.

The transition between these phases was observed in MnF_2 by the Los Alamos group in magnetic fields above 100 T [6.258]. In the visible range MnF_2 is optically transparent and the transition can be conveniently observed by optical experiments. Figure 6.62 shows streak spectra of the magnetooptical absorption in MnF_2 obtained by the Los Alamos group [6.258]. In MnF_2, many absorption lines appear, arising from the optical transition between crystalline field levels. The Los Alamos group found that some of the absorption lines vanish at $H_{c2} = 101$ T. They also observed shifts of some absorption lines.

In ferrimagnetic substances, the situation is a little different because of the difference in the magnetic moments between sublattices [6.259]. In ferrimagnetic substances with two sublattice moments M_a, M_d, ($|M_d| > |M_a|$), for instance, the spin configuration is such that the larger moment M_d is in the direction of H_0 for low fields, as is shown in Fig. 6.63 I. As H_0 is increased, the opposite moment M_a becomes unstable energetically, and the transition occurs at some critical field H_{c1} to the spin canted phase as shown in Fig. 6.63 II. The angle between M_a and M_d is determined by both H_0 and the molecular field, and it decreases as H_0 is increased. At a very high magnetic field above the second critical field H_{c2}, both the sublattice moments become parallel to H_0, as shown in Fig. 6.63 III. The critical fields H_{c1} and H_{c2} at $T=0$ K are given by

$$H_{c1} = \lambda(|M_d| - |M_a|)$$
$$H_{c2} = \lambda(|M_d| + |M_a|), \tag{6.46}$$

where λ is the molecular field coefficient [6.259, 260].

The magnetic phase diagram for YIG ($Y_3Fe_5O_{12}$) is shown in Fig. 6.64 [6.260]. If we take into account the effect of the magnetic anisotropy of the crystal, the phase diagram becomes more complicated [6.261]. In YIG, however, the anisotropy field is much lower than the exchange field and almost negligible. Iron garnet crystals are transparent in the infrared, so that information on magnetization can be obtained by measuring the Faraday rotation, since it is usually proportional to a linear combination of each sublattice moment:

$$\theta_F = \sum_i A_i |M_i|, \tag{6.47}$$

where θ_F is the Faraday rotation angle, M_i is the ith sublattice moment and A_i are coefficients [6.262]. The Faraday rotation in iron garnet crystals was measured in ultrahigh fields both by *Guillot* [6.208] and by the ISSP group [6.263, 264].

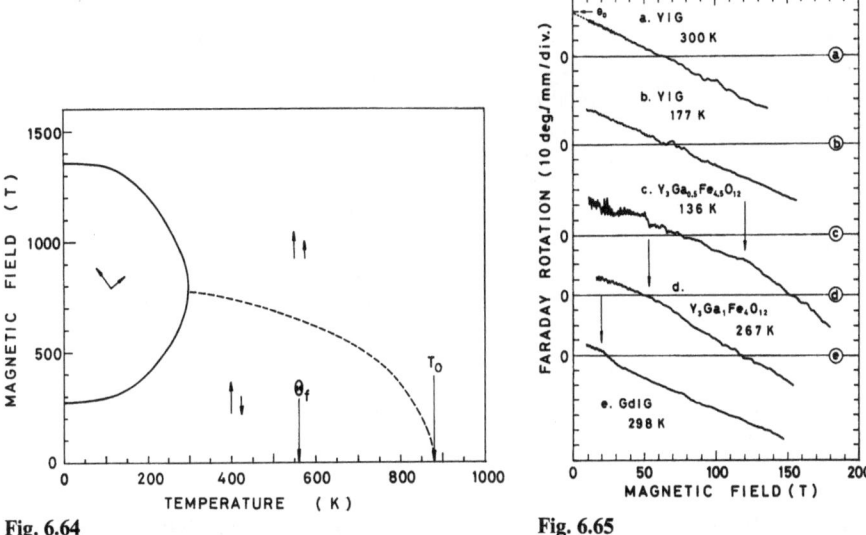

Fig. 6.64

Fig. 6.65

Fig. 6.64. Magnetic phase diagram for YIG [6.260]. The three phases are indicated by arrows as shown in Fig. 6.63

Fig. 6.65. Faraday rotation angle vs magnetic field for YIG ($Y_3Fe_5O_{12}$), $Y_3Ga_xFe_{5-x}O_{12}$ and GdIG [6.263]. The zero of the ordinate is shown for each curve (for details, see text). The transition from the ferrimagnetic phase to the spin cant phase is observed at the fields shown by the arrows

Figure 6.65 shows the Faraday rotation angle against the magnetic field measured by the ISSP group for various iron garnet crystals. In YIG, the critical fields for the spin-flip transition are extremely high and no such transition took place in the available field range. Accordingly, the Faraday rotation is linear as a function of field for various temperatures as shown in Fig. 6.65a,b. If we substitute Ga for some Fe ions, the exchange interaction is effectively weakened and the transition field is greatly reduced [6.258]. In Ga substituted YIG, the transition from the ferrimagnetic phase to the spin canted phase was observed as a kink in the plot of the Faraday rotation versus magnetic field, as shown in Fig. 6.65c,d. In rare earth iron garnets, which have compensation points T_c, the critical field decreases as the temperature approaches T_c [6.259, 260, 265, 266]. Therefore, in the vicinity of T_c, the spin-flip transition can be observed at moderate magnetic fields. For example, Fig. 6.65e shows the transition in GdIG which has the compensation point at 285 K.

The ISSP group found that the experimentally observed critical fields do not seem to agree with those expected from the molecular field theory [6.260, 263]. The large magnetic field coefficient of the Faraday rotation in YIG as shown in Fig. 6.65a,b is also an interesting problem. The ISSP group also measured the

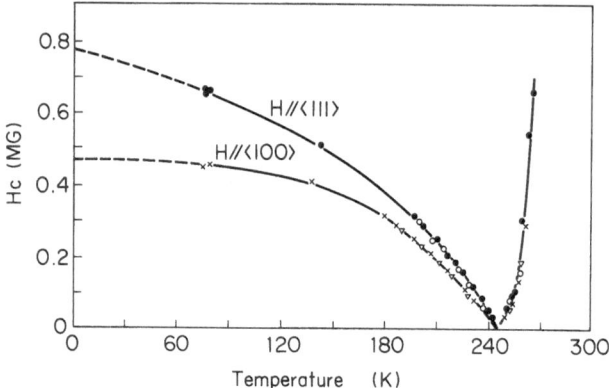

Fig. 6.66. Magnetic phase diagram for TbIG, determined by the Faraday rotation and the magnetization [6.267]

transition fields in TbIG and constructed the magnetic phase diagram for TbIG as shown in Fig. 6.66 [6.267]. Reflecting the large anisotropy of Tb ion, a large anisotropy is seen in the phase diagram.

6.5.8 Other Topics

With all these experiments, the surface has still barely been scratched. There is much room for improvement of the techniques, for an extension to higher fields and lower temperatures, and for exciting discoveries with particular samples. With sufficiently high fields, it can be expected that drastic changes in the electronic properties of solids will be observed.

One such example is a semimetal-semiconductor transition in Bi. Bismuth is a semimetal which has a small energy overlap (38.5 meV) between the electron band and the hole band. When a very high magnetic field is applied to it, the energy overlap may disappear, and Bi may change into a semiconductor [6.99, 134, 268, 269]. At the transition point from a semimetal to semiconductor, Bi takes a zero-gap state, which is an interesting situation in relation to the "excitonic phase" [6.270, 271], or the "gas-liquid type phase transition" [6.272]. The ISSP group performed preliminary Shubnikov-de Haas experiments in nondestructive fields up to 42 T in Bi and Bi-Sb alloys, and predicted that the semimetal to semiconductor transition would occur in Bi at about 100 T [6.99, 129, 134]. The Los Alamos group actually measured the magnetoresistance in Bi in megagauss fields [6.205]. Figure 6.67 shows the plot of the magneto-resistance in Bi. This first experiment was performed at room temperature and for a polycrystalline sample; it would be worthwhile to extend the measurements to low temperatures and to a single crystal.

Very recently, the ISSP group succeeded in observing the magnetic field induced semimetal-semiconductor transition in Bi using megagauss fields

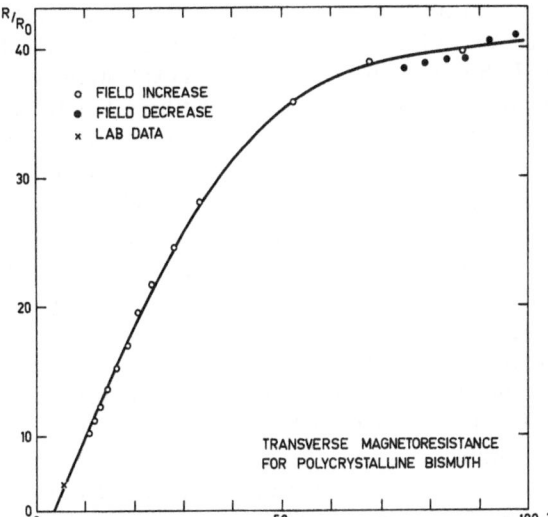

Fig. 6.67. Magnetoresistance in polycrystalline Bi [6.205]

[6.137]. For the magnetic field parallel to the binary axis, the transmission of the far infrared radiation with a wavelength of 337 μm abruptly increased above 88 T at low temperatures. In this experiment, the magnetotransmission is used to monitor the magnetoresistance which is difficult to measure in the megagauss range. The increase of the magnetotransmission at 88 T is interpreted as an indication of the magnetic field induced semimetal-semiconductor transition. The observed critical field agrees well with the theoretically predicted transition field [6.138].

Application of ultrahigh magnetic fields to a substance such as Bi would also be interesting in connection with the "excitonic matter" whose existence on the surface of a neutron star is discussed, where the field is estimated to be 10^8 T [6.273,274].

There are many possibilities that new phenomena in solids may be observed in ultrahigh magnetic fields. An electronic phase transition such as "Wigner crystallization" or "charge density wave transition" [6.275,276] may be observed as a result of enhancing the many-body effect in ultrahigh magnetic fields. When the cyclotron orbit becomes extremely small, crystals may be transformed into a different structure.

Another interesting area concerning the application of ultrahigh magnetic fields is the extremely high pressure associated with these magnetic fields. The Maxwell stress reaches 40 kbar at 100 T and 4 Mbar at 1000 T. Pressures of this magnitude can be obtained in shock waves generated by the impact of high-speed projectiles on the target material [Fig. 6.24, 6.26]. However, this is an adiabatic compression on a very short time scale and thus results in heating, which can become so great as to impede effectively the volume compression of soft materials. In a magnetic compression experiment, the magnetic field acts as a cushion between the target and the projectile and conditions can be adjusted for obtaining an isentropic compression with quasi-hydrostatic pressure. *Hawke* et al. at the Lawrence Livermore Laboratory [6.57,277] used an explosive-driven flux compression system with a peak field of 1000 T. Among others, Lucite was isentropically compressed to a specific volume of about 0.35 under a pressure of about 4 Mbar [6.57], and the electrical conductivity of hydrogen and neon under high pressures was measured [6.277]. It was found that hydrogen becomes conductive at a pressure of about 2 Mbar, but that neon remained an insulator to pressures of more than 5 Mbar. Following a suggestion of the late Francis Bitter, *Bless* at MIT used the more direct method of an electromagnetic linear pinch [6.278]. For a given current, the magnetic stress at the surface of a conducting wire is inversely proportional to the radius and thus can be made arbitrarily high, at least in principle. The time scale of the experiment is limited by the electrical explosion of the wire (6.2) which calls for an extremely fast capacitor bank as the energy source. Nonconducting samples can be subjected to the pressure if they are enclosed in a hollow conductor. Bless studied various samples of geological interest, such as boron nitride, graphite, iron and mullite. It was found that the mullite decomposed into alumina and amorphous silica, and that polymorphous boron nitride, carbon and iron were formed.

Depending on the conductivity and design of the sample assembly, allowing partial penetration of the magnetic field, experiments can be performed with a combination of extreme values for pressure, magnetic field and temperature. As high temperatures can be reached in very short times, the high-speed dynamics of phase transitions, e.g., melting and vaporization, can be studied.

At the present, hydrostatic pressures of the order of 1 Mb are becoming available. Although this is in very small volumes, it is more convenient for most experiments. Meanwhile, magnetically driven compressions are being pushed towards the 100 Mb range [6.70,178].

6.6 Outlook

Solid-state physics is the only area of research where strong pulsed magnetic fields have been used extensively. The research reported in the preceding sections now provides a good basis for further development. In particular, the combination of low temperatures and high magnetic fields, along with improved techniques for pulsed experimentation, will open up new frontiers. Pulsed magnetic fields will be more widely used as a research tool when the large nondestructive pulsed magnets become operational, with extended pulse duration and larger volumes. These will demonstrate the practical value and economical advantages of pulsed field techniques on a larger scale and will clear the path for developments extending to other areas of research and ultimately leading into megagauss physics, where new effects are waiting to be discovered.

Besides in the solid state, Zeeman and Faraday effects can be studied in gases, both in transmission and in absorption. This may be of interest in astrophysics where multimegagauss fields (10^3–10^4 T or more) have been discovered in white dwarfs, and fields of the order of 10^8 T are inferred for neutron stars [6.279].

Plasma physics and nuclear fusion research, although directed mainly at steady-state reactors, have almost exclusively relied on pulsed magnetic fields, mostly of moderate strength but in large volumes. On the one hand, this is for economical reasons because of the high power required by most experiments. On the other hand, the early confinement configurations were all unstable and thus no longer times were needed to conduct experiments. Many of the early difficulties have now been overcome and as the confinement time as well as the size of the machines increases, the capacitor banks are replaced by rotating machines and superconducting magnets. At the time when the instabilities were most troublesome, it had been proposed to circumvent this difficulty by developing fusion reactors with high density plasmas and very strong magnetic fields. This reduces the required confinement time for efficient burning of the nuclear fuel to a time interval comparable to the onset of the instabilities. In fact, the early flux compression research had been inspired by this application [6.280]. However, this approach did not lead to the quick realization of practical fusion reactors, and the different research projects were abandoned. None of the experiments came even close to ignition, perhaps the scale was simply too small.

Recently, this idea was revived in the LINUS project at NRL [6.76, 281], which involves an original and attractive reactor concept: in a volume of several cubic metres, magnetic flux is compressed by the reversible implosion of a liner of liquid metal and in turn compresses a preheated plasma. Unusual as this scheme may appear, it has been shown by means of successful experiments and additional theoretical work that many of the difficult experimental problems related to this type of reactor can be solved [6.76, 282]. In particular, it was demonstrated that a piston-driven liquid metal implosion can be fully reversible if it is stabilized by rotation. In a recent conceptual reactor design (Fig. 6.68), the rotation is

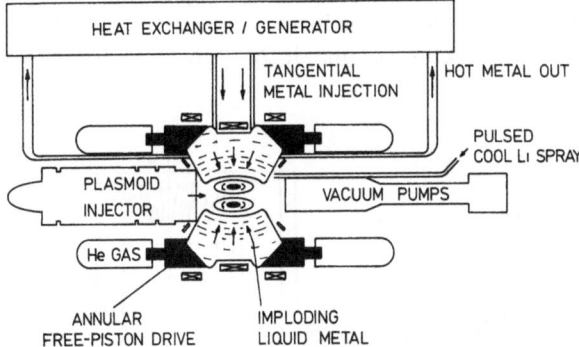

Fig. 6.68. Scheme of a 1800/500 MW LINUS fusion reactor [6.76] with a fully reversible implosion of liquid metal to compress magnetically self-confined plasmoids at the rate of 1 Hz. The initial size of the plasmoid is 1.9 m radius and 7.8 m length, the peak magnetic field is 54 T, the compression ratio 10 and the compressed plasma temperature 15 keV (initially 380 eV). The total reactor radius is 5.1 m and the helium drive pressure is 200 bar

provided by tangential injection of a liquid mixture of lithium and lead. The implosion is driven by the axial displacement of two annular pistons pushed by compressed helium at 200 bar. The plasma is in the shape of a torus with initial radius 1.9 m and 7.8 m length, injected along the axis and carrying its self-contained magnetic field. The compression is three-dimensional, reducing the length to 3.1 m. At a repetition frequency of 1 Hz, the thermal power is estimated at 1800 MW, resulting in a net electrical output of 500 MW, while 100 MW are needed to keep the reactor going. Many of the severe technical problems inherent in all types of fusion reactors are elegantly taken care of by the renewable liner of liquid metal. It performs the multiple functions of reactor first wall, magnet coil, reactor blanket (breeding tritium if it contains lithium), high energy neutron shield, and heat transfer medium. To this end, the liquid metal is axially recirculated and passed through heat exchangers and chemical processing. A small part of the fusion energy is directly converted by doing work against the expanding liner, recompressing the helium for the next implosion. Apart from its potential usefulness as an economically interesting reactor, there is some stark beauty in this concept of a giant diesel-type engine pulsating at the rhythm of the human heart.

Another interesting concept in nuclear fusion research, the Filippov plasma focus, is related to ultrastrong magnetic fields in more than one way. It is a simple device (Fig. 6.69) with a surprising effect: a radial discharge is started at one end of a coaxial electrode structure, it propagates along the axis and collapses onto a focal point as it emerges at the end. Very high plasma density and temperatures are documented by the emission of neutrons and x-rays. Explosive-driven generators have been considered for scaling this device to larger dimensions, for the possible use as the source of a strong pulse of neutrons, if not fusion for energy production [6.283]. More related to our topic is the application as a generator for extremely high magnetic fields. In a research project continued over many years, *Bostick* et al. found some indirect evidence for the occurence of small filaments containing fields of the order of 100 MG [6.61]. At the ISSP, a plasma focus is used to compress an initially introduced axial magnetic field for obtaining a megagauss field in a useful volume [6.284]. In a small pilot

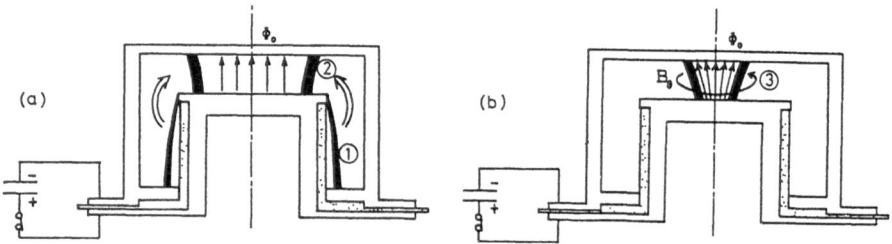

Fig. 6.69a, b. Schematic of a Filippov-type plasma focus device. (a) *Stage* 1 : A current sheet is formed at the start of the discharge. *Stage* 2 : Collapse phase of the current sheet. (b) *Stage* 3 : Maximum compression of the current sheet. Φ_0 is the initial magnetic flux generated near the axis of the cylinder, and B_θ is the azimuthal magnetic field associated with the discharge current [6.284]

experiment with a 55.7 µF capacitor bank at 20 kV, magnetic flux was compressed from 1 T to a measured peak field of 14 T. Experiments with a larger device are in progress with the goal of exceeding the megagauss limit without damage to the sample space.

In elementary particle physics, megagauss fields can be used as a target for high-energy particle beams to measure the radiation reaction which is of interest in astrophysics, and to observe effects directly related to quantum electro-dynamics [6.75]. The feasibility of such experiments was demonstrated in the electron beam of the Stanford Linear Accelerator [6.156]. In the same experiment, it was shown that the exposure of nuclear emulsions in a megagauss field can be used to determine the momentum of charged particles over a short length of the track [6.157]. The megagauss target is an ideal application of ultrastrong fields because the elementary particles do not pick up any disturbances from the megagauss environment, and detectors can be placed at a comfortable distance. It would certainly be worthwhile to provide a multimega-gauss target at one of the new particle accelerators. More on the technical side of high-energy physics, ultrastrong fields can be used on a limited scale for critical beam transport problems and to increase the reaction rate strongly in colliding beam experiments.

Ultrastrong magnetic fields are now available from reliable devices, and experimental techniques for their exploitation have been developed. In closing this chapter, we repeat our invitation to the physics community for making good use of these facilities. Theoreticians are invited to venture into the domain of nonlinear magnetic effects and high field quantum phenomena, and thus provide the needed stimulation for research with megagauss fields. In particular, the megagauss laboratory at the ISSP shall be happy to accommodate guest experiments at the new facilities which are now under development.

Given sufficient support, there is a good chance that magnetic fields well in excess of 10 MG will become available for experimentation within this decade. The field is open for discovery!

References

6.1   D. ter Haar: *The Collected Papers of P. Kapitza* (Pergamon, Oxford 1964)
6.2   G.M.Spruch: Phys. Today **32**, 34 (September 1979)
6.3   W.J. de Haas, J.B.Westerdijk: Nature **158**, 271 (1946)
6.4   T.F.Wall: J. Inst. Electr. Eng. (London) **64**, 745 (1926)
6.5   G.Raoult: Ann. Phys. Paris **4**, 369 (1949)
6.6   K.S.W.Champion: Proc. Roy. Soc. **B63**, 795 (1950)
6.7   W.R.Myers: J. Sci. Instr. **30**, 237 (1953)
6.8   D.Shoenberg: In *Progress in Low Temperature Physics*, ed. by C.J.Gorter (North-Holland, Amsterdam 1957) p. 226
6.9   J.L.Olsen: Helv. Phys. Acta **26**, 798 (1953)
6.10   J.L.Olsen, L.Rinderer: Nature **173**, 682 (1954)
6.11   B.Lüthi: Phys. Rev. Lett. **2**, 503 (1959)
6.12   B.Lüthi: Helv. Phys. Acta **33**, 161 (1960)

6.13   P.Cotti, B.Lüthi, J.L.Olsen: In *Proc. Int. Conf. on High Magnetic Fields*, ed. by H.Kolm, B.Lax, F.Bitter, R.Mills (MIT Press and Wiley, New York 1961) p. 523

6.14   C.Froideveaux, J.R.Keyston, P.Cotti, J.L.Olsen: *Proc. VII Int. Conf. Low Temp. Phys.* (University Press, Toronto 1960) p. 1265

6.15   P.Cotti: In *Proc. Int. Conf. on High Magnetic Fields*, ed. by H.Kolm, B.Lax, F.Bitter, R.Mills (MIT Press and Wiley, New York 1961) p. 539

6.16   H.Benz, R.Fasel, E.Fischer: Rev. Sci. Instrum. **36**, 562 (1965)

6.17   I.S.Jacobs: J. Appl. Phys. **32**, 62 (1962);
       I.S.Jacobs, P.E.Lawrence: Rev. Sci. Instrum. **29**, 713 (1958)

6.18   R.W. de Blois: Rev. Sci. Instrum. **32**, 816 (1961)

6.19   R.B.Flippen: J. Appl. Phys. **34**, 2026 (1963)

6.20   K.W.Blazey, H.Rohrer: Phys. Rev. **173**, 574 (1968);
       K.W.Blazey, H.Rohrer, R.Webster: Phys. Rev. **B4**, 2287 (1971)

6.21   H.Rohrer: Phys. Rev. **174**, 583 (1968);
       H.Rohrer: J. Appl. Phys. **40**, 1472 (1969)

6.22   H.P.Furth, R.W.Waniek: Rev. Sci. Instrum. **27**, 195 (1956)

6.23   S.Foner, H.H.Kolm: Rev. Sci. Instrum. **28**, 799 (1957);
       S.Foner, H.H.Kolm: Rev. Sci. Instrum. **27**, 547 (1956)

6.24   R.J.Keyes, S.Zwerdling, S.Foner, H.H.Kolm, B.Lax: Phys. Rev. **104**, 1804 (1956);
       B.Lax, J.G.Mavroides, H.J.Zeiger, R.J.Keyes: Phys. Rev. **122**, 31 (1961)

6.25   Proc. Conf. Les champs magnétiques intenses, leur production et leurs applications, Grenoble, 12–14 Sept. 1966 (CNRS, Paris 1967)

6.26   H.A.Jordaan, R.Wolf, D. de Klerk: Phys. Lett. **44A**, 381 (1973)

6.27   R.Gersdorf, F.A.Muller, L.W.Roeland: Rev. Sci. Instrum. **36**, 1100 (1965)

6.28   K.Kamigaki, A.Hoshi, M.Fuse, M.Ohashi, T.Kaneko: J. Magn. & Magn. Mater. **11**, 328 (1979)

6.29   H.C.Praddaude, S.Foner: Rev. Sci. Instrum. **10**, 1183 (1979)

6.30   S.Askenazy: Proc. Conf. The Application of High Magnetic Fields in Semiconductor Physics, ed. by J.F.Ryan, (Clarendon Laboratory, Oxford 1978) p. 101

6.31   M.Date: IEEE Trans. **MAG-12**, 1024 (1976)

6.32   Ø.Fischer: *Proc. Conf. Physique sous champs magnétiques intenses, Grenoble*, 1974 (CNRS, Paris 1975) p. 79

6.33   D.Schneider, J.Salge: Z. Angew. Phys. **31**, 346 (1971)

6.34   International Laboratory of High Magnetic Fields and Low Temperatures, Wroclaw, Poland, (descriptive brochure available on request)

6.35   J.A.Lira, W.Boon, F.Herlach, L.Janssens, J.Kuppens: Proc. 5th Int. Conf. on Magnetic Technology, Frascati 21–25 April 1975 (CNEN, Roma 1975) p. 264

6.36   H.P.Furth, M.A.Levine, R.W.Waniek: Rev. Sci. Instrum. **28**, 949 (1957)

6.37   C.M.Fowler, W.B.Garn, R.S.Caird: J. Appl. Phys. **31**, 588 (1960)

6.38   H.Knoepfel, F.Herlach (eds.): Proc. Conf. Megagauss Magnetic Field Generation by Explosives and Related Experiments, Frascati 1965 (Euratom, Brussels 1966)

6.39   C.M.Fowler: Science **180**, 261 (1973);
       C.M.Fowler, R.S.Caird, W.B.Garn, D.J.Erickson: IEEE Trans. **MAG-12**, 1018 (1976);
       also in Los Alamos Internal Report LA-5065-MS (1972)

6.40   F.Herlach, R.McBroom: J. Phys. **E6**, 652 (1973)

6.41   E.C.Cnare: J. Appl. Phys. **37**, 3812 (1966)

6.42   N.Miura, G.Kido, M.Akihiro, S.Chikazumi: J. Magn. & Magn. Mater. **11**, 275 (1979)

6.43   G.A.Shneerson: Z. Eksp. Teor. Fiz. Pis'ma **32**, 1153 (1962); Sov. Phys.-Tech. Phys. **7**, 848

6.44   F.Herlach, H.Knoepfel: Rev. Sci. Instrum. **36**, 1088 (1965)

6.45   F.Herlach, H.Knoepfel, R.Luppi: In Ref. 6.38, p. 287

6.46   R.S.Caird, W.B.Garn, D.B.Thomson, C.M.Fowler: J. Appl. Phys. **35**, 781 (1964); and in Ref. 6.38, p. 101

6.47   A.D.Sakharov, R.Z.Lyudaev, E.N.Smirnov, Yu.I.Plyushchev, A.I.Pavlovskii, V.K.Chernyshev, E.A.Feoktistova, E.I.Zharinov, Yu.A.Zysin: Dokl. Akad. Nauk SSSR **165**, 65 (1965); Sov. Phys.-Dokl. **10**, 1045 (1966)

6.48   E.I.Bichenkov: Dokl. Akad. Nauk SSSR **174**, 779 (1967); Sov. Phys.-Dokl. **12**, 567 (1967);
       E.I.Bichenkov, A.E.Voitenko, A.F.Demchuk, A.A.Deribas, B.I.Kulikov, Yu.E.Nesterikhin,
       O.P.Sobolev: Dokl. Akad. Nauk SSSR **183**, 1289 (1968); Sov. Phys.-Dokl. **13**, 1256 (1968)
6.49   D.W.Forster, C.J.Martin: In Ref. 6.25, p. 361
6.50   S.G.Alikhanov, V.G.Belan, A.I.Ivanchenko, V.N.Karasjuk, G.N.Kichigin: J. Phys. **E1**, 543
       (1968)
6.51   L.V.Babarina, O.P.Sobolev, A.E.Voitenko: Acta Astronaut. **15**, 297 (1969)
6.52   J.W.Shearer: J. Appl. Phys. **40**, 4490 (1969)
6.53   G.A.Shneerson: JETP Lett. **12**, 315 (1970)
6.54   A.M.Andrianov, V.F.Demichev, G.A.Eliseev, P.A.Levit: Zh. Eksp. & Fiz. Pis'ma **11**, 582
       (1970); JETP Lett. **11**, 402 (1970)
6.55   A.Brin, J.E.Besançon, J.L.Champetier, J.P.Plantevin, J.Vedel: In Ref. 6.38, p. 21
6.56   F.Herlach, J.E.Kennedy: J. Phys. **D6**, 661 (1973)
6.57   R.S.Hawke, D.E.Duerre, J.G.Huebel, H.Klapper, D.J.Steinberg, R.N.Keeler: J. Appl. Phys.
       43, 2734 (1972)
6.58   N.Miura, S.Chikazumi: Jpn. J. Appl. Phys. **18**, 553 (1979)
6.59   A.I.Pavlovskii, V.V.Druzhinin, O.M.Tatsenko, R.V.Pisarev: Zh. Eksp. Teor. Fiz. Pis'ma **20**,
       561 (1974); JETP Lett. **20**, 256 (1974)
6.60   V.T.Mikhkel'soo, G.A.Shneerson, A.P.Shcherbakov: Prib. Tekh. Eksp. **17**, 212 (1974);
       Instrum. Eksp. Tech. **17**, 551 (1974)
6.61   W.H.Bostick, V.Nardi, J.Feugeas, L.Grunberger, W.Prior, C.Cortese, F.Mezzetti,
       A.Pedrielli: In Ref. 6.68, p. 533
6.62   C.M.Fowler, R.S.Caird, W.B.Garn, D.J.Erickson: IEEE Trans. **MAG-12**, 1018 (1976)
6.63   R.S.Caird, J.H.Brownell, D.J.Erickson, C.M.Fowler, B.L.Freeman, T.Oliphant: In Ref.
       6.68, p. 461
6.64   N.N.Gennadiev, V.F.Demichev, P.A.Levit: In Ref. 6.68, p. 27
6.65   E.I.Bitchenkov, V.A.Lobanov: In Ref. 6.68, p. 181
6.66   A.M.Andrianov, Yu.A.Alekseev, V.L.Baryshev, V.I.Vasil'ev, M.N.Kazeev, V.V.Kisula: In
       Ref. 6.68, p. 479
6.67   U.N.Botcharov, A.I.Krutchinin, S.I.Krivosheev, A.N.Chetchel, G.A.Shneerson: In Ref.
       6.68, p. 485
6.68   P.J.Turchi (ed.): *Megagauss Physics and Technology*, Proc. of the Second Internat. Conf.
       on Megagauss Magnetic Field Generation and Related Topics, Washington, D.C., May
       30 – June 1, 1979 (Plenum, New York 1980)
6.69   A.I.Pavlovskii, R.Z.Lyudaev, V.A.Zolotov, A.S.Seryoghin, A.S.Yuryzhev, M.M.Kharlamov,
       A.M.Shuvalov, V.Ye.Gurin, G.M.Spirov, B.S.Makaev: In Ref. 6.68, p. 557
6.70   S.G.Alikhanov, V.P.Bakhtin, A.D.Muzychenko, V.P.Novikov: In *High Pressure Science and
       Technology*, Vol. 2 (Plenum, New York 1979, p. 974)
6.71   S.G.Alikhanov, V.P.Novikov: In Ref. 6.68, p. 505
6.72   D.H.Parkinson, B.E.Mulhall: *The Generation of High Magnetic Fields*, (Plenum, New York
       1967)
6.73   D.B.Montgomery: *Solenoid Magnet Design*, (Wiley, New York 1969)
6.74   H.Knoepfel: *Pulsed High Magnetic Fields* (North-Holland, Amsterdam 1970)
6.75   T.Erber: Rev. Mod. Phys. **38**, 626 (1966)
6.76   P.J.Turchi, R.L.Burton, A.L.Cooper, R.D.Ford, D.J.Jenkins: In Ref. 6.68, p. 375
6.77   W.Kaiser, A.Penzkofer, A.Lauberau: In *Proc. Conf. Lasers in Physical Chemistry and
       Biophysics*, ed. by J. Joussot-Dubien (Elsevier, Amsterdam 1975) p. 323;
       A.Lauberau, W.Kaiser: Rev. Mod. Phys. **50**, 607 (1978)
6.78   V.G.Welsby: *The Theory and Design of Inductance Coils* (McDonald, London 1964)
6.79   F.W.Grover: *Inductance Calculations* (Dover, New York 1962)
6.80   D.Melville, D.L.Rayner, W.I.Khan, P.G.Mattocks, K.M.A.-Rawi: J. Appl. Phys. **50**, 7771
       (1979) (abstract only, complete preprint available)
6.81   V.G.Kotenko, Yu.A.Litvinenko: Sov. Phys.-Tech. Phys. **17**, 144 (1972)
6.82   V.I.Ozhogin, K.G.Gurtovoj, A.S.Lagutin: In *High Field Magnetism*, ed. by M.Date (North-
       Holland, Amsterdam 1983) p. 267

6.83   D.Melville, P.G.Mattocks: J. Phys. **D5**, 1745 (1972)

6.84   J.Witters, F.Herlach: J. Appl. Phys. **D16**, 255 (1983)

6.85   F.Herlach, G. De Vos, J.Witters: Proc. 8th Int. Conf. on Magnet Technology 1983, J. Physique **45**, Supplément C1–915 (1984)

6.86   B.E.Mulhall, D.H.Prothero: J. Phys. **D6**, 1973 (1973)

6.87   D.M.Binnie, J.Carr, R.T.Elliott, W.M.Evans, P.S.Flower, D.G.Miller: Proc. 6th Int. Conf. on Magnet Technology, Bratislava (1977)

6.88   G.V.Brown, L.Flax, E.C.Itean, J.C.Laurence: NASA Tech. Rep., NASA TR R-170

6.89   G.Dworschak, F.Haberey, P.Hildebrand, E.Kneller, D.Schreiber: Rev. Sci. Instrum. **45**, 243 (1974)

6.90   A.H.Guenther, M.Kristiansen (eds.): *Proc. 2nd Int. Pulsed Power Conf.*, Lubbock, TX (Texas Tech Univ. 1979)

6.91   D.B.Montgomery: J. Magn. & Magn. Mater. **11**, 293 (1979)

6.92   N.T.Olson, J.Bandas, A.C.Kolb: Proc. Joint Intermag-MMM Conf., New York (1979)

6.93   B.Howland, S.Foner: In *Proc. Int. Conf. on High Magnetic Fields*, Cambridge, Mass., 1961 (MIT Press, Cambridge, Mass. 1962) p. 249

6.94   H.Brechna, D.A.Hill, B.M.Bailey: Rev. Sci. Instrum. **36**, 1529 (1965)

6.95   K.Suzuki, N.Miura: J. Phys. Soc. Jpn. **39**, 148 (1975)

6.96   O.K.Mawardi: In Ref. 6.68, p. 497

6.97   L.W.Roeland, F.A.Muller, R.Gersdorf: In Ref. 6.25, p. 175

6.98   S.Foner: Proc. Conf. Les champs magnétiques intenses, leur production et leurs applications, Grenoble, Sept. 12–14, 1966 (CNRS, Paris 1967) p. 385

6.99   K.Hiruma, G.Kido, N.Miura: Solid State Commun. **31**, 1019 (1979)

6.100   N.Miura, G.Kido, H.Miyajima, K.Nakao, S.Chikazumi: In *Physics in High Magnetic Fields*, ed. by S.Chikazumi, N.Miura, Springer Ser. Solid-State Sci., Vol. 24 (Springer, Berlin, Heidelberg 1981) p. 64

6.101   L.Eaves, R.A.Hoult, R.A.Stradling, S.Askenazy, R.Barbaste, G.Carrère, J.Leotin, J.C.Portal, J.P.Ulmet: J. Phys. **C10**, 2831 (1977)

6.102   J.C.Portal, L.Eaves, S.Askenazy, R.A.Stradling: Solid State Commun. **14**, 1241 (1974)

6.103   J.Leotin, R.Barbaste, S.Askenazy, M.S.Skolnick, R.A.Stradling, J.Tuchendler: Solid State Commun. **15**, 693 (1974)

6.104   M.S.Skolnick, A.K.Jain, R.A.Stradling, J.Leotin, J.C.Ousset, S.Askenazy: J. Phys. **C9**, 2809 (1976)

6.105   C.C.Bradley, P.E.Simmonds, J.R.Stockton, R.A.Stradling: Solid State Commun. **12**, 413 (1973)

6.106   J.Leotin, J.C.Ousset, R.Barbaste, S.Askenazy, M.S.Skolnick, R.A.Stradling, G.Poiblaud: Solid State Commun. **16**, 363 (1975)

6.107   K.Suzuki, N.Miura: Solid State Commun. **18**, 233 (1976)

6.108   J.C.Portal et al.: Proc. 15th Int. Conf. Physics of Semiconductors, Kyoto, J. Phys. Soc. Jpn. **49**, Suppl. A (1980) p. 113 (GaSb), p. 879 (InSe), p. 1167 (Cd_3As_2 thin film)

6.109   R.J.Nicholas, R.A.Stradling, S.Askenazy, P.Perrier, J.C.Portal: Surf. Sci. **73**, 106 (1978)

6.110   J.C.Ousset, S.Askenazy: In *Physics in High Magnetic Fields*, ed. by S.Chikazumi, N.Miura, Springer Ser. Solid-State Sci., Vol. 24 (Springer, Berlin, Heidelberg 1981) p. 161; J.C.Ousset, G.Carrère, J.P.Ulmet, S.Askenazy, G.Creuzet, A.Fert: J.Magn. Magn. Mater. **24**, 7 (1981)

6.111   M.Motokawa, S.Kuroda, M.Date: J. Appl. Phys. **50**, 7762 (1979)

6.112   H.Hori, H.Mollymoto, M.Date: J. Phys. Soc. Jpn. **46**, 908 (1979)

6.113   M.Date, M.Motokawa, K.Okuda, H.Hori, H.Mollymoto, T.Sakakibara: J. Magn. & Magn. Mater. **15–18**, 1559 (1980)

6.114   Y.Sasaki, N.Kuroda, Y.Nishina: In *Physics in High Magnetic Fields*, ed. by S.Chikazumi, N.Miura, Springer Ser. Solid-State Sci., Vol. 24 (Springer, Berlin, Heidelberg 1981) p. 161

6.115   M.Date, M.Motokawa, K.Okuda, H.Hori, T.Sakakibara: In *Physics in High Magnetic Fields*, Springer Ser. Solid-State Sci., Vol. 24 (Springer, Berlin, Heidelberg 1981) p. 44

6.116   K.Okuda, M.Kitagawa, T.Sakakibara, M.Date: J. Phys. Soc. Jpn. **48**, 2157 (1980)

6.117   Ø.Fischer: Appl. Phys. **16**, 1 (1978)

6.118 S.Foner: Proc. Conf. Physique sous champs magnétiques intenses, Grenoble Sept. 18–20, 1974 (CNRS, Paris 1975) p. 423
S.Foner, E.J.McNiff, E.J.Alexander: Phys. Lett. **49A**, 269 (1974)
S.Foner: J. Phys. Soc. Jpn. **50**, 2595 (1981)
6.119 S.Foner, L.R.Momo, A.Mayer: Phys. Rev. Lett. **3**, 36 (1959)
S.Foner, W.Low: Phys. Rev. **120**, 1585 (1960)
S.Foner, H.Meyer, W.H.Kleiner: J. Phys. Chem. Solids **18**, 273 (1961)
S.Foner: Proc. Int. Conf. on Magnetism (The Institute of Physics and the Physical Society, London 1965) p. 438
6.120 D.Schneider: Wiss. Berichte HMFA Braunschweig **1**, 31 (1973)
6.121 J.C.A. van der Sluis, H.R. de Beun, B.A.Zweers, D. de Klerk: Proc. Int. Conf. Les champs magnétiques intenses, leur production et leurs applications, Grenoble, Sept. 12–14, 1966 (CNRS, Paris 1967) p. 429
6.122 J.C.A. van der Sluis: Ph. D. Thesis, University of Leiden (1967)
6.123 R.Weggel, C.Weggel, M.Leupold: Francis Bitter National Magnet Laboratory Annual Report 1979/80 (Nov. 1980) p. 90
6.124 A.Hairie, A.Fortini, M.Huissier, M.Sauzade: Rev. Sci. Instrum. **44**, 1464 (1973)
6.125 D.Schneider: Proc. Int. Conf. "The Application of High Magnetic Fields in Semiconductor Physics", ed. by G.Landwehr, Würzburg, July 24–Aug. 2, 1974, p. 615; Z. Naturforsch. **27a**, 250 (1972)
6.126 W.H.Bergmann: In Ref. 6.25, p. 371
6.127 K.Suzuki, N.Miura, S.Uchida, S.Tanaka: Phys. Status Solidi (B) **76**, 787 (1976)
6.128 G.Kido, N.Miura, M.Akihiro, H.Katayama, S.Chikazumi: In *Physics in High Magnetic Fields*, Springer Ser. Solid-State Sci., Vol. 24 (Springer, Berlin, Heidelberg 1981) p. 72
6.129 K.Hiruma, G.Kido, N.Miura: Solid State Commun. **38**, 859 (1981)
6.130 K.Hiruma, G.Kido, N.Miura: J. Phys. Soc. Jpn. **52**, 2550 (1983)
6.131 K.Nakamura, G.Kido, N.Miura: Solid State Commun. **47**, 349 (1983)
6.132 K.Nakamura, G.Kido, K.Nakao, N.Miura: J. Phys. Soc. Jpn. **53**, 1164 (1984)
6.133 K.Nakamura, T.Osada, G.Kido, N.Miura, S.Tanuma: J. Phys. Soc. Jpn. **52**, 2875 (1983)
·6.134 K.Hiruma, G.Kido, K.Kawauchi, N.Miura: Solid State Commun. **33**, 275 (1980)
6.135 N.Miura, G.Kido, M.Suekane, S.Chikazumi: Physica **117, 118, B**, C66 (1983)
6.136 N.Miura, G.Kido, M.Suekane, S.Chikazumi: J. Phys. Soc. Jpn. **52**, 2838 (1983)
6.137 N.Miura, K.Hiruma, G.Kido, S.Chikazumi: Phys. Rev. Lett. **49**, 1339 (1982)
6.138 K.Hiruma, N.Miura: J. Phys. Soc. Jpn. **52**, 2118 (1983)
6.139 M.P.Vecchi, J.R.Pereira, M.S.Dresselhaus: Phys. Rev. **B14**, 298 (1976)
6.140 Y.Iye, J.Heremans, K.Nakamura, G.Kido, N.Miura, J.P.Michenaud, S.Tanuma: J. Phys. Soc. Jpn. **52**, 1962 (1983)
6.141 J.Heremans, J.P.Michenaud, Y.Iye, N.Miura, K.Nakamura, G.Kido, S.Tanuma: J. Phys. **E16**, 382 (1983)
6.142 S.Tanuma, R.Inada, A.Furukawa, O.Takahashi, Y.Iye: In *Physics in High Magnetic Fields*, ed. by S.Chikazumi, N.Miura, Springer Ser. Solid-State Sci., Vol. 24 (Springer, Berlin, Heidelberg 1981) p. 316
6.143 D.Yoshioka, H.Fukuyama: J. Phys. Soc. Jpn. **50**, 725 (1981)
6.144 Y.Iye, P.Tedrow, G.Timp, M.S.Dresselhaus, G.Dresselhaus, A.Furukawa, S.Tanuma: Phys. Rev. **B25**, 5478 (1982)
6.145 N.Miura, Y.Iwasa, T.Itakura, G.Kido: J. Phys. Soc. Jpn. **51**, 1228 (1982)
6.146 G.Kido, N.Miura, H.Ohno, H.Sakaki: J. Phys. Soc. Jpn. **51**, 2168 (1982)
6.147 A.R.Bryant: In Ref. 6.38, p. 183
6.148 J.D.Lewin, P.F.Smith: Rev. Sci. Instrum. **35**, 541 (1964)
6.149 F.Herlach: J. Appl. Phys. **39**, 5191 (1968)
6.150 F.Herlach: Helv. Phys. Acta **44**, 308 (1971)
6.151 M.Roux: "Analytical and Numerical Treatment of Nonlinear Magnetic Difusion", M. Sc. thesis, Illinois Institute of Technology, Chicago (1972);
R.D.Richtmyer, K.W.Morton: *Difference Methods for Initial Value Problems*, 2nd ed. (Interscience, New York 1967) p. 17;

B.Carnahan, H.A.Luther, J.O.Wilkes: *Applied Numerical Methods* (Wiley, New York 1969) p.441
6.152 R.E.Kidder: In Ref. 6.38, p. 37
6.153 P.W.Bridgman: *The Physics of High Pressure* (Bell, London 1949)
6.154 J.W.Shearer, F.F.Abraham, C.M.Aplin, B.P.Benham, J.E.Faulkner, F.C.Ford, M.M.Hill, C.A.McDonald, W.H.Stephens, D.J.Steinberg, J.R.Wilson: J. Appl. Phys. **39**, 2102 (1968)
6.155 Group GMX-6, Los Alamos Scientific Laboratory, Selected Hugoniots, Los Alamos Scientific Laboratory Rep. LA-4167-MS (1969)
6.156 F.Herlach, R.McBroom, T.Erber, J.Murray, R.Gearhart: IEEE Trans. **NS-18**, 809 (1971)
6.157 H.H.Heckman, F.Herlach: Nucl. Instrum. & Methods **106**, 269 (1973)
6.158 F.Herlach, J.Davis, R.Schmidt, H.Spector: Phys. Rev. **B10**, 682 (1974)
6.159 H.Knoepfel, R.Luppi: J. Phys. **E5**, 1133 (1972)
6.160 Yu.N.Bocharov, S.I.Krivosheyev, G.A.Shneyerson: In Ref. 6.169, p. 77
6.161 F.Herlach: In Ref. 6.169, p. 58
6.162 C. Di Gregorio, F.Herlach, H.Knoepfel: In Ref. 6.38, p. 421
6.163 C.M.Fowler, R.S.Caird, W.B.Garn: An Introduction to Explosive Magnetic Flux Compression Generators, Los Alamos Scientific Laboratory Rpt. LA-5890-MS (1975)
6.164 M.Cowan: In Ref. 6.38, p. 167
6.165 A.K.Aziz, H.Hurwitz, H.M.Sternberg: Phys. Fluids **4**, 380 (1961)
6.166 E.G.Harris: Phys. Fluids **5**, 1057 (1962)
6.167 J.G.Linhart: J. Appl. Phys. **32**, 500 (1961)
6.168 J.E.Besançon: Thèse de doctorat es-sciences physiques, Université Paris-Sud, Centre d'Orsay (1971)
6.169 V.M.Titov, G.A.Shvetsov (eds.): *Ultrahigh Magnetic Fields*. Proceedings of the Third International Conference on Megagauss Magnetic Field Generation and Related Topics, Novosibirsk 1983 (NAUKA, Moscow 1984)
6.170 J.E.Besançon, M.Guillot, F.Herlach: Rev. Phys. Appl. **12**, 573 (1977)
6.171 F.Herlach: Rep. Progr. Phys. **31**, 341 (1968)
6.172 J.E.Besançon, J.Morin, J.M.Vedel: C. R. Acad. Sci. Ser. **B271**, 397 (1970)
6.173 A.F.Demtshuk, V.V.Poljudov, V.M.Titov, G.A.Shvetsov: In Ref. 6.68, p. 55
6.174 K.Nagayama, T.Mashimo: In *High Field Magnetism*, ed. by M.Date (North-Holland, Amsterdam 1983) p. 319
6.175 D.Kachilla, F.Herlach, T. Erber: Rev. Sci. Instrum. **41**, 1 (1970)
6.176 D.Shoenberg: In *Progress in Low Temperature Physics*, ed. by C.J.Gorter (North-Holland, Amsterdam 1957) p. 226
6.177 S.Chikazumi, N.Miura, G.Kido, M.Akihiro: IEEE Trans. **MAG-14**, 577 (1978)
6.178 C.M.Fowler, R.S.Caird, R.S.Hawke, T.J.Burgess: In *High Pressure Science and Technology*, Vol. 2, ed. by K.D.Timmerhaus, M.S.Barber (Plenum, New York 1979) p. 981
6.179 Proc. 1st IEEE International Pulsed Power Conf., Dept. of Electrical Engineering, Texas Tech Univ., Lubbock, Texas, Nov. 9–11, 1976
6.180 R.A.Nuttelman, J.H.Degnan, G.F.Kiuttu, R.E.Reinovsky, W.L.Baker: In Ref. 6.68, p. 37
6.181 D.H.McDaniel, R.W.Stinnett, R.J.Leeper, L.P.Mix: In Ref. 6.68 (not published in the proceedings)
6.182 A.R.Sherwood, E.L.Cantrell, C.A.Ekdahl, I.Henins, H.W.Hoida, T.R.Jarboe, P.L.Klingner, R.C.Malone, J.Marshall, G.A.Sawyer: In Ref. 6.68, p. 391
6.183 C.M.Fowler, R.S.Caird, W.B.Garn, D.B.Thomson: In *High Magnetic Fields*, Proc. Int. Conf., Nov. 1—4, 1961 ed. by H.Kolm, B.Lax, F.Bitter, R.Mills (MIT Press, Cambridge, Mass. 1962) p. 269
6.184 J.C.Crawford, R.A.Damerow: J.Appl. Phys. **39**, 5224 (1968)
6.185 E.I.Bitchenkov, V.A.Lobanov, V.I.Telenkov, A.M.Trubatshev: In Ref. 6.68, p. 471
6.186 F.Herlach: J. Phys. **E12**, 421 (1979)
6.187 J.Morin, J.Vedel: C.R.Acad. Sci. **B272**, 1232 (1971)
6.188 H.Knoepfel, H.Kroegler, R.Luppi, J.E.Van Montfoort: Rev. Sci. Instrum. **40**, 60 (1969)
6.189 A.D.Sakharov: Usp. Fiz. Nauk **88**, 725 (1966); Sov. Phys.-Usp. **9**, 294 (1966)
6.190 I.I.Divnov, Yu.A.Guśkov, N.I.Zotov, O.P.Karpov, B.D.Khristoforov: Fiz. Goreniya Vzryva **12**, 959 (1976)

6.191 E.I.Azarkevich, A.E.Voitenko, V.P.Isakov, Yu.A.Kotov: Zh. Tekh. Fiz. **46**, 1957 (1976); Sov. Phys.-Tech. Phys. **21**, 1141 (1976)

6.192 R.S.Hawke, J.K.Scudder: In Ref. 6.68, p. 297
R.S.Hawke:; Lawrence Livermore Lab. Report UCRL-52778, Part 2 (1979)

6.193 C.M.Fowler, D.R.Peterson, R.S.Hawke, A.L.Brooks, 1981 Topical Conf. Shock Waves in Condensed Matter, AIP Conf. Proc. **78**, 686 (1981); see also R.Hawke et al. in Ref. 6.169

6.194 C.M.Fowler, D.B.Thomson, W.B.Garn, R.S.Caird: LASL group M6 summary report: The Birdseed Program, Los Alamos Scientific Laboratory Rpt. LA-5141-MS (1973)

6.195 E.P.Velikhov, A.A.Vedenov, A.D.Bogdanets, V.S.Golubev, E.G.Kasharskii, A.A.Kiselev, F.G.Rutberg, V.V.Chernukha: Zh. Tekh. Fiz. **43**, 429 (1973) Sov. Phys.-Tech. Phys. **18**, 274 (1973)

6.196 R.Hahn, B.Antoni, J.Lucidarme, C.Rioux, F.Rioux-Damidau: Rev. Phys. Appl. **11**, 409 (1976)

6.197 M.Cowan, E.C.Cnare, W.B.Leisher, W.K.Tucker, D.L.Wesenberg: Cryogenics **16**, 699 (1976)

6.198 F.Herlach, H.Knoepfel, R.Luppi, J.E.Van Montfoort: In Ref. 6.38, p. 471

6.199 W.B.Garn, R.S.Caird, D.B.Thomson, C.M.Fowler: Rev. Sci. Instrum. **37**, 762 (1966)

6.200 W.B.Garn, R.S.Caird, C.M.Fowler, D.B.Thomson: Rev. Sci. Instrum. **39**, 1313 (1968)

6.201 S.C.Rashleigh, R.Ulrich: Appl. Phys. Lett. **34**, 768-70 (1979)

6.202 F.Herlach, H.Knoepfel, R.Luppi, J.E.Van Montfoort: Helv. Phys. Acta **38**, 363 (1965); in Ref. 6.198

6.203 C.M.Fowler, R.S.Caird, W.B.Garn, D.J.Erickson, B.L.Freeman: J. Less Common Met. **62**, 397 (1978)

6.204 C.M.Fowler, R.S.Caird, D.J.Erickson, B.L.Freeman, W.B.Garn: *Physics in High Magnetic Fields*, Proc. Oji Int. Seminar, Hakone, Japan, Sept. 10-13, 1980 ed. by S.Chikazumi, N.Miura (Springer, Berlin, Heidelberg 1981) p. 54

6.205 C.H.Aldrich, C.M.Fowler, R.S.Caird, W.B.Garn, W.G.Witteman: Phys. Rev. **B23**, 3970 (1981)

6.206 G.Kido, N.Miura, S.Chikazumi: Proc. 13th Int. Congress on Highspeed Photography and Photonics, Tokyo, ed. by S.Hyodo (Japan Society of Precision Engineering, Tokyo 1978) p. 552

6.207 V.V.Druzhinin, O.M.Tatsenko: Opt. Spectrosc. **36**, 426 (1974)

6.208 M.Guillot: Proc. Int. Conf. Magnetism, Moscow, 1, 1973 (Nauka, Moscow 1974) p. 268

6.209 N.Miura, G.Kido, I.Oguro, S.Chikazumi: Proc. Int. Colloq. on Physics in High Magnetic Fields, Grenoble 1974 (CNRS 1975) p. 345

6.210 G.Kido, N.Miura, K.Kawauchi, I.Oguro, S.Chikazumi: J. Phys. **E9**, 587 (1976)

6.211 J.A.Davis, F.Herlach: Phys. Lett. **43A**, 303 (1973)

6.212 F.Herlach: Proc. Int. Conf. on Application of High Magnetic Fields in Semiconductor Physics, Würzburg 1974 p. 84

6.213 N.Miura, G.Kido, S.Chikazumi: Solid State Commun. **18**, 885 (1976)

6.214 N.Miura, G.Kido: Proc. 13th Int. Conf. on Phys. Semiconductors, Rome 1976, ed. by F.G.Fumi, p. 1149

6.215 N.Miura, G.Kido, K.Suzuki, S.Chikazumi: Proc. Int. Conf. on Application of High Magnetic Fields in Semiconductor Physics, Würzburg 1976, p.441

6.216 N.Miura, G.Kido, S.Chikazumi: Proc. Int. Conf. on Application of High Magnetic Fields in Physics of Semiconductors, Oxford 1978, ed. by J.F.Ryan, p. 233

6.217 N.Miura, G.Kido, S.Chikazumi: Proc. 14th Int. Conf. Phys. Semiconductors, Edinburgh 1978, ed. by B.L.H.Wilson (Institute of Physics, London 1978) Conf. Ser. **43**, p. 1109

6.218 I.M.Boswarva, R.E.Howard, A.B.Lidiard: Proc. Roy. Soc. **A269**, 125 (1962)

6.219 I.M.Boswarva, A.B.Lidiard: Proc. R. **A278**, 588 (1964)

6.220 V.V.Druzhinin, A.I.Pavlovskii, A.A.Samokhvalov, O.M.Tatsenko: Sov. Phys. JETP Lett. **23**, 233 (1976)

6.221 A.I.Pavlovskii, N.P.Kolokol'chikov, V.V.Druzhinin, O.M.Tatsenko, A.I.Boykov, M.I. Dolotenko: Sov. Phys. JETP Lett. **30**, 193 (1979)

6.222 B.Lax, J.G.Mavroides, H.J.Zeiger, R.J.Keyes: Phys. Rev. **122**, 31 (1961)

6.223 E.D.Palik, G.B.Wright: In *Semiconductors and Semimetals*, 3, 421 (Academic Press, New York 1967)

6.224 E.O.Kane: J. Phys. Chem. Solids 1, 249 (1957)

6.225 L.M.Roth, B.Lax, S.Zwerdling: Phys. Rev. 114, 90 (1959)

6.226 C.R.Pidgeon, R.N.Brown: Phys. Rev. 146, 575 (1966)

6,227 N. Miura, G.Kido, S. Chikazumi: In *The Physics of Selenium and Tellurium*, Springer Ser. Solid-State Sci., Vol. 13, ed. by E.Gerlach, P.Grosse (Springer, Berlin, Heidelberg 1979) p. 110

6.228 M.Hulin: Proc. 10th Int. Conf. on Phys. Semiconductors, Cambridge 1970, ed. by S.P.Keller et al. p. 329

6.229 H.J.G.Meyer: Phys. Lett. 2, 259 (1962)

6.230 D.M.Larsen, E.J.Johnson: J. Phys. Soc. Jpn. Suppl. 21, 443 (1966)

6.231 E.J.Johnson, D.M.Larsen: Phys. Rev. Lett. 16, 655 (1966)

6.232 D.H.Dickey, E.J.Johnson, D.M.Larsen: Phys. Rev. Lett. 18, 599 (1967)

6.233 D.M.Larsen: Proc. 10th Int. Conf. on Phys. Semiconductors, Cambridge 1970, ed. by S.P.Keller et al., p. 145

6.234 W.S.Baer, R.N.Dexter: Phys. Rev. 135, A1388 (1964)

6.235 K.J.Button, B.Lax, D.R.Cohn, W.Dreybrodt: Proc. 10th Int. Conf. on Phys. Semiconductors, Cambridge, 1970, ed. by S.P.Keller et al., p. 153

6.236 K.Nagasaka, G.Kido, S.Narita: Phys. Lett. 45A, 485 (1973)

6.237 K. Nagasaka: Phys. Rev. B15, 2273 (1977)

6.238 K. Sawamoto: J. Phys. Soc. Jpn. 18, 1224 (1963)

6.239 K.J. Button, B.Lax: Bull. Am. Phys. Soc. 15, 365 (1970)

6.240 D.M.Larsen: Phys. Rev. 135, A419 (1964)

6.241 J.Waldman, D.M.Larsen, P.E.Tannenwald, C.C.Bradley, D.R.Sohn, B.Lax: Phys. Rev. Lett. 23, 1033 (1969)

6.242 M.Shinada, O.Akimoto, H.Hasegawa, K. Tanaka: J. Phys. Soc. Jpn. 28, 975 (1970)

6.243 N.Lee, D.M.Larsen, B.Lax: J. Phys. Chem. Solids 34, 1059 (1973)

6.244 Y.Yafet, R.W.Keyes, E.N.Adams: J. Phys. Chem. Solids 1, 137 (1956)

6.245 R.F.Wallis, H.J.Bowlden: J. Phys. Chem. Solids 7, 78 (1958)

6.246 D.Cabib, E.Fabri, G.Fiorio: Nuovo Cimento 10B, 185 (1972)

6.247 A.Baldereschi, F.Bassani: Proc. 10th Int. Conf. on Phys. Semiconductors, Cambridge 1970, ed. by S.P.Keller et al., p. 191

6.248 N.Miura, G.Kido, H.Katayama. S.Chikazumi: J. Phys. Soc. Jpn. 49, Suppl. A, 409 (1980)

6.249 K.Aoyagi, A.Misu, G.Kuwabara, Y.Nishina, S.Kurita, T.Fukuroi, O.Akimoto, H.Hasegawa, M.Shinada, S.Sugano: J. Phys. Soc. Jpn. 21, Suppl. 174 (1966)

6.250 E.Mooser, M.Schlüter: Nuovo Cimento 18B, 164 (1973)

6.251 G.Harbeke, E.Tosatti: Phys. Rev. Lett. 28, 1567 (1977)

6.252 M.S.Skolnick, L.C.Thanh, F.Levy, G.Harbeke: Physica 89B, 143 (1977)

6.253 S.Sugano, Y.Tanabe: J. Phys. Soc. Jpn. 13, 880 (1958)

6.254 S.Sugano, I.Tsujikawa: J. Phys. Soc. Jpn. 13, 889 (1958)

6.255 K.Aoyagi, A.Misu, S.Sugano: J. Phys. Soc. Jpn. 18, 1448 (1963)

6.256 G.Kido, N.Miura: Appl. Phys. Lett. 41, 569 (1982)

6.257 Y.Shapira: J. Appl. Phys. 42, 1588 (1971)

6.258 R.S.Caird, W.B.Garn, C.M.Fowler, D.B.Thomson: J. Appl. Phys. 42, 1651 (1971)

6.259 A.E.Clark, E.Callen: J. Appl. Phys. 39, 5972 (1964)

6.260 N.Miura, I.Oguro, S.Chikazumi: J. Phys. Soc. Jpn. 45, 1534 (1978)

6.261 N.F.Kharchenko, V.V.Eremenko, S.L.Gnatchenko, L.I.Belyi, E.M.Kabanova: Sov. Phys. JETP 41, 531 (1976)

6.262 W.A.Crossley, R.W.Cooper, J.L.Page: Phys. Rev. 181, 896 (1969)

6.263 N.Miura, G.Kido, I.Oguro, K.Kawauchi, S.Chikazumi, J.F.Dillon, Jr., L.S.van Uitert: Physica 86–88B, 1219 (1977)

6.264 G.Kido, N.Miura, K.Kawauchi, I.Oguro, J.F.Dillon, Jr., S.Chikazumi: Physica 89B, 147 (1977)

6.265 J.Bernasconi, D.Kuse: Phys. Rev. B3, 811 (1971)

6.266 J.L.Féron, G.Fillion, G.Hug: J. Physique **34**, 247 (1973)
6.267 F.M.Yang, N.Miura, G.Kido, S.Chikazumi: J. Phys. Soc. Jpn. **48**, 71 (1980)
6.268 N.B.Brandt, E.A.Svistova, Yu.G.Kashirskii: Sov. Phys. JETP Lett. **9**, 136 (1969)
6.269 M.P.Vecchi, J.R.Pareira, M.S.Dresselhaus: Proc. 12th Int. Conf. on Phys. Semiconductors, 1974, ed. by M.H.Pilkuhn (Teubner, Stuttgart 1974) p. 1181
6.270 T.Sakai, N.Goto, S.Mase: J. Phys. Soc. Jpn. **35**, 1064 (1973)
6.271 N.B.Brandt, S.M.Chudinov: J. Low Temp. Phys. **8**, 339 (1972)
6.272 D.Yoshioka: J. Phys. Soc. Jpn. **45**, 1165 (1978)
6.273 S.T.Chui: Phys. Rev. **B9**, 3438 (1974)
6.274 M.Ruderman: Phys. Rev. Lett. **27**, 1306 (1971)
6.275 H.Fukuyama: Solid State Commun. **19**, 551 (1976)
6.276 H.Fukuyama, P.M.Platzman, P.W.Anderson: Phys. Rev. **B19**, 5211 (1979)
6.277 R.S.Hawke, T.J. Burgess, D.E.Duerre, J.G.Huebel, R.N.Keeler, H.Klapper, W.C.Wallace: Phys. Rev. Lett. **41**, 994 (1978)
6.278 S.J.Bless: J.Appl. Phys. **43**, 1580 (1972)
6.279 M.Ruderman: J. Magn. & Magn. Mater. **11**, 269 (1979)
6.280 C.M.Fowler, R.S.Caird, W.B.Garn, D.B.Thomson: In Ref. 6.38, p. 1
6.281 A.E.Robson: In Ref. 6.68, p.425
6.282 Y.Itoh, Y.Fujii-E: In Ref. 6.68, p. 437
6.283 C.M.Fowler, R.S.Caird, D.J.Erickson, B.L.Freeman: Proc. 3rd IEEE Pulsed Power Conf., Albuquerque, June 1–3, 1981
6.284 K.Hirano, M.N.Hattori, K.Shimoda, G.Kido, N.Miura, S.Chikazumi: J. Phys. Soc. Jpn. **51**, 297 (1982)

Additional References

Arakawa, Y., Nishioka, M., Miura, N.: Use of high magnetic fields to estimate carrier leakage current in GaInAsP-InP double heterostructure lasers. Appl. Phys. Lett. **45**, 7 (1984)
Arakawa, Y., Sakaki, H., Nishioka, M., Kido, G., Miura, N.: Characteristics of double-heterostructure lasers in strong magnetic fields. Jap. J. Appl. Phys. **22**, Suppl. 22-1, 283 (1983)
Arakawa, Y., Sakaki, H., Nishioka, M., Miura, N.: Two-dimensional quantum-mechanical confinement of electrons in LED by strong magnetic fields. IEEE Trans. **ED-30**, 330 (1983)
Arakawa, Y., Sakaki, H., Nishioka, M., Miura, N.: Two-dimensional quantum-mechanical confinement of electrons in DH lasers by strong magnetic fields. IEEE J. Quant. Electron. **QE-19**, 1255 (1983)
Arakawa, Y., Sakaki, H., Nishioka, M., Okamoto, H., Miura, N.: Spontaneous emission characteristics of quantum well lasers in strong magnetic fields – an approach to quantum-well-box light source. Jap. J. Appl. Phys. Lett. **22**, L804 (1983)
Bieri, J.B., Fert, A., Creuzet, G., Ousset, J,C.: Magnetoresistance of amorphous CuZr: weak localization in a three dimensional system. Solid State Commun. **49**, 849 (1984)
van Deursen, A.P.J., Buiting, J.J.M., Weger, M., Moses, D.: Magnetic breakdown in incommensurate $Hg_{3-\delta}AsF_6$. J. Phys. **F14**, L101 (1984)
van Deursen, A.P.J., de Vroomen, A.R.: A 40T pulsed field de Haas-van Alphen spectrometer. J. Phys. **E17**, 155 (1984)
Devillers, M.A.C., van Ruitenbeek, J.M., Schreurs, L.W.M., de Vroomen, A.R.: The Fermi surface of In_2Bi. Solid State Commun. **49**, 613 (1984)
Horstman, R.E., van den Broek, E.J., Wolter, J., van Deursen, A.P.J., André, J.P.: Fractional quantum Hall effect in MOCVD-grown GaAs/AlGaAs heterostructures at pulsed high magnetic fields. Proc. 17th Int. Conf. on Physics of Semiconductors, San Francisco 1984
Leotin, J., Goiran, M., Askenazy, S.: Cyclotron resonance of holes in cadmium antimonide. Phys. Status Solidi (b) **123**, K43 (1984); Proc. 17th Int. Conf. on Physics of Semiconductors (San Francisco 1984)

Leotin, J., Poirier, A., Carrère, G., Askenazy, S., Rössler, U., Braun, M.: Quantum cyclotron resonance in silicon. Int. J. Infrared and Millimeter Waves **4**, 575 (1983)

Maeda, Y., Miura, N., Sakata, M., Ohta, E.: Transverse magnetophonon resonance in p-Te in high pulsed magnetic fields. J. Phys. Soc. Jpn. **53**, 3120 (1984)

Maeda, Y., Taki, H., Sakata, M., Ohta, E., Yamada, S., Fukui, T., Miura, N.: Magnetophonon resonance in epitaxial n-InP in high pulsed magnetic fields. J. Phys. Soc. Jpn. **53**, 3553 (1984)

Marquez, J., Ulmet, J.P., Leotin, J.: Déclenchement par fibres optiques d'un commutateur de puissance à thyristors. Proc. Conf. Optical Fibers in Broadband Networks, Instrumentation, and Urban and Industrial Environments, Paris 1983 (The International Society for Optical Engineering, P.O. Box 10, Bellingham, Washington 98227-0010 USA) p. 206

Miura, N.: "Infrared Magneto-optical Spectroscopy of Semiconductors and Magnetic Materials in Pulsed High Magnetic Fields", in *Infrared and Millimeter Waves*, Vol. 12, ed. by K.J. Button (Academic, New York 1983); Tech. Rept. ISSP Ser. A, No. 1349 (1983)

Miura, N., Iwasa, Y., Tarucha, S., Okamoto, H.: Magneto-optics of two-dimensional excitons in GaAs-AlAs heterostructures in high magnetic fields. Proc. 17th Int. Conf. on Physics of Semiconductors (San Francisco 1984)

Miura, N., Osada, T., Goto, T.: Magneto-transport and electronic phase transition in graphite in high magnetic fields. Proc. 17th Int. Conf. on Physics of Semiconductors (San Francisco 1984)

Nakao, K., Herlach, F., Goto, T., Takeyama, S., Sakakibara, T., Miura, N.: A laboratory instrument for generating magnetic fields over 200 tesla with single turn coils. Submitted to J. Phys. E: Sci. Instrum. (1985)

Ousset, J.C., Cantaloup, S., Durand, J., Bertrand, D., Fert, A.: Crystal-field effects in amorphous alloys containing praseodymium. J. Non-Crystalline Solids **61** and **62**, 385 (1984)

van Ruitenbeek, J.M., van Deursen, A.P.J., Schreurs, L.W.M., de Groot, R.A., de Vroomen, A.R., Fisk, Z., Smith, J.L.: Geometry and field dependence of the Fermi surface in $TiBe_2$ studied with the DHVA effect in fields up to 35 T. J. Phys. **F14**, 2555 (1984)

Tarucha, S., Okamoto, H., Iwasa, Y., Miura, N.: Exciton binding energy in GaAs quantum wells deduced from magneto-optical absorption measurement. Solid State Commun. **52**, 815 (1984)

Ulmet, J.P., Auban, P., Askenazy, S.: High field Shubnikov–de Haas effect and magnetoresistance in the organic metal $(TMTSF)_2ClO_4$. Solid State Commun. **52**, 547 (1984)

Editor's Appendix:
New Developments Regarding Electron Correlation Effects in $Hg_{0.8}Cd_{0.2}Te$

Fritz Herlach

With 1 Figure

Since the publication of the data shown in Fig. 2.16, a great deal more experimental data has become available and this has resulted in profound changes in the interpretation of the results. The breaks in the log $(\sigma_\perp)$ vs log (B) plots in Fig. 2.16 are no longer regarded as indicating a transition related to Wigner condensation of the electrons or charge density waves. It was shown that these breaks depend on the current passed through the sample [A.1]. Recently it has been suggested that the breaks are due to a surface layer which provides a shunt resistance [A.2]. Extensive measurements as a function of temperature and magnetic field have revealed that the conductivity is thermally activated. This was first shown for the low field range [A.1] and later extended to magnetic fields up to 35 T [A.3]. It is possible to give a consistent interpretation of these results by assuming "magnetic freeze-out" of conduction electrons which occurs when in a strong magnetic field the distance between the conduction band and the impurity levels becomes of the order kT. This analysis is based on both the Hall effect and the magnetoresistance; the carrier density in the conduction band is determined from the σ_{xy} component of the conductivity tensor which is related to the measured quantities $\varrho_\perp$ (transversal magnetoresistance) and $\varrho_H = R_H B$ (Hall resistivity; $R_H =$ Hall constant, $B =$ magnetic induction) by [A.4]

$$\sigma_{xy} = \varrho_H / (\varrho_H^2 + \varrho_\perp^2). \tag{A.1}$$

In the high field limit $\omega_c \tau \gg 1$, the charge carrier density is then given by

$$n = |\sigma_{xy}| B / e. \tag{A.2}$$

The activation energies determined from σ_{zz}, n and $\sigma_{xx} = \varrho_\perp / (\varrho_\perp^2 + \varrho_H^2)$ are given in Table A.1 (see also Fig. A.1). The consistency of the analysis is strengthened by finding the same values for the activation energy from both σ_{xy} and $\sigma_{zz} = 1/\varrho_\parallel$ ($\varrho_\parallel$: longitudinal magnetoresistance).

While experimental data obtained at Leuven/Nijmegen and Köln are in complete agreement, at Köln a different interpretation is given to these and some additional results. The idea of an electron correlation effect is maintained by interpreting the magnetotransport data as if the effect were caused by changes in the mobility rather than the carrier concentration. The thermally activated conductivity is interpreted as resulting from a transition of the electron gas to a viscous liquid [A.5]. Several arguments are given against magnetic freeze-out: it

Table A.1. The activation energies of thermally activated conduction in $Hg_{0.8}Cd_{0.2}Te$ at different magnetic fields as determined in Fig. A.1

B	Activation energy		
[T]	from σ_{zz} [meV]	from $\sigma_{xy}B/e$ [meV]	from σ_{xx} [meV]
4	0.33	0.40	
6	0.50	0.51	0.28
8	0.68	0.75	0.39
10	0.82	0.85	0.48
12	0.94	1.05	0.53
14	1.03	1.14	0.56
	±0.05	±0.1	±0.1

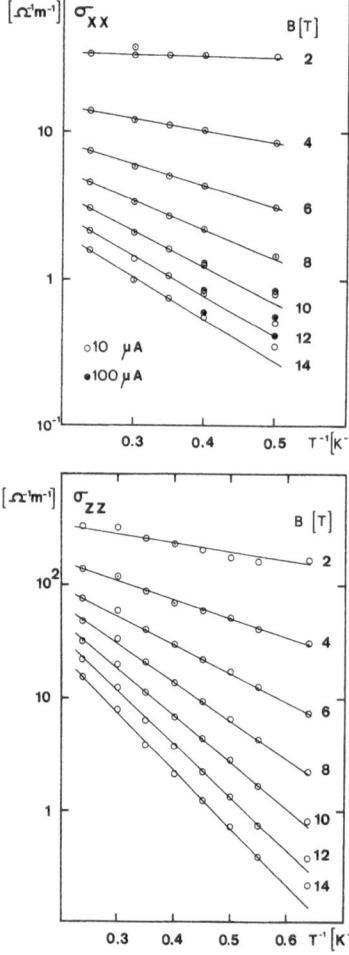

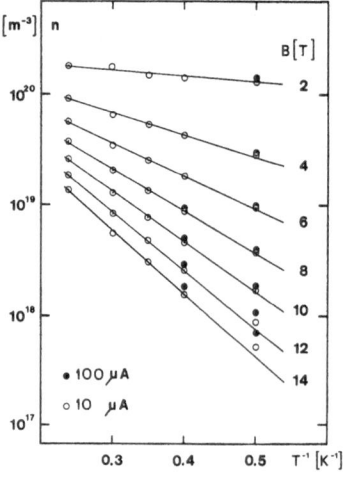

Fig. A.1. Components of the conductivity tensor plotted logarithmically as a function of $1/T$ for different fields, showing that electrical conductivity is thermally activated in $Hg_{0.8}Cd_{0.2}Te$ [A.3]. σ_{xy} has been transformed into the charge carrier density according to (A.2). At low temperatures and high magnetic fields, deviations due to the sample current (10 μA and 100 μA) become evident

is pointed out that earlier experiments in zero magnetic field have not revealed any impurity levels (i. e. no ionized impurity scattering) to which freeze-out could occur. A complete survey of earlier work on HgCdTe and other narrow gap semiconductors is given in [A.6]. The transient response of electrical current to an applied voltage step does not show – within the range of magnetic field, temperature and applied electric field of these experiments – the typical delay depending on the applied voltage as is seen in InSb and which is indicative of impact ionization. In the voltage-current characteristics, a negative resistance and instabilities due to avalanche formation have been seen in InSb but not yet in HgCdTe. Finally, measurements of the differential photoconductivity and the specific heat have been carried out in strong magnetic fields and the results are reported to support the assumption of a thermally activated mobility and electron correlation effects [A.7].

References

A.1 G. De Vos, F.Herlach: In *Application of High Magnetic Fields in Semiconductor Physics*, ed by
     G. Landwehr, Lecture Notes in Physics, Vol. 177 (Springer, Berlin, Heidelberg 1983) p. 378
A.2 J.P.Stadler, G.Nimtz, B.Schlicht, G.Remenyi: Solid State Commun. **52**, 67 (1984)
A.3 G. De Vos: Ph. D. Thesis, Katholieke Universiteit Leuven (1984);
     G. De Vos, F.Herlach, H.Myron: In preparation
A.4 R.Mansfield, L.Kusztelan: J. Phys. **C11**, 4157 (1978)
     A.Raymond, J.L.Robert, R.L.Aulombard, C.Bousquet, O.Valassiades, M.Royer: In *Physics
     of Narrow Gap Semiconductors*, ed. by E.Gornik, H.Heinrich, L.Palmetshofer, Lecture Notes in
     Physics, Vol. 152 (Springer, Berlin, Heidelberg 1982) p. 387
A.5 B.Schlicht: Ph. D. Thesis, Universität Köln (1983)
A.6 R.Dornhaus, G.Nimtz, B.Schlicht: *Narrow-Gap Semiconductors*, Springer Tracts Mod. Phys.,
     Vol. 98 (Springer, Berlin, Heidelberg 1983)
A.7 J.Gebhardt, G.Nimtz, B.Schlicht, J.P.Stadler: Submitted to J. Phys. C

Subject Index

Superconductivity in Magnetic and Exotic Materials

Proceedings of the Sixth Taniguchi International Symposium, Kashikojima, Japan, November 14–18, 1983

Editors: **T. Matsubara, A. Kotani**
1984. 106 figures. XII, 211 pages
(Springer Series in Solid-State Sciences, Volume 52)
ISBN 3-540-13324-0

Contents: Introduction. – Ferromagnetic Superconductors. – Antiferromagnetic Superconductors. – Organic Superconductors. – $BaPb_{1-x}Bi_xO_3$ and Some Metal Hydrides. – Appendix. – Index of Contributors.

Physics in High Magnetic Fields

Proceedings of the Oji International Seminar, Hakone, Japan, September 10–13, 1980

Editors: **S. Chikazumi, N. Miura**
1981. 257 figures. X, 358 pages
(Springer Series in Solid-State Sciences, Volume 24)
ISBN 3-540-10587-5

Contents: Recent Progress in the Steady High Field System. – Physics in Pulsed High Magnetic Fields. – Cyclotron Resonance and Laser Spectroscopy of Semiconductors. – Impurity States in High Magnetic Fields. – Magneto-Transport Phenomena. – Excitons and Magneto-Optics. – Electron-Hole Drops and Semimetals. – Narrow Gap Semiconductors. – Space Charge Layer and Superlattice. – Layered Materials and Intercalation. – Magnetism and Magnetic Semiconductors. – Photograph of the Participants of the Seminar. – List of Persons in the Photograph. – Index of Contributors.

Springer-Verlag
Berlin
Heidelberg
New York
Tokyo

Light Scattering in Solids I

Introductory Concepts

Editor: **M. Cardona**

2nd corrected and updated edition. 1983. 111 figures.
XV, 363 pages. (Topics in Applied Physics, Volume 8)
ISBN 3-540-11913-2

Contents: *M. Cardona:* Introduction. – *A. Pinczuk,
E. Burstein:* Fundamentals of Inelastic Light Scattering in
Semiconductors and Insulators. – *R. M. Martin,
L. M. Falicov:* Resonant Raman Scattering. – *M. V. Klein:*
Electronic Raman Scattering. – *M. H. Brodsky:* Raman
Scattering in Amorphous Semiconductors. – *A. S. Pine:*
Brillouin Scattering in Semiconductors. – *Y.-R. Shen:*
Stimulated Raman Scattering. – Overview. – Additional
References with Titles. – Subject Index. – Contents of
Light Scattering in Solids II, III and IV.

Application of High Magnetic Fields in Semiconductor Physics

Proceedings of the International Conference Held in
Grenoble, France, September 13–17, 1982

Editor: **G. Landwehr**

1983. XII, 552 pages. (Lecture Notes in Physics,
Volume 177). ISBN 3-540-11996-5

Contents: Quantized Hall Effect. – Localization in 2D
Systems. – Magneto-Transport in 2D- and Intercalated
Systems. – Magneto-Optics and Electronic Structure of
2D-Systems. – Magneto-Optics in 3D-Systems, Except
InSb. – Properties of Indium-Antimonide in High
Magnetic Fields. – Narrow Gap Materials. – Magneto-
Transport. – Magnetic Semiconductors. – General. –
Reports by High Magnetic Field Facilities, Technique. –
Historical. – Author Index.

Springer-Verlag
Berlin
Heidelberg
New York
Tokyo

Topics in Applied Physics Founded by Helmut K. V. Lotsch